W9-CSF-216

Combustion Residues

Combustion Residues

Current, Novel and Renewable Applications

Edited by

Michael Cox
University of Hertfordshire, UK

Henk Nugteren
Delft University of Technology, The Netherlands

Mária Janssen-Jurkovičová
KEMA Nederland BV, The Netherlands

John Wiley & Sons, Ltd

Library of Congress Cataloging in Publication Data

Cox, Michael, 1933–
    Combustion residues : current, novel and renewable applications / Michael Cox, Maria
Janssen-Jurkovicova, Henk Nugteren.
        p. cm.
    Includes bibliographical references and index.
    ISBN 978-0-470-09442-6 (cloth : alk. paper)
    1. Combustion product—Research.   2. Materials science—Research.   I. Janssen-Jurkovicova, Maria.
II. Nugteren, Henk.   III. Title.
    QD516.C69 2008
    621.31′21320286—dc22

                                                                                        2007046640

British Library Cataloguing in Publication Data

A catalogue record for this book is available from the British Library

ISBN 978-0-470-09442-6

Typeset in 10.5/13.5pt Sabon by Integra Software Services Pvt. Ltd, Pondicherry, India
Printed and bound in Great Britain by TJ International Ltd, Padstow, Cornwall
This book is printed on acid-free paper

Contents

Preface

In 1998, an EU Thematic Network (PROGRES),* consisting of 16 partners drawn from academia, research institutes and industry, was formed under the guidance of Dr Mária Janssen-Jurkovičová, with the aim of investigating newer uses of combustion residues largely arising from coal-burning power stations. During the project, the Network partners collaborated in a series of meetings and workshops to present their studies on current and future uses of combustion residues, to which other researchers from around the world were invited. The highlight of these was a three-day international meeting in 2001, held in Morella, Spain, where 75 participants from China, Indonesia, Israel, Japan, Poland, Russia, South Africa and the USA joined members of the Network and other workers in Europe. Selected papers from this meeting were published in a Special Issue of the *Journal of Chemical Technology and Biotechnology*.†

Over the four years, the partners produced a number of individual reports and scientific papers on various research topics, and it was decided that it would be useful to industry and other researchers if these could be correlated and expanded and published separately in the form of a book to supplement the final Network report to the EU.

The world-wide production of combustion residues from electric power production and waste incineration in 1999 was estimated at 500 Mt/y, of which the EU contribution was 60 Mt. Since then power generation has increased significantly, with coal still the major energy source. Thus, for the foreseeable future, the disposal of combustion

---

* PROducts from Glassy combustion RESidues.
† *Journal of Chemical Technology and Biotechnology*, volume 77, March 2002.

ashes will continue to be a major concern world-wide. While in the EU on average about 50 % of these residues are reused, mainly in production of low-value bulky building materials, such as cement, concrete and fillings in road construction, world-wide the figure is considerably lower. Thus the majority of such residues are land-filled or stored for potential future use.

As shown in the following chapters, combustion residues, as well as being used in the above traditional outlets, can also be a useful source of primary feed materials for a range of industrial high-value products, such as zeolites, ceramics, glass fibres, glass polyalkenoate cements for biomedical applications, fire-resistant materials, fillers for rubber and plastics, etc.

The introduction of newer materials into established industrial processes can pose difficulties, especially where existing materials are subjected to precise specifications. This has caused problems for the application of combustion residues in some areas, and if industries are to take advantage of these relatively low-cost resources, then specifications may have to be amended to accommodate them. This topic is addressed in the final chapter, with an example of how one particular industry has approached the problem.

The editors would like to thank the authors for their patience and help in the preparation of the final manuscript. In particular, thanks are due to Dr Alain Adjemian (EC Project Officer), for his advice and assistance; and Dr Ruud Meij (KEMA), for his knowledge on the health and safety issues of combustion residues.

Michael Cox
Mária Janssen-Jurkovičová
Henk Nugteren
September 2007

List of Contributors

Michael Anderson, University of Staffordshire, UK

Ian Barnes, Hattersall Associates, UK

Flavio Cioffi, Contento Trade srl, Terenzano (UD), Italy

Michael Cox, University of Hertfordshire, Hatfield, UK

Constantino Fernández-Pereira, Universidad de Sevilla, Spain

Robert Hill, Imperial College, London, UK

Rod Jones, University of Dundee, Scotland

Richard Kruger, Ash Resources (Pty) Ltd, Republic of South Africa

Mercedes Maroto-Valer, University of Nottingham, UK

Michael McCarthy, University of Dundee, Scotland

Fritz Moedinger, IRSAI srl, Carbonara di Po (MN), Italy

Natalie Moreno, CSIC, Barcelona, Spain

Henk Nugteren, Delft University of Technology, The Netherlands

Ann Sullivan, University of Limerick, Ireland

Alan Thompson, University of Nottingham, UK

Xavier Querol, CSIC, Barcelona, Spain

Luis F. Vilches Arenas, Universidad de Sevilla, Spain

1

The Current and Future Nature of Combustion Ashes

Alan Thompson
University of Nottingham, UK

List of Abbreviations and Acronyms

AFBC   Atmospheric Fluidised Bed Combustion
ABFBC   Atmospheric Bubbling Fluidised Bed Combustion
ACFBG   Atmospheric Circulating Fluidised Bed Gasification
APC   Air Pollution Control
ASR   Automotive Shredder Residue
ASTM   American Society of Testing & Materials
BOD   Biological Oxygen Demand
BFB   Bubbling Fluidised Bed
BGL   British Gas Lurgi
BMD   Building Materials Decree (Dutch)
°C   degrees Centigrade (Celsius)
CFB   Circulating Fluidised Bed
CFBC   Circulating Fluidised Bed Combustion
ESP   Electrostatic Precipitator
EWC   European Waste Catalogue
°F   degrees Fahrenheit
FB   Fluidised Bed
FBA   Furnace Bottom Ash
FBC   Fluidised Bed Combustion

Combustion Residues: Current, Novel and Renewable Applications   Edited by Michael Cox,
Henk Nugteren and Mária Janssen-Jurkovičová   © 2008 John Wiley & Sons, Ltd

FBG	Fluidised Bed Gasification
FGD	Flue Gas Desulfurisation
g	grams
GCV	Gross Calorific Value
GFT	Groente, Fruit and Tuinavfal (Dutch; category for waste classification = vegetables, fruit and garden waste)
GW_e/GW_{th}	gigawatt electrical thermal
HCOSFA	High-Calcium Oil Shale Fly Ash
HFO	Heavy Fuel Oil
IGCC	Integrated Gasification Combined Cycle
J	joule
JSIM	Japanese Society of Industrial Machinery Manufacturers
kg	kilogram
LOI	Loss On Ignition
MJ	megajoule
MPF	Mixed Plastics Fraction
MSW	Municipal Solid Waste
MW_e	megawatt electrical
MW_{th}	megawatt thermal
NCV	Net Calorific Value
NO_x	nitrogen oxides
OMC	Oil Mill Cake
OMWW	Oil Mill Waste Water
OSA	Oil Shale Ash
OSFA	Oil Shale Fly Ash
PE	polyethylene
PF	Pulverised Fuel
PFA	Pulverised Fuel Ash
PFBC	Pressurised Fluidised Bed Combustion
PPF	Paper and Plastics Fraction
RE	relative enrichment factor
RDF	Refuse-Derived Fuel
SRC	Short Rotation Coppice
TCLP	Toxicity Characteristic Leaching Procedure
TLV	Threshold Limited Value
TRACE	Trace Radioactivity Ash Coal Emissions
TSPM	Total Suspended Matter
WID	Waste Incineration Directive (EU2000/76/EC)
WTE	Waste-To-Energy
WWTP	Waste Water Treatment Plant

1.1 Introduction

Coal ash has a long and venerable history of use within concrete. It is often argued that the use of ash can be traced back to at least Roman times and perhaps beyond, but it is worthwhile mentioning that these were typically naturally occurring pozzolanic materials or were created by burning clay minerals. Whilst these were clearly pozzolanic in nature, this is about the only similarity with modern coal fly ash from the high-temperature combustion of pulverised coals. Although the word 'fly ash' is traceable to about 1914,[1] the classic work identifying the use of coal fly ash in concrete was carried out by R.A. Davies and co-workers in a series of classical papers published in the 1930s.[2-4] These provided the fundamental underpinning of all modern specifications and identified the need to assess the quality of fly ash using fineness, loss on ignition and moisture content. These key characteristics still form the basis for many ash-based product specifications, as will be shown in Chapter 2.

With the advent of pulverised coal firing systems, significant quantities of combustion residues became available. Amounts have increased dramatically during the past few decades, such that by 1999 world production of ash and slag was 480 million tonnes, with the main producers being China, North America, India, Russia and Europe. The quantity produced in the European Union in 1999 was 55 million tonnes, rising in 2000 to 59 million tonnes.

It was evident that simply discarding such quantities of material would be unacceptable on environmental grounds and methods for their reuse have been sought for a number of years, particularly in Europe, the USA and Japan. However, there is a wide variation in usage throughout the world, ranging from total reuse to total disposal. Where usage is high, it is a result of a changed perception towards combustion residues. Far from it being an unwanted and unwelcome by-product from the use of coal, it has been recognised as a valuable resource, with a range of uses that are continuing to increase daily. However, the use of such residues is not simply the result of enlightened thinking; it is also driven by the realisation that, because of their impact on the environment, they must be disposed of in a responsible manner. Landfill taxes, already high in some European countries, and likely to continue to increase, will ensure that producers of combustion residues will, of necessity, seek to reuse and thus generate a stream of income from what are unavoidable by-products.

Whilst significant progress has been made in terms of reusing combustion residues – for example, the Netherlands 100 % reuse,[5] Western

Europe 90 %,[6] the UK around 50 %[5] and the USA 33 % (2001)[7] – there are now new challenges to be faced. Until the early 1990s, the majority of ash produced across the European Union arose from conventionally configured and operated pulverised coal-fired power generating plant. Ash quality was generally predictable and reasonably consistent with regard to its properties. Since that time, a number of changes in the utility marketplace, growing pressure for an increased use of renewable energy sources and increasingly stringent environmental legislation have led to changes in the nature of utility ashes that require new approaches and that open new opportunities for utilisation.[8]

Specific topics that need to be addressed by producers and users of combustion residues include the following:

- The widespread use of NO_x reduction technologies, which can lead to the formation of fly ash containing unacceptably high levels of unburned carbon and, in some cases, adsorbed ammonia.
- The increase in world-traded coals, which has resulted in the combustion of, in some cases, less familiar coals with the production of fly ash whose properties are less well understood.
- Co-combustion of coal with minor amounts of biomass, such as wood and so on, or waste materials such as petroleum coke in a pulverised fuel fired boiler. This may lead to changes in fly ash quality and, as a result, its value as a marketable product.
- Combustion of single-source noncoal fuels or wastes, such as municipal solid waste (MSW) or sewage sludge. The nature and composition of the residues and their leachability may hinder or prevent their reuse.
- The use of cleaner coal combustion technologies such as fluidised bed combustion (FBC), where the fuel ash is produced together with an emission-trapping sorbent, often calcium salts. This also applies to flue gas desulfurisation (FGD). These residues and ashes have to be stabilised before reuse.
- The use of newer noncombustion technologies, such as gasification of coal and wastes. The process produces mainly glassy, hard residues, for which applications have been found, and dusts, whose compositions are quite different and for which uses are being sought.[8]

This book aims to provide a view of the current status of combustion residues in terms of what residues are available, possibilities for their uses and any limitations. This chapter will describe the nature of coal and the origins of ash and residues produced from the minerals in coal using traditional methods of combustion. More recent developments into the use of blends of coal and renewable materials and wastes are described, together with more advanced combustion systems. The latest technological developments in the reuse of these residues are described in later chapters.

1.2   Coal: the Principal Source of Combustion Residues

1.2.1   Background

Coal has been used as a source of fuel since early Greek and Roman times. The Greek philosopher Theophratus, in his 4th century *Treatise on Stones*, describes a fossil substance used as a fuel. Coal cinders in Roman ruins in Britain indicate that coal was used during the period of Roman occupation, from approximately AD 50 to 450. Despite this, there was little incentive to find and use coal while wood was so plentiful. However, as industry grew, a shortage of wood occurred and a substitute was needed. To fuel its industrial expansion, Britain began extensively mining coal in the 13th century.[9] It was coal that provided the energy to fuel the Industrial Revolution of the 19th century and, in turn, enabled the proliferation of electric power in the early 20th century.

There are a number of reasons why the use of coal has increased in recent years. These include its abundance (proven coal reserves are estimated to last over 200 years), wide geographical distribution (present in over 100 countries), politically stable locations (Australia, the USA and China), and cost.[10]

Coal is now the most widely used fuel for generating electricity, accounting for about 38 % of the total production world-wide. In some countries, coal is almost the only fuel used for electricity generation. For example, in Poland, 96 % of the electricity is obtained from coal firing, in South Africa 90 % and in Australia 84 %.[11]

Total hard coal production for 1999 was 3466 million tonnes.[11] If one assumes that the average ash content of the coal is 10 %, then approximately 350 million tonnes of ash will be formed. Whilst not all of the coal that is produced will be used in combustion, nevertheless this quantity of ash represents a huge challenge to the power generator and much effort, particularly in recent years, has gone into its resolution.

The use of such residues is not simply the result of forward thinking; it is partly driven by the recognition that such residues can have a major impact on the environment and as such must not be disposed of irresponsibly. As noted above, landfill taxes will act as a driver for the development of new uses.

1.2.2 Terminology and Classification

Due to the great importance coal has had for the world economy since the start of the Industrial Revolution, much research has been dedicated to its origin, appearance, chemical and physical properties, extraction, preparation and use. From such work, a vast amount of nomenclature and terminology and many classification systems have evolved, all of which has been thoroughly reviewed by Speight.[12]

The origin of coal lies in the simple photosynthetic reaction of carbon dioxide and water, using sunlight as energy source and chlorophyll as a catalyst to form carbohydrates. Large accumulations of carbohydrates are found in plants, and if sufficiently concentrated under favourable anaerobic conditions, coal can result from the decay and maturation of floral remains over geological time. So, coal can be regarded as an organic sediment, which has hardened through burial and exposure to elevated pressures and temperatures, to become an organic rock-like natural product.

The vast terminology that developed arose mainly from the need for classification tools; so, in contrast to many other areas, terminology cannot be viewed separately from classification systems. Terminology used to describe carbon content and the constituent parts of coal (petrography) is clearly related to classification.

The three most frequently used general terms to describe particular characteristics of coal are as follows:

- *type* – concerned mostly with the nature of the coal material;
- *rank* – which refers to the degree of metamorphism to which the coal has been subjected; and
- *grade* – which relates best to the amount of inorganic (mineral) matter within the coal matrix.

Different classification systems have been devised using the geological age of coal, its (banded) structure and the type and grade, but the most common are for coal rank. Rank refers to the degree of maturation of the coal and, in general, higher-rank coals have higher carbon contents. In some systems, the rank of coal is equated directly with carbon content. Therefore, some relationship, although not simple, must exist between coal composition and its behavioural properties, such as calorific value and volatile matter. Seyler's coal classification was one of the first schemes that attempted to quantify this relationship.[13]

Proximate analysis of coal involves the determination of moisture, ash, volatile matter and fixed carbon content. The American Society for Testing and Materials (ASTM) has adopted a coal classification system that uses the fixed carbon content of coal as well as other physical properties.[14] The fixed carbon value, by definition, is calculated by subtracting from 100 % the percentages of moisture, volatile matter and ash.[12] It thus is determined by methods of proximate analysis, as opposed to the elemental carbon content that evolves from ultimate analysis. Higher-rank coals (having less than 31 % w/w volatile matter on a dry, ash-free basis) are classified on the basis of their fixed carbon content. However, coals containing more than this figure are classified according to their calorific value.

The International system of classification is mainly used in Europe. It separates coals into two types, with the division occurring at a calorific value of $23.9 \, MJ \, kg^{-1}$ or $10\,260 \, BTU \, lb^{-1}$. Coals with a higher calorific value are called *hard coal*, and those with a lower calorific value *brown coal*. The coals are further divided into classes on their ash-free calorific values. Hard coals are divided into groups and subgroups according to their caking properties (swelling test) and coking properties (slow heating test) respectively. A three-digit code number (indicating class, group and subgroup number) is employed to identify the coal. Brown coals are subdivided into groups according to their yield of tar on a dry, ash-free basis.

The ASTM rank system uses a mineral-matter-free calorific value, whereas the International system employs an ash-free calorific value for class divisions, which introduces a slight discrepancy. However, the ASTM ranks can be approximated to the International system classes (Table 1.1).[15] For further subdivision into groups and subgroups, reference should be made to the overview given by Speight, who also presents tables of other local systems such as the British system (mainly based on coking behaviour) and the systems used in Belgium, Germany, Poland and other countries.[12]

Current coal classification systems, although successful in classifying coal, do not provide any tools by which the potential environmental liability of coal usage can be determined. Classification data is usually reported on a mineral-free or ash-free basis and, as a result, the effects

Table 1.1 An overview of the simplified ASTM (USA) and International coal classification systems (Reproduced by permission of Bohn Stafleu van Loghum BV).

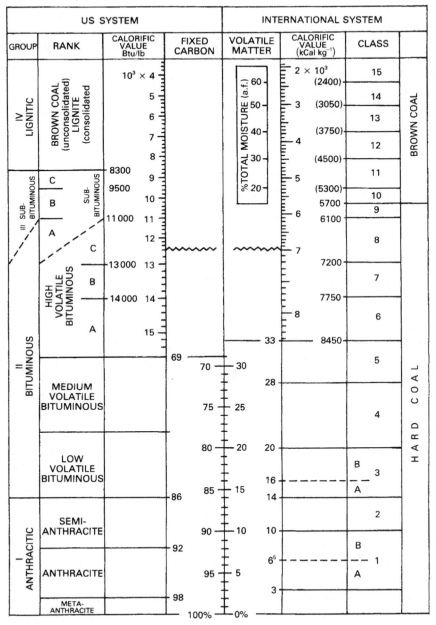

of ash content and thus ash production and ash characteristics cannot be ascertained. Therefore, the mineral content of coal requires further discussion.

1.2.3 Mineral Composition

The phrase 'minerals in coal' is usually meant to include all of the inorganic noncoal material found in coal as mineral phases and the elements in coal that are considered inorganic. The definition thus includes all elements in coal except carbon, hydrogen, oxygen, nitrogen and sulfur. However, carbon is also found in inorganic carbonates such as calcite, hydrogen is found in water bound to clay material and sulfur occurs in minerals such as pyrite.

Mineral matter occurs in coal in two general forms: as elements required for the growth of the original plant life from which the coal was formed and as inorganic material, typically sand, silt or clay, deposited in the accumulating plant debris as discrete particles of foreign matter. Inorganic material also may have been deposited later by mineral-laden water percolating through the coal seam. The former is known as 'inherent' and the latter 'extraneous' mineral matter. The key difference between these mineral types is that extraneous mineral matter may be removed by beneficiation, at least in part, whereas inherent mineral matter cannot be removed by normal coal cleaning processes. This can have important implications, not only for boiler plant design but also for ash production rates. Coals with up to 48 % ash, most of it from inherent matter, have been identified for use in Brazil and in Uttar Pradesh, India.[16]

The mineral matter found in coal consists of a large number of different compounds. However, by far the most abundant noncombustible species in bituminous coals are those of the aluminosilicate clay minerals. Together with quartz, they account for between 60 % and 90 % of the total mineral matter in coal. The most common species of clay minerals are muscovite–illites (potassium aluminosilicates), kaolinites (aluminosilicates) and mixed-layer illite–montmorillonites of variable composition.[17]

The nature of mineral matter in coal does not appear to change very much with coal rank. However, there is a noticeable difference in the ash chemistry of low-rank coals such as lignites and sub-bituminous coals compared to higher rank coals. Low-rank coals, unlike bituminous coals, have a high oxygen content. Some of this oxygen is present as carboxylic groups, which can act as ion exchange sites. This results in the incorporation of significant quantities of sodium, calcium and magnesium into such coals. These high concentrations of alkali and alkaline earth components are due to groundwater containing these substances filtering through the low-rank coal seams.[18] For this reason, the fly ashes from low- and high-rank coals are given two different designations, depending on their composition.[19]

Certain geological/geographical differences in coal mineralogy have been noted.[20] For example, the mineral composition of Australian coals differs substantially from those of North American and European coals. The Black coals contain from 10 to 17 % ash and are low in moisture. The Victorian Brown coals contain less than 5 % ash and more than 60 % moisture. Virtually all Australian coals contain less than 1.3 % sulfur and very little chlorine. Unlike Northern Hemisphere coals, in which iron appears primarily as pyrites, the Australian coals contain siderite (iron carbonate) and ankerite (calcium iron carbonate). In the low-rank Brown coals, iron is also found as organically-bound mineral matter and frequently occurs in concentrations exceeding that of aluminium, not a trend usually found in Northern Hemisphere coals. Calcium is generally low and usually appears as a carbonate in older coals and organically-bound in low-rank Brown coals. Magnesium, sodium and significant portions of potassium also appear as organically-bound mineral matter. The highest concentration of sodium is found in the low-rank coals.

Despite the apparent order in mineral formation by geological origin, anomalies do occur. Some lignites have been found enriched with pyrites, which could be responsible for otherwise unpredicted slagging. In some instances, minerals are reported that are peculiar to a particular coal source. For example, zeolites are the leading source of sodium in Texas lignite; and the mineral analcime, a complex sodium aluminium silicate, is present in high concentration in certain Utah coals.[20]

1.3   Coal Ash: the Principal Combustion Residue

1.3.1   Origin

Pulverised coal firing is by far the commonest form of coal combustion used today for power generation. This system uses finely-ground coal, which is introduced via burners into water tube-walled furnaces. In the following sections, 'coal combustion' will refer to pulverised coal combustion. Other types of coal firing systems will be discussed separately in later parts of this chapter.

The bulk of the ash (~90%) that is formed when pulverised coal burns is converted into fine dry powder usually known as fly ash or pulverised fuel ash (PFA). Other terminology, used to differentiate between fuel types and whether it has been collected by boiler equipment or emitted from stacks into the atmosphere, has been introduced in the Netherlands. Other countries, such as the UK and the USA, tend not to make this distinction.

A smaller proportion (~10%) of the coal ash exits the boiler as furnace bottom ash (FBA). This is much denser and larger in size. Effectively, this is fused ash that has agglomerated and fallen to the base of the furnace.

1.3.2 Ash Composition

The composition of the ash depends on numerous parameters, summarised in Table 1.2.[21] While the term 'ash' is applied to all the materials produced, these materials are in fact quite different from one another, with different physical, chemical and, consequently, toxicological properties.

Considerable differences can occur between the fly ashes produced by the combustion of coal. All coal-fired power stations in the Netherlands are based on the following combination:

Table 1.2 Parameters influencing the composition of ash (Reproduced by permission of KEMA).

Primary parameter	Secondary parameter
Fuel	Coal (subdivided according to type, origin and ash content), brown coal, peat, industrial waste, domestic waste, paper, etc.
Combustion technique	Pulverised coal boilers with dry ash removal; pulverised coal boilers with wet ash removal; cyclone boilers; stoker-fired boilers; gasification process; fluidised bed (atmospheric, pressurised, bubbling, circulating, combustion, gasification); etc.
Temperature and period of residence	During combustion and in the flue gas ducts
Type of particle filter	ESP (high-temperature, low-temperature); cyclone; baghouse; wet systems; also filter efficiency
Other flue gas clean-up systems	De-NO_x; FGD system; etc.

- a pulverised coal boiler with dry ash removal;
- a high-efficiency E-filter, at approximately 120–140 °C; and
- a wet FGD system of the calcium/limestone/gypsum type.

Additionally, there is one gasification unit based on the Shell process.

Since different combinations are used in other countries, it cannot be assumed that ashes produced in the Netherlands have a similar composition to the ashes described in the academic literature. Even in cases where the plant is comparable, the ash produced may differ on account of the type of coal used. Coal can be categorised on the basis of volatility due to variations in the degree of carbonisation. Most of the coal fired in the Netherlands is bituminous and comes from Australia, Colombia, Indonesia, Poland, South Africa, the USA (particularly the Eastern states) and, to a lesser extent, from China, Russia and Venezuela.

The ash content of this coal varies between 5 and 20 %, with the average in the Netherlands being around 14 % (2002). Different types of coal, and even coals of the same type from geographically different sources, can differ considerably in composition. For this reason, KEMA has been conducting a systematic study on behalf of the electricity sector since 1978, to determine the composition of coals fired in the Netherlands and the associated ash types and flue gases.

The various ash flows in Dutch power stations are listed in Table 1.3, together with their principal properties. Distribution between the different flows is indicated in both Table 1.3 and Figure 1.1.

Table 1.3 Nomenclature of ash flows from coal-fired power stations with their relative distribution and particle size (as applied in the Netherlands) (Reproduced by permission of KEMA).

Term	Definition	Percentage of all ash produced	Particle size
Bottom ash	Collected at bottom of boiler	12 (10–20)	Coarse
Pulverised fuel ash (ESP ash)	Removed from flue gases by ESP	~ 87.8	50 % of the overall mass made up of particles under 30 μm
Fly ash	In flue gases after passing through ESP and (in plants without FGD system) in the stack	~ 0.25	50 % of the overall mass made up of particles under 3 μm
Fly dust	In flue gases in the stack (after passing through FGD system)	~ 0.05	50 % of the overall mass made up of particles under 0.3 μm

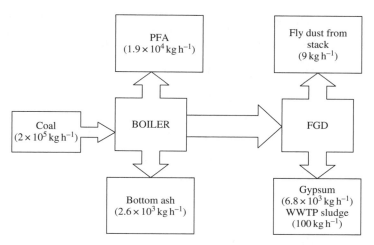

Figure 1.1 Materials flows associated with a coal-fired power station with a net capacity of 600 MW$_e$ (Reproduced by permission of KEMA).

The largest ash flow is that from the electrostatic precipitator (ESP). In the Netherlands, this material is referred to as pulverised fuel ash (PFA), partly to make it clear that it comes from pulverised coal-fired boilers and partly to distinguish it from that released into the atmosphere from the stack. The latter is referred to as fly ash or fly dust. The term 'fly ash' is used to refer to material from plants without FGD systems (of which there are no longer any in the Netherlands), while the term 'fly dust' is applied to material from plants with FGD systems.

The captured material reaches temperatures of between 1 000 and 1500 °C, causing it to melt. As a result, the ash forms glassy globules, which are then largely removed from the flue gas using filters. All the flue gas filters used in the Netherlands are high-efficiency ESPs. By installing several such filters in series, a very high proportion of the fly ash (average 99.75 %) can be removed. The particle in associated flue gas is on average 25 mg m^{-3}.

It will be apparent from Table 1.3 that the various ash types differ greatly from one another in particle size. The concentrations of certain elements in the various ash types also differ considerably; this issue is considered in more detail below.

Ash produced during co-combustion of secondary fuels has similar physical and chemical properties, and so the same terminology as for 100 % co-firing is used.

Details of the typical composition of bituminous coal, mostly used in coal-fired power stations, are shown in Table 1.4.[22] For comparison, in

Table 1.4 A comparison of element concentrations for coal firing in the Netherlands with the corresponding concentrations reported for coal in general, soil and the Earth's crust (Reproduced by permission of KEMA).

Element	NL mean[21]	Reported date for global coal use[22]			
		Typical conc.	Range	Soil	Earth's crust
Macro-elements, concentrations in %					
Cl	0.04	0.10	0.01−0.20	0.01	0.02
P	0.04	0.02	0.001−0.30	0.05	0.10
Ti	0.10	0.06	0.001−0.20	0.30	0.44
Trace- and micro-elements, concentrations in $mg\,kg^{-1}$					
As	3.1	10	0.5−80	7	1.0
B	63	50	5−400	30	10
Ba	181	200	20−1000	500	425
Be	1.2	2	0.1−15	1.0	3
Br	4.4	20	0.5−90	10	0.8
Cd	0.12	0.5	0.1−3	0.6	0.2
Ce	18	20	2−70		60
Co	5	5	0.5−30	10	25
Cr	17	20	0.5−60	55	100
Cs	1.0	1.0	0.3−5	4	3
Cu	12	15	0.5−50	25	55
Eu	0.4	0.5	0.1−2		1.2
F	106	150	20−500	300	544
Ge	1.7	5	0.5−50	1.0	2
Hf	1.2	1.0	0.4−5		3
Hg	0.11	0.1	0.02−1	0.10	0.08
I	2.8	5	0.5−15	7	0.3
La	6.9	10	1−40		30
Mn	43	70	5−300	550	950
Mo	1.9	3	0.1−10	1	2
Ni	14	20	0.5−50	20	75
Pb	8	40	2−80	20	13
Rb	9	15	2−50		90
Sb	0.55	1.0	0.1−10	0.7	0.2
Sc	3.5	4	1−10	9	22
Se	1.9	1.0	0.2−10	0.4	0.1
Sm	1.8	2.0	0.5−6		6
Sr	225	200	15−500	240	375
Th	3.4	4	0.5−10	9	7
Tl	<1			0.2	0.5
U	1.3	2	0.5−10	3	2
V	26	40	2−100	80	135
W	1.2	1.0	0.5−5	300	2
Zn	25	50	5	70	70

addition to element concentration data for global coal supplies, the table contains corresponding figures for the coal fired in the Netherlands, all of which is imported.[21] In the past 25 years, more than 150 coals have been analysed by KEMA for about 45 elements. Table 1.4 also provides data on the concentrations in which the elements generally occur in common soil and in the Earth's crust.

It also appears that elemental concentrations in coal used in the Netherlands are generally lower than the corresponding concentrations reported for global coal use.[22] Nevertheless, the concentrations of certain elements found in the coal are higher than the corresponding concentrations in the Earth's crust. This is particularly true for selenium (concentration in the coal 50 times higher) and, to a lesser extent, for iodine, bromine > boron, arsenic > antimony and chlorine > molybdenum.

1.3.3 The Behaviour of Elements in the Boiler and ESP

The ash remaining after combustion generally contains the same elements present in the coal, but enriched in the ash by a factor known as the 'coal ash ratio'. However, it has been found that enrichment also depends on the type of ash and the particular element. The term 'relative enrichment' was introduced by Meij to describe properly the observed behaviour.[23] The relative enrichment factor (RE) is defined as follows:

$$RE_X = \frac{[X]_{ash}}{[X]_{coal}} \frac{\% \text{ ash content of coal}}{100} \tag{1.1}$$

Thirty-three mass balance studies were carried out at 11 different coal-fired units in the Netherlands between 1980 and 1991. The measurements were performed on 22 different bituminous coals from seven countries. In all cases, the RE factors for each type of ash (furnace bottom ash, PFA and emitted fly ash) were calculated. Meij used these results to divide the elements into three classes (Table 1.5), which are discussed below.[23]

Class I Elements

Class I elements are those elements that do not vaporise during combustion and their concentration in all ash types is the same. The RE factor is approximately one and is independent of size (Figure 1.2).

Table 1.5 A classification of elements by volatility and behaviour in pulverised coal fired installations with their relative enrichment factor (RE) according to Meij[23] (Reproduced by permission of KEMA).

Class	Relative enrichment factor			Behaviour in installation	Classified elements
	Bottom ash	PFA	Fly ash[a]		
I	≈ 1	≈ 1	≈ 1	Not volatile	Al, Ca, Ce, Cs, Eu, Fe, Hf, K, La, Mg, Sc, Sm, Si, Sr, Th, Ti
IIc	< 0.7	≈ 1	$1.3 < \ldots \leq 2$	Volatile, but condensation within the installation on the ash particles	Ba, Cr, Mn, Na, Rb
IIb	< 0.7	≈ 1	$2 < \ldots \leq 4$		Be, Co, Cu, Ni, P, U, V, W
IIa	< 0.7	≈ 1	> 4		As, Cd, Ge, Mo, Pb, Sb, Tl, Zn
III	$\ll 1$	$\ll 1$		Very volatile, moderate to little condensation in the installation	B, Br, C, Cl, F, Hg, I, N, S, Se

Volatility→

[a] Emitted fly ash and PFA from last hopper of ESP (finest fraction).

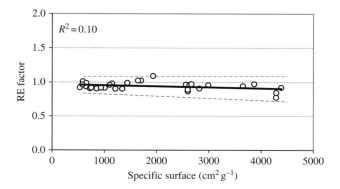

Figure 1.2 The relation between the RE factor of aluminium, a Class I element, and specific surface with the 95 % confidence interval (Reproduced by permission of KEMA).

Class II Elements

In addition to Class I elements, there are those that redistribute among the various ash types; that is, bottom ash, pulverised-fuel ash (collected) and fly ash (in the flue gases downstream of the ESP). These elements are vaporised in the boiler during the combustion process. Concomitant with the route of the flue gases through the boiler, ducts, air preheater and ESPs, the temperature decreases from about 1600 °C to about 120–140 °C. Depending on the chemical compound, somewhere on this route the dew point will be passed and condensation will start on the surface of the fly ash particles. Also, particles can form through nucleation of vaporised material and grow through coagulation and heterogeneous condensation. The smallest particles have the largest specific areas. Therefore, on a weight basis, the condensing elements are found in greatest concentrations on the smallest particles. All elements that condense within the installation are grouped as Class II. The RE factor of the bottom ash is less than 0.7, because elements originally present in the vapour phase have no opportunity to condense on the bottom ash particles. The RE factor of the PFA from the collection tank is approximately one for elements of Class II; the factor for the smaller particles exceeds 1.3. The smaller particles are found in the last two hoppers of the ESP and in the flue gases downstream of the ESP. Class II is subdivided into three classes; a, b and c. These subclasses refer to the degree of volatility (Table 1.5). The RE factors and the elements in question are given in Table 1.5. An example of the RE factor of lead as a function of the specific surface is given in Figure 1.3, where there is a clear relationship between the RE factor and specific surface, or particle size.

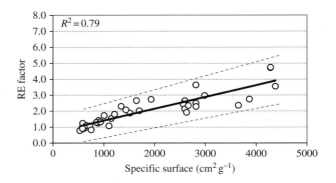

Figure 1.3 The relation between the RE factor of lead, a Class IIa element, and specific surface with the 95 % confidence interval (Reproduced by permission of KEMA).

Class III Elements

Elements that occur in compounds with a low dew point condense only partly within the installation and, if FGD is absent, they are totally or partly emitted in the vapour phase. These are grouped as Class III. Their RE factor can be very small ($\ll 1$), especially in the bottom ash and to a lesser extent in the PFA in the silo. The RE factor of the smallest fly ash particles found in flue gas downstream of the ESP can also be high (Table 1.5).

The distinction between Classes II and III depends on the RE factor of the PFA (hopper 1 or silo). If the RE factor equals one, with a margin of ± 0.3, then it is a Class II element. If the RE factor is significantly less than one, then it falls into Class III.

1.3.4 Modelling and Prediction of Ash Composition

The studies mentioned in previous sections yield typical parameters, such as the RE factor, that provide relationships between the streams. These parameters are independent of the situation at a particular moment and can be used in a general way in models for predicting the chemical composition of the streams. Such a model (the KEMA TRACE MODEL®) was developed by Meij.[24] It is an empirical and statistical model based on extensive studies, in which the behaviour and the fate of (trace) elements in power stations is studied (Figure 1.4). The name

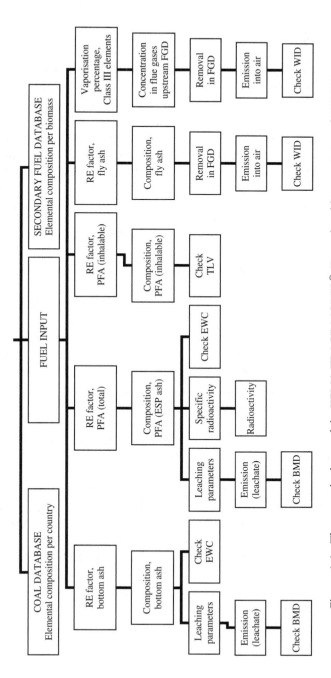

Figure 1.4 The general scheme of the KEMA TRACE MODEL® (Reproduced by permission of KEMA).

TRACE implies that the model traces the trace elements, originating in the fuel, (Trace Radioactivity Ash Coal Emissions into air).

The basis for the model is the composition of the fuel. If the coal composition is known, then the composition of various ashes and flue gases can be calculated. If the coal composition is unknown but the origin of the coal is known, the coal composition can be obtained from a database. This latter approach yields a fairly good prediction, is also valid for blends of coal and was tested in practice with 21 samples from six different Dutch power stations in 1993 and 1994. A comparison between the measured and the predicted values of the coal composition (42 elements) yielded a mean R^2 value of $93 \pm 7\%$. The same procedure was followed for the composition of the pulverised fuel ash, yielding a mean R^2 value of $93 \pm 8\%$. The prediction for some individual elements, such as (in decreasing order) Hg > Br > Be, Se > Cd and Zn, can be less accurate.[24] In particular, Class III elements show the largest deviations.

The origin of the coal in a particular year is well recorded, so for that year the averaged composition together with the standard deviation of the various streams can be calculated. Hence, a representative figure for the year is obtained. An example for the coal fired in the Netherlands in 1999 using yearly average figures has been calculated and shown in Table 1.4 and expanded data is shown in Table 1.6. The maximum value is the mean value with twice the standard deviation. It does not indicate the maximum value of the fired coal in 1999, but the uncertainty of the calculation of the mean value.

In addition to the coal data, concentrations in the various ash flows – that is, bottom ash, pulverised fuel ash and fly ash (fly dust) – are given.[21] These were also calculated with the KEMA TRACE MODEL®. When coal is fired a process of concentration takes place, so that in most cases the ash element concentrations are higher than those in the coal. The ash is found to consist principally of silicon, aluminium and a small amount of iron, in compounds such as aluminium silicates and iron oxides. Even after this process of concentration, the amounts of these macro-elements contained in the ash are broadly comparable with the amounts in sand or clay (Table 1.4).

From Table 1.6, it will be apparent that the four types of ash differ in composition according to the elements involved. For instance, in general terms, the macro-elements occur in similar concentrations in all three ash flows. By contrast, the concentrations of some trace elements – for example, arsenic, cadmium, molybdenum, nickel, lead, antimony, selenium and zinc – increase markedly from furnace bottom ash to

Table 1.6 The average elemental composition of coal and associated ash composition (furnace bottom ash, pulverised fuel ash, fly ash and fly dust), for Dutch power stations in 1999 (Reproduced by permission of KEMA).

	Coal (1999)		Bottom ash		PFA$_{total}$		PFA$_{inhalable}$		Fly dust	
	Mean	Max.	Mean	Max.	Mean	Max.	Mean	Max.	Mean	Max.
Macro- and main elements, concentrations in %										
Al	2.0	3.3	14.0	16.0	14.0	16.0	14.0	16.0	14.0	16.0
C	70		2		5					
Ca	0.32	0.65	2.3	3.2	2.3	3.2	2.3	3.2	2.3	3.2
Cl	0.04	0.09	0.002	0.008	0.003	0.008	0.003	0.008	0.091	
Fe	0.63	1.28	4.6	6.2	4.6	6.2	4.6	6.2	4.6	6.2
K	0.14	0.3	1	1.4	1	1.4	1	1.4	1	1.4
Mg	0.1	0.22	0.8	1.1	0.8	1.1	0.8	1.1	0.8	1.1
Na	0.05	0.1	0.3	0.5	0.4	0.5	0.4	0.6	0.6	
P	0.05	0.1	0.1	0.5	0.4	0.6	0.5	1.1	1	1.1
S	0.9	1.7	0.14		0.14					
Si	3.2	5.3	24	25	24	25	24	25	24	25
Ti	0.12	0.21	0.9	1	0.9	1	0.9	1	0.9	1
Trace and micro-elements, concentrations in mg kg^{-1}										
As	3.1	6.3	1.6	12	23	43	39	100	246	555
B	63	92	184	266	253	444	321	533	459	621
Ba	181	357	1060	1721	1325	2582	1457	3443	2374	
Be	1.2	2.2	6	11	9	15	11	21	18	
Br	4.4	13.5	0.3	2	3.3	16	0.3	3.9	39	254
Cd	0.12	0.41	0.1	0.8	0.9	4	1	5	7	25
Ce	18	30	130	188	130	145	130	145	130	145
Co	5.2	8.8	27	43	38	60	49	72	89	
Cr	17.3	29.3	114	141	127	155	127	183	199	451
Cs	1	2	8	10	8	10	8	10	8	10

Table 1.6 (Continued)

	Coal (1999)		Bottom ash		PFA$_{total}$		PFA$_{inhalable}$		Fly dust	
	Mean	Max.	Mean	Max.	Mean	Max.	Mean	Max.	Mean	Max.
Cu	12	23	45	76	91	153	100	175	215	404
Eu	0.4	0.7	2.7	3.2	2.7	3.2	2.7	3.2	2.7	3.2
F	106	213	62	205	125	246	311	1231	1167	
Ge	1.7	3	6	10	13	22			99	143
Hf	1.2	2.3	9	11	9	11	9	11	9	11
Hg	0.11	0.48	0.06	0.23	0.43	2.6	0.3	1.9	1.6	8.1
I	2.8	7.6	0.2	0.4	0.6					
La	6.9	12.9	50	62	50	62	50	62	50	62
Mn	43	94	317	453	317	589	348	770	528	
Mo	1.9	4	7	13	14	29	21	58	61	112
Ni	14.2	30.2	73	146	104	175	135	291	333	845
Pb	7.7	15.2	23	58	57	110	102	197	261	643
Rb	9.4	22.2	69	107	69	107	69	107	69	107
Sb	0.6	1.1	0.8	2.1	4.1	7.3	7.3	14	23	60
Sc	3.5	6.3	26	30	26	30	26	30	26	30
Se	1.9	3.5	0.1	0.5	10	20	19	50	848	1472
Sm	1.8	2.9	13	14	13	14	13	14	13	14
Sn	2.3		9		17		17		77	
Sr	22.5	426	1646	2055	1646	2055	1646	2055	1646	2055
Te	<0.4		<1.5		<2.9		<2.9		<13	
Th	3.4	6.5	25	31	25	31	25	31	25	31
Tl	<1		<4		<7		<7		<33	
U	1.3	2.1	6	9	9	13	10	19	26	70
V	26	46	134	221	192	265	249	397	580	
W	1.2	6.2	6	30	9	45	13	74	27	
Zn	25	101	92	291	184	679	258	1164	1025	4172

pulverised fuel ash and from pulverised fuel ash to the inhalable fraction and fly dust. This is one of the main reasons for applying different names to the various types of ash. As previously reported, there is a shift to smaller sizes when going from FBA to PFA_{total} to $PFA_{inhalable\ or\ PM_{50}}$ and finally to fly dust. From Table 1.6, it can be seen that the concentration of Class I elements remain the same, whereas the concentrations of Class II and III elements increase.

Based on the KEMA coal database and with the help of the KEMA TRACE MODEL®, the composition of furnace bottom ash and pulverised fuel ash from coal of various origins was calculated, assuming the layout of Dutch power stations. The results are given in Tables 1.7 and 1.8, together with their standard deviations.

The standard deviation refers to the standard deviation of the coal composition and the standard deviation of the RE factor. For the calculations, an average figure for the coal for a specific geographical region is used and the standard deviation applies to differences between the individual samples within that geographical region.

1.3.5 The Particle Size Distribution of PFA

The particle size distribution of PFA was historically determined by sieving, but now many different instrumental methods are available for particle size distribution measurements, of which laser diffraction methods are currently the most widely used. The advantage of these is a wider operational range than for sieving (1–500 microns). Samples can be presented to the instrument as a dry powder or suspended in a liquid (alcohol or water). The averaged results for 17 representative PFA produced in the Netherlands are given in Table 1.9.[25–27] The results are obtained by the dry method and expressed as their geometric or projected diameter.

Particles < 4.5 μm in diameter account for 10 % of pulverised fuel ash, with half consisting of particles < 21.4 μm, and 90 % of particles < 90.4 μm. For scientific purposes and for applications in wet conditions, a determination under wet conditions is preferable. The results obtained using the wet methodology were about 55 % of the figures obtained in dry conditions,[27] the differences being ascribed to agglomeration. However, if the results are to be used to describe the behaviour in air – for example, in health and dispersion studies – the dry method is recommended, as it simulates what happens by dispersion in the air, such as coagulation. It is then

Table 1.7 The elemental composition of furnace bottom ash originating from coal of various origins (according to Meij and Te Winkel) (Reproduced by permission of KEMA).

Origin of coal	Australia		USA		Colombia		Poland		Indonesia		South Africa	
Number of samples	24		25		11		8		11		16	
Macro- and main elements (%)												
Al	14.7	5.4	14.2	6.0	11.3	3.7	14.7	5.0	15.6	10.1	15.9	2.4
Ca	1.6	1.7	1.6	2.0	2.3	2.0	3.1	2.0	2.0	1.4	5.1	1.5
Cl	0.002	0.002	0.006	0.006	0.002	0.002	0.007	0.007	0.001	0.002	0.000	0.000
Fe	3.7	2.1	6.7	6.0	5.5	2.1	6.1	1.7	5.5	2.8	2.7	0.7
K	0.8	0.4	2.1	1.3	1.5	0.7	2.2	0.9	1.0	0.5	0.5	0.1
Mg	0.4	0.2	0.7	0.3	1.0	0.4	1.6	1.1	0.9	0.5	1.1	0.3
Na	0.2	0.1	0.3	0.1	0.5	0.4	0.5	0.2	0.3	0.1	0.1	0.1
P	0.2	0.2	0.1	0.0	0.1	0.2	0.2	0.1	0.1	0.1	0.3	0.2
S	0.08	0.03	0.18	0.15	0.15	0.03	0.11	0.02	0.21	0.19	0.08	0.02
Si	27.2	9.1	23.9	11.3	25.9	7.7	20.4	10.1	23.1	11.6	20.8	4.7
Trace- and micro-elements (mg kg^{-1})												
As	1.2	3.1	5.0	12.1	2.2	5.3	2.0	4.7	2.6	6.3	1.0	2.5
B	73	46	118	82	176	66	77	35	483	194	112	60
Ba	996	1088	1078	415	1026	359	1528	393	488	142	1916	385
Be	6.3	3.2	11.2	5.8	4.4	2.5	9.4	2.3	5.1	3.3	8.3	3.4
Cd	0.10	0.23	0.08	0.14	0.20	0.33	0.12	0.18	0.04	0.07	0.05	0.09
Co	36	18	40	15	21	12	39	12	32	20	34	12
Cr	79	33	151	49	160	100	190	67	115	58	180	27
Cu	52	21	80	31	46	34	87	22	41	31	36	12
F	61	58	56	50	52	57	75	60	46	45	132	111

Ge	7.4	1.5	6.8	1.6	8.4	3.3	6.5	4.0	9.3	1.9	5.3	1.1
Hg	0.04	0.03	0.08	0.07	0.05	0.03	0.19	0.29	0.06	0.05	0.05	0.02
Mn	291	313	301	225	389	219	893	395	148	58	389	126
Mo	5.8	2.5	11.9	6.2	15.5	9.4	6.7	1.8	5.4	3.4	6.9	5.5
Ni	75	50	78	34	80	37	145	87	103	70	78	32
Pb	27	21	26	18	13	8	55	33	15	11	29	16
Sb	0.9	0.7	1.4	0.8	1.6	1.3	2.1	1.1	0.5	0.5	0.4	0.3
Se	0.09	0.11	0.32	0.34	0.49	0.52	0.08	0.08	0.10	0.14	0.06	0.07
Sn	8.6	2.6	4.2	1.2	5.7	1.7	4.1	1.4	6.2	1.9	6.7	2.8
Sr	1492	785	997	565	1252	1447	1189	191	1082	550	2906	1131
Te	4.2	1.3	2.1	0.6	12.6	3.8	1.9	0.6	3.1	0.9	1.8	0.5
Th	34.1	20.5	28.2	12.5	16.1	15.2	26.9	8.1	17.5	8.7	52.5	7.9
Tl	3.5	1.2	15.0	5.0	5.1	1.7			8.9	3.0	4.2	1.4
U	6.1	3.0	7.5	3.5	5.0	3.4	8.3	2.8	2.3	1.2	9.7	2.9
V	138	52	185	74	175	71	206	55	294	319	128	42
W	6.1	4.1	4.4	1.7	2.9	2.3	21	38	5.0	7.0	6.0	2.0
Zn	167	239	75	38	92	46	134	58	99	48	50	18

Table 1.8 The elemental composition of pulverised fuel ash originating from coal from various origins (according to Meij and Te Winkel) (Reproduced by permission of KEMA).

Origin of coal	Australia		USA		Colombia		Poland		Indonesia		South Africa	
Number of samples	24		25		11		8		11		16	
Macro- and main elements (%)												
Al	14.7	5.4	14.2	6.0	11.3	3.7	14.7	5.0	15.6	10.1	15.9	2.4
Ca	1.6	1.7	1.6	2.0	2.3	2.0	3.1	2.0	2.0	1.4	5.1	1.5
Cl	0.003	0.002	0.008	0.005	0.002	0.002	0.010	0.005	0.001	0.002	0.001	0.001
Fe	3.7	2.1	6.7	6.0	5.5	2.1	6.1	1.7	5.5	2.8	2.7	0.7
K	0.8	0.4	2.1	1.3	1.5	0.7	2.2	0.9	1.0	0.5	0.5	0.1
Mg	0.4	0.2	0.7	0.3	1.0	0.4	1.6	1.1	0.9	0.5	1.1	0.3
Na	0.2	0.1	0.3	0.1	0.6	0.5	0.6	0.2	0.4	0.2	0.2	0.1
P	0.4	0.3	0.1	0.1	0.3	0.4	0.4	0.2	0.2	0.1	0.7	0.3
S	0.08		0.18		0.15		0.11		0.21		0.08	
Si	27.2	9.1	23.9	11.3	25.9	7.7	20.4	10.1	23.1	11.6	20.8	4.7
Trace- and micro-elements (mg kg^{-1})												
As	17.6	17.5	72.1	38.0	31.2	19.9	28.0	13.9	37.5	18.7	14.7	5.9
B	100	71	163	124	242	119	105	60	665	343	154	97
Ba	1245	1386	1348	595	1282	528	1910	642	610	222	2395	707
Be	9.0	4.5	16.0	8.2	6.2	3.5	13.4	3.2	7.2	4.7	11.8	4.8
Cd	1.0	1.7	0.8	0.7	2.0	1.6	1.2	0.7	0.4	0.3	0.5	0.4
Co	52	26	57	22	30	17	56	17	46	28	48	17
Cr	88	37	168	55	178	111	211	74	128	65	200	30
Cu	104	43	159	62	91	69	175	43	83	62	73	23
F	122	76	113	61	104	87	149	58	91	62	264	119
Ge	14.9	3.7	13.5	3.8	16.8	7.1	12.9	8.3	18.5	4.6	10.5	2.6
Hg	0.31	0.26	0.64	0.61	0.40	0.28	1.46	2.36	0.44	0.43	0.36	0.20

Mn	291	316	301	230	389	226	893	417	148	62	389	139
Mo	11.7	5.4	23.9	12.9	31.1	19.4	13.4	4.1	10.8	7.0	13.7	11.2
Ni	108	68	112	44	115	48	206	119	147	96	111	40
Pb	68	43	66	36	32	16	139	58	38	21	72	24
Sb	4.6	3.0	6.8	2.7	7.8	5.4	10.7	2.3	2.3	2.1	2.1	1.1
Se	6.53	5.02	22.13	12.00	34.52	16.45	5.28	2.25	6.98	6.98	4.39	2.89
Sn	17.1	1.7	8.3	0.8	11.3	1.1	8.3	1.4	12.4	1.2	13.3	4.2
Te	8.4	0.8	4.2	0.4	25.1	2.5	3.7	0.4	6.2	0.6	3.5	0.4
Th	34.1	20.5	28.2	12.5	16.1	15.2	26.9	8.1	17.5	8.7	52.5	7.9
Tl	5.8	0.6	25.0	2.5	8.6	0.9			14.8	1.5	7.0	0.7
U	10.2	4.6	12.4	5.3	8.3	5.5	13.8	3.7	3.9	1.8	16.2	3.6
V	197	64	264	93	251	90	294	55	421	449	183	50
W	8.7	6.0	6.3	2.5	4.2	3.3	31	55	7.2	10.1	8.6	3.1
Zn	334	482	149	80	185	98	267	126	198	102	99	40

Literature:

1 Meij, R., *Status Report on Health Issues Associated with Pulverised Fuel Ash and Fly Dust. Introduction and Summary.* Revision (version 2.1). KEMA Report no. 50131022-KPS/MEC 01-6032 (2003).

2 Smith, IEA report (1992).

3 Meij, R., Van der Sluys, J.L.G., Siepman, F.G.C. and Van der Sloot, H.A., *The emission of fly ash and trace species from pulverised coal fired utility boilers.* In Proc. VIth World Congress on Air Quality, Paris, France May 1983, part IV, pp. 317–324 (1983).

4 Meij, R., *Trace elements behaviour in coal-fired power plants.* Fuel Processing Technology, 39: 199–217 (1994).

5 Meij, R., *Prediction of environmental quality of by-products of coal-fired power plants; elemental composition and leaching.* In Proc. Int. Conf. WASCON '97, Houthem St. Gerlach, the Netherlands, 4–7 June, 1997, published in '*Studies in Environmental Science 71, Waste Materials in Construction; Putting Theory into Practice*'. J.J.J.M. Goumans, G.J. Senden and H.A. van der Sloot (eds), Elsevier, Amsterdam (1997).

Table 1.9 The particle size distribution of PFA produced in the Netherlands, determined by laser diffractometry in the dry form (Reproduced by permission of KEMA).

Geometric diameter (μm)	Particle size distribution of PFA		
	D_{10}	D_{50}	D_{90}
Mean	4.5	21.4	90.4
SD	0.2	3.3	10.8
V (%)	5	16	12
Minimal	4.0	16.0	72.3
Maximal	4.8	29.7	109
n	17	17	17
Range according to 95 % confidence interval:			
Mean − 2 SD	4.1	14.7	68.8
Mean + 2 SD	5.0	28.0	112

common to express the particle size as its aerodynamic diameter; that is, the diameter of a spherical particle with a density of 1000 kg m^{-3} that has a drop velocity equal to that of the particle in question. The aerodynamic diameter can be calculated from the Stokes diameter according to the following formula:

$$D_{ae} = D_s \sqrt{\frac{\rho_p}{\rho_0}}, \qquad (1.2)$$

where D_{ae} is the aerodynamic diameter, D_s is the Stokes diameter (identical to the geometrical diameter for PFA), ρ_p is the density of the particle in question (PFA), with a value of 2.1×10^3 kg m^{-3}, and, ρ_0 is the density of the particle to which it will be normalised, in this case 1×10^3 kg m^{-3}. PFA particles are generally spherical, so in this case the Stokes diameter is equal to the geometric diameter.

The results of the calculations are given in Table 1.10, together with the proportion of inhalable (PM$_{50}$), respirable (PM$_4$), fine dust (PM$_{10}$) and ultra-respirable (PM$_{2.5}$) fractions of the PFA calculated according to the definitions of the CEN/ISO conventions.

In Figure 1.5, the average size distribution of PFA is given, together with the CEN/ISO conventions of the different health related fractions of the total suspended matter (TSPM).[26] These curves have been used for the calculation of the results in Table 1.10.

Table 1.10 The particle size distribution of PFA, as produced in the Netherlands, expressed in aerodynamic diameter and their relative contributions of various particle size conventions (Reproduced by permission of KEMA).

Aerodynamic diameter	Particle size distribution			Part of the total PFA (%)			
	D_{10} (µm)	D_{50} (µm)	D_{90} (µm)	$PM_{2.5}$	PM_4	PM_{10}	PM_{50}
Mean	6.5	31.0	131.0	1.3	4.6	19.9	54.7
SD	0.3	4.8	15.6	0.2	0.5	1.8	2.8
V (%)	5	16	12	12	11	9	5
Minimal	5.8	23.2	104.8	1.2	4.1	17.1	49.1
Maximal	7.0	43.0	157.5	1.7	5.9	24.0	59.9
n	17	17	17	17	17	17	17

Range according to 95 % confidence interval:

Mean − 2 SD	5.9	21.3	99.7	1.0	3.6	16.2	49.2
Mean + 2 SD	7.2	40.6	162.3	1.7	5.6	23.5	60.3

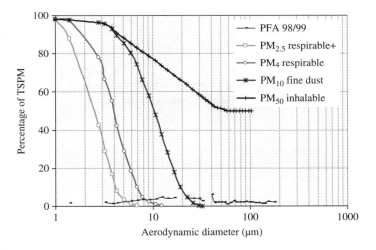

Figure 1.5 The particle size distribution of coal fly ash and various health-related CEN/ISO conventions (Reproduced by permission of KEMA).

1.4 Other Forms of Combustion

1.4.1 Stoker Firing

Stokers are forms of fuel bed firing, where the coal is pushed, dropped or thrown onto a grate by a mechanical device. Air is blown through the fuel bed, and part of the fuel is volatilised as a combustible gas and burns above the bed. Additional combustion air is supplied above the fuel bed and the remaining charred material burns on the grate. Ash remaining after combustion is removed from the furnace continuously by movement of the grate.

Stokers can be divided into two general classes depending on the direction from which raw coal reaches the fuel bed: overfeed stokers, in which the coal comes from above, and underfeed stokers, in which it comes from below. The overfeed group includes spreader and mass-burning stokers. The latter are more commonly referred to by specific design features; that is, chain-grate, travelling-grate or water-cooled vibrating-grate. Types of underfeed stokers are single-retort and multiple-retort. A schematic diagram of a spreader stoker is shown in Figure 1.6.

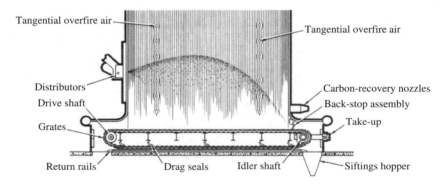

Figure 1.6 A schematic of a spreader stoker with a continuous ash discharge grate. Reproduced with the permission of ALSTOM Power Inc., Windsor, CT, from *Combustion Fossil Power Systems* (copyright © 1981).

Whilst stoker firing was originally the main method used to produce electricity from coal, this is no longer the case. Larger PF-fired units are now the commonest forms of combustion plant for power generation. The use of stoker-fired plant has, however, continued on a smaller scale mainly in industrial boilers. In recent years, the opportunity has been taken to develop stoker firing for use with biomass, either alone or as a blend with coal (Section 1.5).

Ash from Stokers

Although there are a number of different types of stoker, their ashes may be considered similar in nature. Grate or bottom ash, which is carried off the end of the stoker, is usually a lightly sintered mass and falls into a water-filled trough, where it is quenched, crushed and removed for disposal or reuse in another application. Most of the ash from a stoker is produced in this way, and in large stokers accounts for 75–85 % of all the ash produced. This material can be considered similar to FBA from PF firing, although the unburned carbon content will be higher. This material could probably be used in some of the uses of PF-derived FBA, but it is more variable in quality and is available in smaller amounts.

Introducing primary combustion air beneath the grate and secondary air over the grate results in a fine material consisting of ash and partly burnt coal particles being carried over from the bed. Most stokers have mechanical dust collectors that re-inject the fly ash into the grate to improve overall combustion efficiency. The fly ash is collected in cyclones, bag filters or ESPs. The variable quality and relatively small amounts of stoker fly ash mean that although similar to PF-derived ash, it is less attractive as a material for reuse.

1.4.2   Wet-bottom Boilers

The term 'wet-bottom' in the context of a boiler is used to describe the molten condition of the ash as it is drawn from the bottom of the furnace. One of the advantages of this type of boiler is that it can use low ash fusion coals that might otherwise cause problems in PF-firing systems. It also results in low ash carry over, although the high gas temperatures of this system produce very high levels of NO_x. There are two types of wet-bottom boilers: the slag-tap boiler and the cyclone boiler (Figure 1.7).

The slag-tap boiler burns pulverised coal and the cyclone boiler crushed coal. Pulverised coal is approximately 70 % < 75 µm, whereas that used on cyclone burners is approximately 95 % < 4.7 mm (4 mesh).[28] In each boiler, the bottom ash is kept in a molten state and tapped off as a liquid. Both have a solid base with an orifice that can be opened to permit the molten ash that has collected at the base to flow into the ash hopper below, which contains quenching water. When the molten slag contacts the quenching water, it instantly fractures, solidifies and forms pellets. When pulverised coal is burned in a slag-tap furnace, as much as 50 % of the ash is retained in the furnace as boiler slag.

In a cyclone-fired system, 80–90 % of the ash leaves the bottom of the boiler as a molten slag, thus reducing the load of fly ash passing through the heat transfer sections to the ESP or fabric filter. Cyclone-fired units operate at close to atmospheric pressure, simplifying

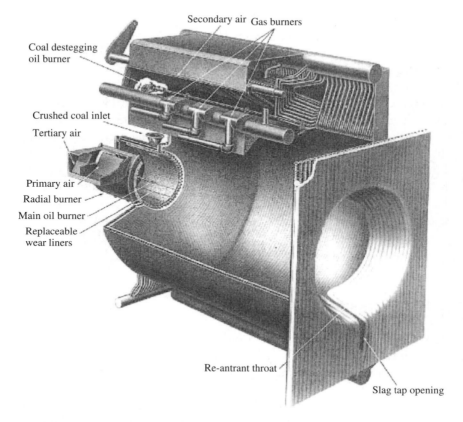

Figure 1.7   The cyclone furnace (Reproduced by permission of Babcock & Wilcox).

the passage of coal and air through the plant. Combustion temperatures in the external cyclone furnaces range from 1650 to > 2000 °C,
which helps carbon burnout but produces very high NO$_x$ levels. Molten
ash flows by gravity from the base of the cyclone furnaces, and is
removed from the system at the bottom of the boiler. It drops into
a quench tank, thus losing a substantial amount of heat. Precautions
against gas build-up and explosions are essential in and around the slag
quench tank.

Cyclone-fired boilers are suitable for coals with:

• Volatile matter content > 15 % (dry basis).
• Ash contents between 6 and 25 % for bituminous, or 4 and 25 % for sub-
bituminous, coals. The ash must have particular slag viscosity characteristics, and the maximum temperature at which the slag has a viscosity

of 250 cp is 1340 °C for bituminous and 1260 °C for sub-bituminous coals. Ash slag behaviour is critical to satisfactory operation.
- Moisture content < 20 % for bituminous and < 30 % for sub-bituminous coals.

Slag from Wet-bottom Boilers

Wet-bottom boiler slag, often termed 'black beauty' in the USA, is a coarse, hard, black, angular, glassy material. It is predominantly single-sized and, after quenching, exhibits a size range of 0.5–5.0 mm. Ordinarily, it has a smooth surface texture, but if gases are trapped in the slag as it is tapped from the furnace, the quenched slag will become somewhat vesicular or porous. Boiler slag from the burning of lignite or sub-bituminous coal tends to be more porous than that from higher rank bituminous coals. Boiler slag is essentially a coarse to medium sand, with 90–100 % passing a 4.75 mm sieve, 40–60 % a 2.0 mm sieve, ≤ 10 % passing a 0.42 mm sieve and ≤ 5 % passing a 0.075 mm sieve. The specific gravity of the dry bottom ash is a function of chemical composition, with higher carbon content resulting in lower specific gravity.

Boiler slag is composed principally of silica, alumina and iron, with smaller percentages of calcium, magnesium, sulfates and other compounds. The composition of the boiler slag particles is controlled primarily by the source of the coal and not by the type of furnace.

1.4.3 Fluidised Bed Combustion (FBC)

When a fluid is passed upwards through a bed of particles, a point is reached when the upward drag force exerted by the fluid on the particles is equal to the apparent weight of particles in the bed. At this point the particles are lifted by the fluid, the separation of the particles increases and the bed becomes fluidised. The status of any particle fluid system is limited by the minimum fluidisation velocity and the entrainment velocity. When a system is operated between these two velocities, it is known as a fluidised bed; and if a combustion reaction is involved, the process is known as atmospheric fluidised bed combustion (AFBC) (Figure 1.8).

In fluidised bed combustion systems, the particles are usually a mixture of sand, fuel and fuel ash. Large ash granules are removed

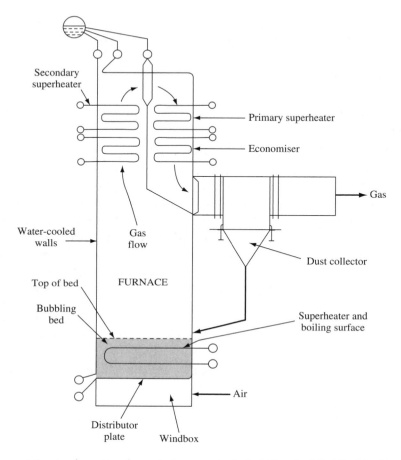

Figure 1.8 A schematic of a typical atmospheric bubbling bed fluidised bed boiler (Reproduced by permission of Babcock & Wilcox).

from the base of the fluidised bed and fly ash by entrainment with the flue gas once the ash particles have been sufficiently reduced in size by the eroding action of the fluidised sand. The combustion heat is recovered via in-bed heat exchangers and modified boiler equipment. Fly ash can be recycled using the fuel feeding system. This type of system can also be operated at pressures above atmospheric, for example 12 atmospheres – where it is known as PFBC, or pressurised fluidised bed combustion.[29]

When a fluidised bed combustion system is designed to operate above the entrainment velocity, it is known as a circulating fluidised bed combustor, or CFBC (Figure 1.9). A high-duty cyclone is used to

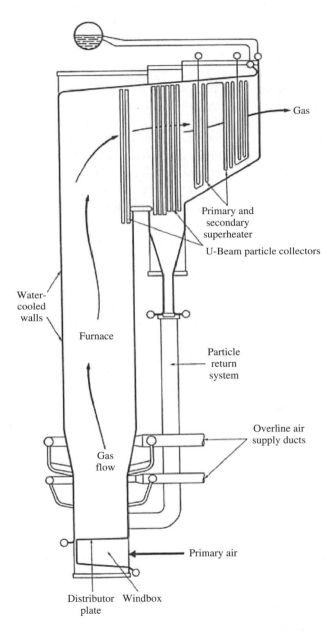

Figure 1.9 A schematic of a typical circulating fluidised bed boiler (Reproduced by permission of Babcock & Wilcox).

separate flue gas from the entrained particles, which are continuously recycled to the bottom of the fluidised bed. Individual particles may be recycled anything from 10 to 50 times, depending on their size and on how quickly the char burns away. Combustion conditions are relatively uniform throughout the combustor, although the bed is somewhat denser near the bottom of the combustion chamber. There is a great deal of mixing, and residence time during one pass is very short. The recycled particles often pass through heat exchangers before passing into the bed and the heat recovered is used to augment steam raising. For this reason, CFBCs can usually dispense with in-bed heat exchangers, which can suffer erosion and eventual failure.

CFBCs are usually designed for a specific quality of coal, principally low-grade, as high-ash coals are difficult to pulverise and may have variable combustion characteristics. It is also suitable for co-firing coal with low-grade fuels, including some waste materials. The direct injection of limestone into the bed offers the possibility of economic sulfur dioxide removal, as the coal sulfur is chemically combined with the sorbent and a solid by-product consisting of calcium sulfate dihydrate (gypsum), coal ash, and unreacted calcium oxide is produced. However, the presence of partly reacted sorbent results in high disposal costs for this type of combustion.

The boiler design must take into account both the ash quantity and properties. While combustion temperatures are low enough to allow much of the mineral matter to retain its original properties, particle surface temperatures can be as much as 200 °C above the nominal bed temperature. So, if any softening takes place on the surface of either the mineral matter or the sorbent, then there is a risk of agglomeration or of fouling. Fly ash leaves the cyclone with the flue gases and is normally separated using an ESP.

Ash from FBC

Most systems produce two types of residue, bed ash and fly ash. According to Robl, in a report by Sloss,[30] bottom ash from FBC systems is coarse and difficult to handle in comparison with the finer fly ash collected in the cyclone or bag-house. The mode of operation of FBC boilers produces larger amounts of combustion residues than similar-sized PF boilers without FGD systems. This is due to the amounts of limestone necessary to remove 90 % of the sulfur oxides from FBC boiler flue gases, calcium : sulfur molar ratios of between 2 : 1 and 5 : 1

usually being required. Table 1.11 shows ash analysis data from the coal and also typical fly ash, circulating ash and bed ash from a Lurgi CFBC plant.[31]

Table 1.11 Comparative ash and residue compositions (in %) from a CFBC boiler.[31]

	Laboratory ash	Fly ash	Circulating ash	Bottom ash
SiO_2	47.1	34.3	22.5	16.2
Al_2O_3	28.8	20.2	10.2	6.8
Fe_2O_3	10.7	8.2	4.3	3.7
CaO (total)	2.7	16.5	33.5	39.7
CaO (free)	0.0	3.0	10.0	12.7
MgO	1.7	1.1	1.4	1.2
Na_2O	1.1	No data	No data	No data
K_2O	3.8	2.3	1.1	0.8
SO_3	1.3	12.2	25.2	29.6

In some, generally smaller, FBC boilers, there may be a need for the provision of bed material that may be silica sand or something similar. In cases where this is used, it has to be separated from the fuel ashes and sorbents and returned to the bed.

The relatively low temperatures in the different FBC systems mean that ash melting is not common and the ash components are therefore mainly crystalline. The key fuel properties that can influence the residue characteristics from FBC boilers are the sulfur and ash contents of coal, ash composition, and the size and friability of the coal minerals. The chemistry of the sorbent is also important, as this also ends up in the residue.[32]

Despite the heterogeneous nature of this ash, attempts have been made to convert FBC residues into a synthetic aggregate for use in highway construction.[33]

1.5 Co-combustion Ashes

It was recently reported that combustion of biomass or wastes as the *sole* source of fuel had met with variable success.[34] The limitations were primarily due to a failure to appreciate the highly inconsistent properties of biomass and wastes. The high ash and moisture contents, together with low heat content present in certain fuels, have caused ignition and combustion problems. In addition, low-melting ash components have also resulted in unacceptable levels of slagging and fouling.[34]

Many of the problems associated with mono-fuel firing of biomass and wastes can be avoided or ameliorated by co-firing with coal. Indeed, there are some synergistic effects relating to emission reduction of NO_x by ammonia in certain wastes and for trapping of sulfur oxides by calcium in biomass ashes. Other nontechnical issues relating to the replacement of coal by renewables such as biomass are being driven by concerns for the environment (carbon dioxide, long and short cycles) and legislative pressures (such as carbon taxes). The downside of the use of co-firing is the effect on ash quality. Markets for PFA have been developed and the incorporation of certain biomass or waste ash components may render the PFA unsaleable. An example of such a problem arose in a boiler co-firing coal and straw. The presence of high levels of chlorine and unburned carbon in the co-firing ash meant that a limit of 10 % straw firing had to be imposed to ensure acceptable PFA quality.[35]

1.5.1   The Composition of the Ash from Co-combustion

Since 1993, KEMA has studied the effect of direct co-combustion in 39 test series in the Netherlands (Table 1.12). In the first three years there were 23 tests on full-sized plant in each of the seven Dutch coal-fired power stations. In these tests, secondary fuels were co-combusted in proportions of up to 10 % by dry mass. In 2001, a test series was started in which secondary fuels were co-combusted in proportions containing more than 10 % by dry mass. Initially, three tests were carried out in the KEMA 1 MW_{th} test boiler, followed by six tests at three different power stations. The ashes were used as a raw material to make concrete test cubes. These cubes were subjected to leaching and other tests to determine their technical parameters. These results are discussed in a later chapter.

Analytical data, as well as predicted values, were obtained for some of the ashes and dusts that served to validate the KEMA model and to enable RE factors to be established. These were similar to those obtained for 100 % coal firing. It must be mentioned that this observation is valid for co-combustion of up to 10 % by mass but, because there is little data at more than 10 %, further investigation is still needed. This means that the KEMA TRACE MODEL® can also be used for predicting the ash composition for co-combustion (Table 1.13). This is

Table 1.12 A survey of all test series carried out by KEMA involving co-combustion in the Netherlands[a] (Reproduced by permission of KEMA).

Test series number[a]	Date	Plant and unit	Plant capacity (MW$_e$)	Origin of coal	Secondary fuel	Secondary fuel amount (mass %, dry)
I*	1993	KTB[b]	1	Australia	Demolition wood y1	4.5, 9.2, 12.9
II*	1993	KTB	1	Australia	Demolition wood y2	9.7
III*	1994	KTB	1	Colombia	Demolition wood x1	4.6, 8.6, 13.0
IV*	1994	KTB	1	Colombia	Demolition wood x2	8.1
V*	1994	KTB	1	Colombia	Demolition wood x3	9.2
VI*	1994	KTB	1	Colombia	Sewage sludge	4.8, 7.6, 10.3
VII*	1995	KTB	1	USA	Petcokes	4, 8, 12, 12.2
VIII*	1995	MV-2	518	Blend	Petcokes	5–9
IX*	1995	HW-8	600	Blend	Sewage sludge	3, 6
X*	1995–12	AC-8	600	Blend	Paper sludge	5, 8
XI*	1995	AC-9	600	Blend	Petcokes A	5, 10
XII*	1995	AC-9	600	Blend	Petcokes B	5, 10
XIII	1996	BS-12	403	Colombia	Hydrocarbon gas (P)	3[c]
XIV*	1996	MV-2	518	Blend	Petcokes A	5, 10
XV*	1996	MV-2	518	Blend	Petcokes B	5, 10
XVI	1997	MV-2	518	Blend	Biomass pellets	5
XVII*	1997–3	CG-13	600	Blend	Demolition wood	5, 10
XVIII	1998–3	AC-9	600	Blend	Paper sludge	5
XIX	1998–9	MV-2	518	Blend	Biomass/citrus pellets	8.5
XX*	1998–10	BS-12	403	Poland	Sewage sludge	5, 9
XXI	1998–12	BS-12	403	Blend	Paper sludge	5
XXII*	1998/9/11	KTB	1	Venezuela	Municipal waste (plastic fraction)	5, 10

Table 1.12 (Continued)

Test series number[a]	Date	Plant and unit	Plant capacity (MW$_e$)	Origin of coal	Secondary fuel	Secondary fuel amount (mass %, dry)
XXIII	1999–9	MV-1	518	Blend	Coffee grounds	3
XXIV	1999/11/11	BS-12	403	Blend	Cacao shells	9
XXV	1999/11/11	MV-1	518	Blend	Poultry dung	3
XXVI*	2000–2	CG-13	600	Blend	Demolition wood	6
XXVII*	2001–1	KTB	1	Venezuela	Poultry dung	24, 34, 39
XXVIII*	2001–1	KTB	1	Venezuela	Waste wood	16, 30, 42
XXIX*	2001–1	KTB	1	Venezuela	RDF	14, 28, 34
XXX*	2001–4	MV-1	518	Blend	Meat and bone meal	2.7
XXXI*	2001–6	MV-1	518	Blend	Meat and bone meal	2,2
XXXII*	2001–8	MV-1	518	Blend	Meat and bone meal	5,1
XXXIII*	2001–11	AC-9	600	Blend	Fresh wood	7
XXXIV*	2001–11	MV-1	518	Blend	Biomass blend	16
XXXV*	2002–1	MV-2	518	Blend	Biomass blend	15
XXXVI*	2002–4	BS-12	403	Blend	Wood + paper sludge	±10.9
XXXVII*	2002–11	BS-12	403	Blend	ONF + biomass	11.5
XXXVIII	2002	AC-9	600	Blend	Wood pellets	33
XXXIX*	2003–5	MV-2	518	Blend	M&B meal + biomass	20

[a] In all cases, the ash is tested for application as pozzolanic filler in concrete. Where an asterisk (*) is shown, then mass balance studies for (trace) elements have also been performed.

[b] KTB = KEMA Test Boiler, capacity in MW$_{th}$.

[c] Co-combusted in proportions of up to x % by energy.

Table 1.13 The elemental composition of pulverised fuel ash originating from 10 % co-combustion (calculations with the KEMA TRACE MODEL® according to Meij and Te Winkel) (Reproduced by permission of KEMA).

Fuel	Coal	Paper sludge	MS sludge	Waste wood	Poultry dung	RDF pellets	Petcokes	Biomass mixture
Macro- and main elements (%)								
Al	13.5	12.7	11.6	13.0	11.1	12.2	13.4	13.0
Ca	2.3	10.2	6.8	2.5	6.7	4.2	2.3	5.6
Cl	0.003	0.003	0.003	0.004	0.009	0.006	0.003	0.003
Fe	4.6	3.6	6.7	4.5	3.8	4.2	4.6	4.4
K	1.0	0.9	1.0	1.0	2.9	1.1	1.0	1.0
Mg	0.8	0.8	1.1	0.8	1.1	0.8	0.8	0.9
Na	0.38	0.3	0.6	0.4	0.5	0.6	0.4	0.4
P	0.35	0.3	2.1	0.3	1.9	0.3	0.3	0.8
S	0.14	0.11	0.12	0.13	0.12	0.12	0.22	0.11
Si	22.0	19.3	20.9	21.5	18.4	22.0	22.0	20.2
Ti	0.9	0.7	0.7	0.9	0.7	0.8	0.9	0.8
Trace- and micro-elements ($mg\,kg^{-1}$)								
As	23	19	22	30	19	24	24	19
B	190	143	142	182	155	160	189	145
Ba	1283	1178	1280	1749	1044	1356	1295	1133
Be	8.4	6.3	6.3	8.0	6.8	7.1	8.3	6.5
Br	3.3	2.4	2.4	3.1	2.6	2.7	3.2	3.5
Cd	0.9	1.0	3.0	1.9	0.7	2.6	1.0	3.2
Ce	122	91	91	116	99	103	121	93
Co	39	32	34	42	31	39	40	42
Cr	130	109	166	185	106	199	144	145
Cs	7.5	5.6	5.6	7.1	6.1	6.3	7.4	5.7

Table 1.13 (Continued)

Fuel	Coal	Paper sludge	MS sludge	Waste wood	Poultry dung	RDF pellets	Petcokes	Biomass mixture
Cu	92	129	315	193	120	1683	93	227
Eu	2.7	2.1	2.0	2.6	2.2	2.3	2.7	2.1
F	121	99	105	117	99	256	128	98
Ge	12	9	9	12	10	10	12	9
Hf	9.1	6.9	6.8	8.8	7.4	7.7	9.1	7.0
Hg	0.5	0.4	1.0	0.5	0.4	0.5	0.5	0.8
I	0.4	0.3	0.3	0.4	0.3	0.3	0.4	0.3
La	69	52	51	66	56	58	68	52
Mn	316	276	569	375	440	373	320	486
Mo	14	15	17	19	12	21	20	14
Ni	104	84	128	124	84	127	329	98
Pb	58	63	202	503	47	338	59	85
Rb	68	51	51	65	55	57	68	52
Sb	4.2	3.9	5.7	16.5	3.4	32.2	4.7	4.5
Sc	26	20	19	25	21	22	26	20
Se	9.5	7.3	8.1	11.7	7.7	8.5	9.7	8.2
Sm	13	10	10	12	11	11	13	10
Sn	14	14	34	18	11	39	14	23
Sr	1610	1211	1204	1543	1310	1359	1604	1311
Te	2.9	2.8	2.5	3.7	2.4	3.0	3.4	15.0
Th	25	19	19	24	20	21	25	19
Tl	7.3	5.5	5.5	7.0	6.0	6.2	7.3	5.6
U	9.3	7.0	7.0	9.0	7.6	7.9	9.3	7.1
V	199	153	163	198	162	178	1463	165
W	8.6	6.5	6.4	8.3	7.0	7.3	8.6	6.6
Zn	186	424	961	808	401	437	191	413

based on co-combustion with 10 % by mass of paper sludge, municipal sewage sludge, waste wood, poultry dung, RDF pellets, petroleum cokes and biomass mixture. For ease of comparison, the coal composition has been kept constant at the average coal composition in the Netherlands in 1999 (Table 1.6). It appears that the concentrations of most heavy metals and other trace elements found in PFA originating from co-combustion are similar to 100 % coal firing (Table 1.8), except for cadmium, copper, molybdenum, nickel, lead, antimony, vanadium and zinc.

1.5.2   Ashes from Coal–Biomass Co-firing

Coal and Wood

Waste wood from primary wood processing and demolition presents both a problem and an opportunity. If disposed to landfill, it occupies large volumes and, on decaying, produces methane, carbon dioxide and other greenhouse gases. Nevertheless, as an energy source in a coal-fired power plant it reduces the consumption of fossil fuels and thereby reduces the greenhouse effect. In the Netherlands, this has been successfully demonstrated at the Gelderland power station, where pulverised wood fuel was used. This has the advantage of being a very dry and fine material that is uniform and easy to handle, with a high energy content, that can be burned much like oil or gas. On this basis, a design for a wood powder production plant and burning system with a capacity of 10 tonnes per hour for 6000 hours per year was produced. The aims of the project were:

- substitution of 45 000 tonnes of coal, averaging 4.5 % of the total yearly fuel input;
- reduction of the carbon dioxide emissions by 110 000 tonnes per year;
- the use of 60 000 tonnes of waste and demolition wood per year; and
- partial solving of a landfill problem.

There were several conditions mandatory on the project. Firstly, there must be no risk to power plant availability; secondly, the use of fly ash must not be jeopardised and, finally, all emissions must remain within the limits set by Dutch environmental laws and regulations. The project was successful and has met its projected targets. Since there is no net addition of carbon dioxide when wood is burned and therefore no contribution to the greenhouse effect, this amount can be seen as a reduction. The other outcome was a reduction of 4000 tonnes of fly ash per year, since the ash content of wood is ten times lower than that of coal.[36]

1.5.3   Ashes from Coal–waste Co-firing

Coal and Manure

About 110 million US tons of cattle and poultry manure are produced annually on a dry basis, while the total amount of all animal waste produced in the USA is estimated to be 300 million wet tons per year. The waste, if stockpiled rather than utilised as fertiliser or in other applications, poses economic and environmental liabilities. The stored animal waste anaerobically releases methane, ammonia, hydrogen sulfide, amides, volatile organic acids, mercaptans, esters and other chemicals. Methane is a greenhouse gas and emissions from stored animal waste account for about 8 % of US greenhouse gas methane emissions. Various technologies including anaerobic digestion have been developed to dispose of such wastes. Anaerobic digestion, however, involves the use of scarce water resources and, furthermore, residual process solids incur large capital costs for their disposal. Thus an alternate least-cost technology such as direct firing in existing boiler burners is desirable. Typically, animal-based biomass fuels have high moisture and ash contents that result in low heating values and hence may cause combustion problems if fired separately. For this reason, they are mostly fired as blends with coal.

Blending manure with a coal eliminates the combustion problems from using manure alone, since the high flame temperatures produced by the coal burn the biomass completely. By blending such waste with coal and firing in utility-type boiler burners, the waste becomes an energy source and pollution from stockpiled or over-applied manure can be reduced. A 10 % blend of manure with coal in US electric utilities will require about 80 million tons of manure (as collected), which is approximately the annual production amount of feedlot waste. The blend technology offers the advantages of no major equipment modifications, lower capital costs for the conversion and tolerance of higher moisture content in manure. An economic analysis of using 10 % manure with the balance coal revealed a fuel cost saving of up to US$ 9.3 million.[37]

Coal and MSW (or RDF)

Although it is impractical to burn unprocessed MSW in combustion systems designed to burn coal, this is probably not so for co-firing refuse-derived fuel (RDF) in coal-fired utility boilers. Because utility boilers are generally very large, typically consuming several hundreds of tonnes of coal per hour, large quantities of MSW can be fired while maintaining a relatively small fraction of the boiler fuel input. Demonstrations of RDF firing have been conducted in cyclone-, stoker- and wall-fired boilers.[38,39]

The benefits of lower fuel costs and emissions by using MSW have to be weighed against the likely disadvantages of increased slagging, fouling, corrosion, erosion and trace element release. Thus the quantity of bottom ash formed would probably increase and fly ash resistivity may also be altered. Tests to confirm the continuing acceptability of PFA and FBA as combustion by-products would be required to ensure the viability of such a fuel replacement.

Coal and Sewage Sludge

Co-combustion of sewage sludge and coal is a method pursued in Germany and in other European countries, because it is considered a reasonable and cost-effective solution. A detailed survey of German practice and current status for co-combustion of coal with sewage sludge and other wastes has recently been reported.[40] By 2002, a total of 17 plants in Germany had experience of the co-combustion of coal and sewage sludge, of which ten are currently operating continuously. The destruction of 350 000 tonnes of sludge per year is possible by this disposal route. Most is disposed of in PF-fired plant at a blend ratio of between 4 % and 9 % by weight of sludge in coal. The power output of the plants range from 200 to >900 MW$_{th}$ and the disposal rate at the plants is from 10 000 to 140 000 tonnes per year. The sludge must first be dewatered before burning.[40] The benefit of using sewage sludge in this manner is that it does not suffer from any seasonal availability problems, unlike some of the biomass alternatives; nor does it show much variability of composition. The disadvantages of high moisture and ash content, lower calorific value and ash composition will limit the extent to which sludge can be safely blended with coal. The need to ensure that PFA from this form of co-combustion can continue to be sold is of major importance.

At present, one plant in the UK is developing the combustion of coal/sewage sludge blends. This unit is a wall-fired PF combustion plant in Scotland.[41]

Coal and Plastics

A number of detailed studies into the viability of burning waste plastics in co-combustion with coal have been prepared for, and issued by, the Association of Plastics Manufacturers in Europe. They include the co-combustion of spent life greenhouse polyethylene (PE) with coal in a 550 MW$_e$ tangentially fired boiler in Spain.[42] The PE was added at a rate of 3 t h^{-1} and the normal coal firing rate at full load of 205 t h^{-1} was adjusted accordingly to provide the same thermal input. Analysis of the slag and ESP dusts showed no compositional changes during the plastic firing episodes, thereby demonstrating the sound ecological recovery route offered by this disposal route. In the Netherlands, a plastics/paper fraction (PPF) was separated from household waste. This fraction was then further separated by vigorous stirring in aqueous suspension to produce mixed plastic fractions (MPFs). These fractions have a high calorific value, 32–35 MJ kg^{-1}, with 80–90 % volatile matter content and 6–9 % ash content. Testing of the MPF fractions, (5 and 10 % by weight) with coal on small, pilot and full-scale plant has taken place. No changes in the nature of the ashes produced during MPF blend firing have been detected.[43]

1.5.4   Ashes from Mixed-waste Co-firing

MSW and Plastics

Trials have been carried out co-combusting MSW with additions of end-of-life plastic waste. The study focused on plastic wastes arising from four different market sectors: packaging, automotive, electrical and electronic, and building and construction.[44] In the case of packaging waste, an improvement in burnout was found. The ash quality was improved by a reduction of the heavy metal content and the low elution values obtained should ensure the continuing marketability of the grate ash. In the case of Auto Shredder Residue (ASR), the concentration in ash of certain metals, such as zinc, copper, antimony, cobalt and nickel, increased. The effect of the increases on leachability is subject to different regulations in different countries and so the use of such materials needs to be investigated on a case-by-case basis. The trials using electrical, electronic and construction materials, such as insulating foams, did not deliver clear-cut benefits. The composition of these plastics is quite different to, for example, packaging waste, with chlorine, bromine and antimony being present in electrical waste plastics. Insulating foams contain fluorine and chlorine from blowing agents that were found to cause acid formation during combustion. It is evident that the use of residues from these co-combustion systems will require a more detailed investigation to prove their acceptability as by-products.

1.6 Biomass and Waste

In many cases, the distinction between biomass and wastes is obvious – for example, used solvents, petroleum coke and so on are clearly process wastes. Sawdust, paper waste, olive stones, sewage sludge, bagasse and so on are all of organic origin and could therefore be termed biomass; however, they are also the by-products of other processes. To clarify matters, it has been decided in this context to adopt definitions for each as given in a recent European report.[35] Thus biomass was defined as 'carbonaceous material that is derived from forestry and agricultural operations, including forestry residues, dry agricultural residues (e.g. straw), energy crops (e.g. miscanthus or short rotation coppice (SRC)), wet or semi-dry agricultural residues (e.g. animal slurries or chicken litter) and uncontaminated wood processing residues'. Biomass from forestry operations includes wood fuel chips, whole-wood chips and whole-tree chips. Branches from harvesting and trimming/cleaning operations complete with bark, mostly in chipped form, are commonly referred to as forestry residues. Table 1.14 shows the composition of some typical biomass materials.

Wastes are defined in the same report as 'industrial and municipal wastes, wood processing residues and demolition wood which has been contaminated with coatings or preservatives, MSW and its processed derivatives'. Wood processing wastes may be exemplified by sawmill dust and some specific wastes such as sludge from paper mills, paper recycling and de-inking processes.

Table 1.14 The composition of typical biomass and waste materials.

	Almond shells[53]	Olive stones[53]	Wheat straw[53]	Rice husks[53]	Bagasse[53]	Willow[53]	Poplar[53]	Demolition wood[53]	Mixed paper[53]	RDF[53]	Tyres (steel-free)[83]
Dry basis											
Ash (%)	3.29	1.72	7.02	20.26	2.44	1.71	2.7	13.12	8.33	26.13	8.83
Volatile matter (%)	76.0	82.0	75.27	63.52	85.61	82.22	84.81	74.56	84.25	73.4	68.00
Fixed carbon (%)	20.71	16.28	17.71	16.22	11.95	16.07	12.49	12.32	7.42	0.47	23.17
Carbon (%, as C)	49.3	52.8	44.92	38.83	48.64	49.9	50.18	46.3	47.99	39.7	72.89
Hydrogen (%, as H)	6.0	6.69	5.46	4.75	5.87	5.9	6.06	5.39	6.63	5.78	6.81
Nitrogen (%, as N)	0.76	0.45	0.44	0.52	0.16	0.61	0.6	0.57	0.14	0.8	0.36
Sulfur (%, as S)	0.04	0.05	0.16	0.05	0.04	0.07	0.02	0.12	0.07	0.35	1.24
Oxygen (%, as O)	40.63	38.25	41.77	35.47	42.82	41.8	40.43	34.45	36.84	27.24	9.67
Chlorine (%, as Cl)	< 0.01	0.04	0.23	0.12	0.03	< 0.01	0.01	0.05	nd	nd	nd
Gross calorific value (GCV) (MJ kg^{-1})	19.49	21.59	17.94	15.84	18.99	19.59	19.02	18.41	20.78	15.54	32.92
Ash composition %											
SiO_2	8.71	30.82	55.32	91.42	46.61	2.35	5.9	45.91	28.1	33.81	22.0
Al_2O_3	2.72	8.84	1.88	0.78	17.69	1.41	0.84	15.55	52.56	12.71	9.09
Fe_2O_3	2.3	6.58	0.73	0.14	14.14	0.73	1.4	12.02	0.81	5.47	1.45
CaO	10.5	14.66	6.14	3.21	4.47	41.2	49.92	13.51	7.49	23.44	10.64
MgO	3.19	4.24	1.06	< 0.01	3.33	2.47	18.4	2.55	2.36	5.64	1.35
TiO_2	0.09	0.34	0.08	0.02	2.63	0.05	0.3	2.09	4.29	1.66	2.57
Na_2O	1.6	27.8	1.71	0.21	0.79	0.94	0.13	1.13	0.53	1.19	1.10
K_2O	48.7	4.4	25.6	3.71	0.15	15	9.64	2.14	0.16	0.2	0.92
P_2O_5	4.46	2.46	1.26	0.43	2.72	7.4	1.34	0.94	0.2	0.67	1.03
SO_3	0.88	0.56	4.4	0.72	2.08	1.83	2.04	2.45	1.7	2.63	15.38
ZnO	nd	nd	nd	nd	nd	nd	nd	nd	nd	nd	34.5

nd = Not determined.

1.6.1   Types of Biomass

Of the many possible biomass materials, not all are either available in sufficient quantity or in a suitable form to be easily burnt and thereby produce ash. A short description of several of the more important biomass fuels follows.

Straw

Within Europe, straw burning has stimulated limited interest in Austria, the UK, the Netherlands and Sweden; however, since the late 1980s it has been burned extensively in small combined power and district heating plants in Denmark.[45,46] There are special reasons why straw utilisation in Denmark is much higher than in other countries. Firstly, in the other countries straw has to compete on commercial terms with other fossil fuels and more abundant biomasses. Secondly, it must be recognised that straw is not the simplest of fuels to burn and that, outside Denmark, there is limited experience in its combustion. Finally, Denmark has an efficient infrastructure for the collection of straw.[45] In addition, there are legislative requirements to use large amounts of biomass, and by the year 2000 it was planned that 1 200 000 tonnes of straw and 200 000 tonnes of wood chips would be burnet corresponding to 6 % of the annual energy consumption in power stations in Denmark.[47]

The technical problems associated with straw combustion are related to the fusibility and volatility of its constituents. The ash components usually melt at a rather low temperature and thereby produce cakes of sintered ash, which are difficult to remove. Of the three types of straw studied (barley, rape and wheat), barley ash had the lowest softening temperature, of 750–800 °C.[48] Studies of straw ashes have shown that the predominant elements are silicon, potassium and calcium (Table 1.15).[47] The adverse effects of harmful elements in straw can be alleviated by co-firing with coal, as demonstrated in Denmark and Germany.[49] In addition, recent work has shown that it is possible to extract the water-soluble elements, which cause sintering and deposition. The economics, scale-up and extract disposal issues of such a system have not yet been investigated.[50]

At present, the use of straw ash from mono-fuel firing does not allow it to be used in traditional coal fly ash products. However, co-combustion with coal may result in minimal changes to the nature of the combined ash, allowing it to continue to be saleable. It is likely that other applications of straw ash will be developed.

Table 1.15   Analysis of straw samples (Reproduced by permission of KEMA).

	Wheat straw	Wheat/barley (1:1)	Danish cereal straw	
	Dry basis	As received	Typical	Variation
Moisture	Nil	10.85	14.0	8–23
Ash (%)	7.02	4.18	4.5	2–7
Volatile matter (%)	75.27	81.10	78.0	75–81
Fixed carbon (%)	17.71	3.87	3.5	
		Dry, ash-free		
Carbon (%, as C)	44.92	50.74	47.5	47–48
Hydrogen (%, as H)	5.46	5.36	5.9	5.4–6.4
Nitrogen (%, as N)	0.44	0.58	0.7	0.3–1.5
Sulfur (%, as S)	0.16	0.15	0.15	0.1–0.2
Oxygen (%, as O)	41.77	43.17	45.75	
Chlorine (%, as Cl)	0.23	No data	0.4	0.1–1.1
Potassium (%, as K)	No data	No data	1.0	0.2–1.9
GCV (MJ kg^{-1})	17.94	15.66		
NCV (MJ kg^{-1})			14.9	12.3–16.9
Ash composition %[53]				
SiO_2	55.32			
Al_2O_3	1.88			
Fe_2O_3	0.73			
CaO	6.14			
MgO	1.06			
TiO_2	0.08			
Na_2O	1.71			
K_2O	25.6			
P_2O_5	1.26			
SO_3	4.4			

Poultry Litter

The increased use of battery farming methods has led to the production of higher levels of waste materials. About 1.4 million tonnes of poultry litter (a combination of bird droppings and the wood shavings, straw and shredded paper used to cover the floors of 'chicken houses'), is produced in the UK each year. In some areas the amount produced exceeds the quantities required for use as an agricultural fertiliser, and it is sent to landfill. The combustion of this waste stream has recently been found to be an environmentally sound and profitable activity, with the heat generated from the combustion process used to generate electricity, and the ash residue is a valuable and saleable by-product.

The ash from the burning of poultry litter is a premium fertiliser, free from the disadvantages of using litter directly. It is odour and pathogen free, contains no nitrates and is high in potash and phosphate. The litter itself contains calcium and ammonia, which in the combustion process limits the emission of sulfur dioxide and oxides of nitrogen. The biomass fuel also closes the carbon cycle, emitting negligible net carbon dioxide. Fibrowatt, an independent power developer in the UK, is currently operating Europe's largest biomass power station, a 38.5 MW power station in Thetford, fuelled by chicken litter. Two other smaller plants at Eye and Glanford, each around 12 MW, also burn chicken litter. Fibrowatt also has developments in Minnesota, Maryland and Mississippi, where poultry litter and forestry residues will be burned. A similar plant is also being operated in the Netherlands.[51]

Typically, the plant incorporates a single, conventional boiler design with a feed system and grate specifically designed to combust poultry litter and other biomass fuels. The boiler is equipped with a combustion chamber, a superheater, a generating bank and an economiser. At this point temperatures exceed 850 °C, thereby destroying any odour and bacteria. High-pressure steam produced in the boiler is used to produce electricity in a condensing extraction turbine generator. Bag-filters or ESPs remove dust from the flue gas. Grate ash, boiler bank ash and fly ash are collected in containers and hoppers. The ash by-product is blended and conditioned before being sold under the brand name Fibrophos.

Energy Power Resources Ltd have developed the first fluidised bed combustion system to burn poultry litter. At their Westfield Biomass Plant, a steam turbine feeds electricity to the Scottish grid via a long-term electricity purchase contract. The plant has a net electricity output of 10 MW and converts 115 000 tonnes per year of poultry litter into energy and ash for fertiliser, without producing any waste. Ash is collected from four points in the plant: the fluidised bed, the super-heater, the economiser and the bag-filter unit. The ash is pneumatically transmitted to an ash storage silo, from where it is discharged into bulk vehicles and sold as a high-grade fertiliser under a long-term contract.[52]

Bagasse

Bagasse is a waste fuel produced from the extraction of sugar from cane. The cane stalks, (3–4 m long) are processed by chopping into short lengths and shredding to expose the soft inner fibres in which the sugars are held. Extraction of the sugar solution using hot water and pressure results in the formation of the bagasse residue, typically containing 40 % cellulose fibre, 2.5 % sugar, 55 % moisture and 2.5 % ash. The ash in many bagasse samples is derived from the soil and other inorganic contaminants picked up during harvesting. Details of the analysis of a number of bagasse samples are shown in Table 1.16.[53-55]

So far, there appears to have been little development of bagasse ash into useful products. One exception is an Indian report that bagasse ash has good pozzolanic character. It is claimed that two thirds of the silica in bagasse ash is amorphous. A quick-setting mixture suitable for mortar and plastering is said to be produced from bagasse ash containing 10 % ordinary Portland cement and 4 % finely ground (90 % < 75 μm) gypsum.[56]

Olive Oil Waste

Mediterranean countries produce 95 % of the world production of olive oil, estimated to be 2.4 million tonnes per year, the largest producers being Spain (950 000 tonnes), Italy (450 000 tonnes) and Greece (430 000 tonnes). The environmental impact of its production is considerable because of a need for some 12 million tonnes of water and the production of 8 million tonnes of sludge.

Oil is extracted from olives using two systems. In the first, a traditional process, the olives are ground and the oil separated by pressure. Two types of waste are produced, of which the principal pollutant is olive-mill waste water (OMWW), together with a solid known as oil mill cake (OMC), or sometimes 'foot cake'. Foot cake contains from 3 to 6 % olive oil and from 20 to 25 % water. A more recent process separates the olive oil from the waste products either by centrifugation or decantation, and produces only OMC waste. However, most of the water is retained by the OMC, such that it may contain 50–60 % moisture (Table 1.17).[57,58]

A Spanish company has developed a commercial waste-to-energy system burning powdered olive pits and dry pulp to generate electricity. The use of this by-product removes a contaminating waste that is otherwise hard to eliminate. The Enemansa plant is the first in the world to operate at full capacity employing olive waste as fuel. The result of extracting olive oil in a continuous two-stage centrifugal system is a

Table 1.16 Analytical data on bagasse.

	Unknown[53]	Unknown[54]	Cuba[55]	Hawaii[55]	Java[55]	Mexico[55]	Peru[55]	Puerto Rico[55]
Moisture (%)	10.39	52.0	0	0	0	0	0	0
Ash (%)	2.19	1.7	2.9	1.5	1.68	1.32	1.75	1.35
Volatile matter (%)	76.72	40.2						
Fixed carbon (%)	10.70	6.1						
Carbon (%, as C)	43.59	23.4	43.15	46.2	46.03	47.3	49	44.21
Hydrogen (%, as H)	5.26	2.8	6	6.4	6.56	6.08	5.89	6.31
Nitrogen (%, as N)	0.14	0.1						
Sulfur (%, as S)	0.04	Trace						
Oxygen (%, by difference)	38.39	20.0	47.95	45.9	45.55	45.3	43.36	47.72
GCV (MJ kg^{-1})	17.02	9.30	18.57	18.98	20.19	20.33	19.49	19.51
Ash composition %								
SiO_2	46.61							
Al_2O_3	17.69							
Fe_2O_3	14.14							
CaO	4.47							
MgO	3.33							
TiO_2	2.63							
Na_2O	0.79							
K_2O	4.15							
P_2O_5	2.72							
SO_3	2.08							

Table 1.17   Analysis of olive oil wastes – oil mill cake.

	Spain[57]	Tunisia[58]
Moisture (%, as received)	66.4	9.1
Ash (%, dry)	5.8	
Volatile matter (%, dry)	74.5	
Fixed carbon (%, dry)	19.7	
Carbon (%, as C dry)	52.2	38.6–44.2
Hydrogen (%, as H dry)	6.7	4.8–5.9
Nitrogen (%, as N dry)	1.1	1.7
Sulfur (%, as S dry)	0.1	
Oxygen (%, by difference)	34.0	27.3–33.1
NCV ($MJ\,kg^{-1}$, as received)	6.6	16.5
Ash composition %		
SiO_2	21.2	
Al_2O_3	2.9	
Fe_2O_3	2.7	
CaO	13.8	
MgO	8.4	
Na_2O	0.5	
K_2O	42.5	
P_2O_5	5.5	

semi-solid substance, containing pits, skin and the pressed pulp, known in the Spanish oil industry as *alperujo*. Using an organic solvent, the remaining oil (*orujo* oil) is extracted, leaving a dry, solid by-product called *orujillo*, which has a net calorific value of between 15.5 and 18.0 $MJ\,kg^{-1}$. The fuel is fired in suspension in a water tube boiler and the dust is extracted from the flue gas by a cyclone and sleeve filter. The ash is then transported to a silo. It is not known whether the ash, which probably contains significant amounts of potassium, has any use as a recycled by-product.

The Enemansa plant has an annual treatment capacity of 1 000 000 tonnes of olive biomass. It is expected to be in operation 24 hours a day, with 90 % availability. It is expected to export more than 100 GW_{th} of electricity to the grid, sufficient to meet the electricity demands of 30 000 people, with notably lower CO_2 emissions than conventional systems.[59]

A similar boiler plant near Cordoba, Spain, rated at 25 MW_e, was constructed in 2000 to fire olive oil waste. In this case, the gas cleaning system was an ESP.[59]

1.6.2 The Composition of Ash from Biomass

Table 1.18 gives eight examples of the elemental composition of filter ashes originating from combustion or gasification of biomass. In all cases except one (a grate firing boiler), the process involves fluidised bed (FB), some of combustion type (FBC) whilst others are of gasification type (FBG). In most cases, reference is to circulating bed (CFB) and in one case to bubbling bed (BFB). In all cases, the installations operate under atmospheric conditions (AFBC). The fuel was wood (from pruning to waste wood), refuse-derived fuel (RDF), peat and municipal sewage sludge (MSS). The variation in concentrations is large, as a consequence of the concentration in the fuel of the nonvolatile elements and the presence of flue gas cleaning of the elements partly present in the gaseous phase. Thus, the high copper concentration in the ash originating from RDF comes from the fuel, whereas the low concentration of mercury in the ash form (the MSS) is the result of good mercury removal prior to the filter. It is to be expected that the MSS has the relative highest mercury concentrations of the fuels discussed. Ash originating form waste wood shows relatively the highest concentrations of heavy metals such as cobalt, chromium, molybdenum, nickel, vanadium, lead, tin and zinc. However, the burning of pruning wastes yields ash with higher concentrations of mercury and cadmium than ash originating from waste wood.

Wood represents about 80 % of the biomass consumed for raising steam, with the remainder being straw, manure, shells and so on. Although there is broad similarity in the composition of the combustible fraction of wood, there are differences in the ash make-up. This is believed to be due to differences in plant species, growth conditions, geographical and soil factors, and the method by which the wood is burned. The ash content of wood is generally very low, typically between 1 and 3 %. However, the ash content and composition are affected by the amount of soil removed when the fuel crop is harvested.

The dominant constituents of wood ash are the macronutrients calcium, potassium, magnesium and phosphorus. Calcium, the main constituent of the ash (50–60 %), is usually present as calcite or calcium oxide and results in high ash fusion temperatures. Quartz is present at 10 % or less, and clay minerals are only present in trace amounts. Potassium is also found in amounts that are generally higher than in most coals. In addition, wood ash contains a broad spectrum of trace elements such as manganese, zinc, copper and boron. Most of these elements are essential for plant growth but are toxic in high concentrations.[60] The composition of the ash from wood may be altered slightly during its

Table 1.18 The elemental composition of ash originating from biomass (according to Meij and Te Winkel).

Installation fuel	ABFBC wood, prunings	ACFBG wood	ACFBG waste wood	ACFBG waste wood	Grid wood	ACFBG RDF	ACFBC peat	ACFBC MSS
Macro- and main elements (%)								
Al	1.8	3.9	8.3	1.5	1.6	8.1	6.3	4.9
Ca	11.0	2.9	5.4	7.4	23.5	11.2	7.5	11.3
Cl	0.68	0.08	0.6	1.4	2.8	0.5	0.03	<0.08
Fe	1.3	2.0	3.7	1.1	1.0	3.4	9.0	13.1
K	4.3	1.6	1.1	0.9	8.6	0.8	1.6	1.8
Mg	1.16	0.70	1.1	0.8	3.9	1.5	1.1	1.2
Na	0.4	1.0	0.7	0.7	1.7	1.6	1.3	0.6
P	0.9	0.044		0.1	0.5	0.4	0.7	9.5
S		0.04		0.60	2.5	0.24	0.22	0.53
Si	27.3	19.7		8.3	2.6	22.8	16.0	12.0
Ti	0.12	0.2	1.3	2.1	0.35	0.6	0.29	0.74
Trace- and micro elements (mg kg^{-1})								
As	12	1.9	88	71	42	6.8	66	44
B	177	30		168	395	174	25	85
Ba	438	492		5 032	1 878	1 103	974	981
Be						1.3	2.2	1.0
Br	43	10	34	65		30	87	5.0
Cd	16	<0.17	8	7	10	0.13	1.16	4.5
Ce	6	7.8	43	7		14	39.8	11.8
Co	7	69	35	24	9.0	19	18.5	17.8
Cr	81	140	389	326	311	622	60	135
Cs	2.0	2.6	5.0	1.5		1.1	1.67	2.7
Cu	113	1334	478	419	353	15 393	61	1065
Eu	0.4	0.5	2.0	0.3		0.5	2.3	0.6
F	<222					<50	210	<50

Table 1.18 (Continued)

Installation fuel	ABFBC wood, prunings	ACFBG wood	ACFBG waste wood	ACFBG waste wood	Grid wood	ACFBG RDF	ACFBC peat	ACFBC MSS
Ge	42	56		28	16	4.1	4.1	2.9
Hf	5.3	10	7.9	3.3				
Hg	0.4	0.03		0.02	1.1	0.002	0.10	<0.03
I	<0.6	<2	<5	<4		<3	11.4	<5
La	11	14	66	11	14	22	61.4	20
Mn	2601	579	725	964	5751	603	1009	3472
Mo	12	2.9	21	17	15	7.5	8.5	19
Ni	35	47	245	39	30	102	36	81
Pb	204	303		5044	3505	186	65	407
Rb	104	68	61	37		34	67	41
Sb	3.9	14	60	110	44	218	1.77	14
Sc	2.2	6.1	21	1.8		4.9	12.4	3.6
Se	4.8	<1.7	7.4	0.32	1.6	<3	7.2	<3
Sm	1.8	2.5	11	1.5		3.3	11.6	2.8
Sn	5.0	5.8		69	46		1.4	
Sr	394	136	1433	331	1918	503	726	1217
Te	<0.9	<2	5.2	<6		<3	<0.7	<4
Th	3.1	5.0	25	2.4		6.8	9.5	4.0
U	1.0	1.6	7	0.9		1.6	3.8	5.4
V	29	39	111	26	20	36	98	38
W	23	480	4	78	27	4.2	<0.1	7.8
Zn	1067	773	3087	4840	1807	2022	83	2690

Table 1.19 Wood bark ash compositions[20] (in %).

	Balsam	Black spruce	White spruce	Red spruce	Jack pine	Poplar	Birch, white	Birch, yellow	Maple, hard	Maple, soft	Elm	Beech	Tamarack	Hemlock
SiO_2	24.6	6.4	2.0	7.6	16.0	1.5	3.0	4.1	39.5	6.1	3.6	12.4	7.3	10.0
Al_2O_3	1.8	1.1	0.6	0.0	6.3	0.5	0.6	0.3	3.8	3.1	0.0	0.0	8.4	2.1
Fe_2O_3	2.5	1.1	0.7	3.1	5.0	0.6	2.9	0.8	1.7	0.8	0.3	1.1	3.6	1.3
CaO	43.2	67.6	62.9	58.4	51.6	62.3	58.2	54.2	55.5	60.4	67.1	68.3	50.3	53.6
MgO	2.2	1.7	6.4	4.7	5.5	1.9	4.2	5.4	19.4	2.3	2.0	11.5	8.5	13.1
TiO_2	0.2	No data	No data	0.1	0.2	No data	No data	No data	No data	0.1	0.1	No data	0.1	No data
Na_2O	2.5	2.5	0.8	2.0	3.1	3.9	1.3	1.7	2.2	0.9	0.7	0.9	3.2	1.1
K_2O	10.1	6.2	7.3	5.3	4.1	7.2	6.6	8.0	5.8	6.3	4.4	2.6	5.3	4.6
P_2O_5	4.6	2.2	2.6	2.2	2.8	2.0	2.9	3.8	1.1	0.3	1.3	2.3	4.7	2.1
SO_3	2.7	1.4	2.2	1.3	2.6	0.6	3.2	1.3	1.4	2.0	0.8	0.8	2.6	1.9

transportation prior to combustion. This can be due to sand inclusions or the absorption of sodium chloride from immersion in seawater.

Wood contains very low concentrations of sulfur. Consequently, during combustion the minerals commonly found in wood form a high-melting, calcite-rich ash, slightly contaminated with potassium salts and silica. However, contamination from sand and sodium chloride can lead to low-melting deposits from the reaction between sand particles and the alkalis.[20]

The reuse of wood ash from forestry sources is desirable, as it contains similar amounts of nutrients to the biomass that was harvested. It should be returned to the land, as is the case in many Scandinavian countries. However, not all types of wood ash are suitable for recycling into other locations, as the content of heavy metal and other contaminants may be too high. Also co-combustion of wood with coal will alter the composition of the ash.[61]

The handling of dry, untreated ash should be undertaken with care. In addition to the salt content and high pH, the dust problems during handling and spreading are severe and the ashes may be highly corrosive. The ashes are sometimes reburned to reduce the carbon content to below 20 %. They are then moistened and granulated before being returned to the forest soil.[61] Some examples of typical wood ash compositions are shown in Table 1.19.[20]

1.6.3   Types of Waste

Municipal Solid Waste (MSW)

Electricity can be produced by burning 'municipal solid waste' (MSW). Such power plants, also called waste-to-energy (WTE) plants, are designed to dispose of MSW and to produce electricity as a by-product of the incinerator operation.

The definition of MSW is different in different countries. In the UK and the USA, for example, MSW describes the stream of solid waste generated by households and apartments and some small commercial establishments. In the Netherlands, however, most of what constitutes MSW in the UK is separated by the public (or households) into eight different categories that include:

- GFT (vegetables, fruit, food and garden wastes, grass cuttings);
- small chemical residues (batteries, paint);
- domestic, household and kitchen appliances (appliances);

- paper and cartons;
- glass and bottles;
- textiles (clothing, parts of furniture);
- large items (furniture, wood and metal – separated); and
- grey material (not separable and deemed suitable for incineration).

Many other countries have similar schemes with different categories. Thus MSW consists of everyday household items but does not include medical, commercial and industrial hazardous or radioactive wastes, which must be treated separately. MSW is presently managed by a combination of disposal to landfill, recycling and incineration – the latter often producing electricity in WTE plants.

The particle size and composition of MSW is obviously extremely variable. This variability is made more pronounced by the degree of waste separation in force in different countries. For example, in the case of the Netherlands, the 'grey' material (only suitable for incineration) would be expected to differ in composition dramatically from MSW in a country such as the UK, where waste segregation is minimal or nonexistent. Although very heterogeneous, for the purposes of comparison a 'typical' MSW would be expected to possess a calorific value of around 8–12 MJ kg^{-1}, which is less than one half of that of a bituminous coal. Its ash content would be about 2.5 times that of bituminous coal, at around 25 %, and it could contain significantly more chlorine than many coals (0.2–0.6 %). The chlorine in MSW is mostly derived from plastic materials[62] or foods. Where plastic is one of the major components in MSW, such as in the Netherlands, the chlorine content of the ash may be as high as 10 %. However, in Germany, where plastics are collected separately, the ash of the remaining fraction would be very low in chlorine.[63]

Most MSW is burnt in specifically designed incinerators. Many are of the mass burn type, which are large and can process virtually untreated MSW. In a typical MSW incinerator, waste will normally be tipped into a holding area, from which it will be picked up and dropped into the feed hoppers. The waste will then be mechanically pushed by a hydraulic ram onto the moving grate within the incinerator. This will allow the refuse to gradually pass through the incinerator over a period of about 2.5 hours. After combustion, the ash falls off the end of the grate, is quenched in a water trough and the metal content extracted using an electromagnet. After heat extraction in a boiler, the flue gases are cleaned before being emitted into the atmosphere. The gases first go to a dry scrubber reactor to treat acid pollutants (sulfur dioxide and hydrogen chloride) and active carbon is injected to remove residual organic compounds such as dioxins. The flue gas then passes through a bag-house filter, which removes particulate matter. The majority of toxic and environmentally damaging components have now been removed from the flue gases and the remainder (mostly carbon dioxide and water vapour) is discharged through the stack. A schematic diagram of a mass-burn municipal incinerator is shown in Figure 1.10.

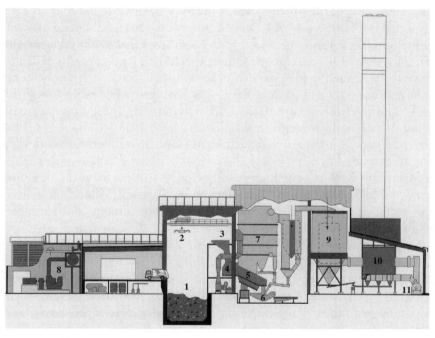

Key:

1	Storage pit	2	Crane	3	Feed hopper
4	Feeder	5	Grate	6	Ash discharger
7	Combustion chamber	8	Turbogenerator	9	Scrubber reactor
10	Baghouse filter	11	Induced draft fan			

Figure 1.10 A schematic of a mass burn incinerator (Reproduced by permission of Martin Engineering Systems Ltd).

Refuse-derived Fuel (RDF)

Refuse-derived fuel (RDF) is a pelletised material that is based on upgraded MSW. The MSW is first shredded and incombustible material such as glass and metal removed. The combustible portion is extruded into pellets and used as a fuel, either alone or blended with coal. The pellets can be conveniently stored and transported. The calorific value of RDF is about 8–15 MJ kg^{-1}, depending upon the content of combustible organic materials in the waste, additives and binder materials, if any, used in the process. By comparison, the calorific value of MSW is typically 8–12 MJ kg^{-1}.

Although RDF is a higher-quality fuel than MSW, a major disadvantage is the cost of the process, which needs to be operated at a significant capacity so that the sale of RDF and by-product materials, such as steel, contribute to revenue generation. Also, a guaranteed market for the product has to be developed. Often, the RDF plant operator will not have the revenue (or even expertise) to build an accompanying incineration plant and so will have to develop a business partnership to sell the fuel. The payback period is long, as costs are high; therefore revenue from sales has to be guaranteed. For the RDF plant to make a profit, this necessitates a guaranteed partnership with an incineration company for an extended period.[64]

Ash from RDF is similar in many respects to that from the combustion of MSW and in most cases has to be disposed to landfill. This is due to the unacceptable nature of some of its constituents; for example, chlorine. However, in Japan, where landfill is scarce and costly, a washing process for RDF incinerator fly ash has been developed. This will allow the use of such 'cleaned' ash in what is termed 'Ecocement' and the first commercial plant commenced operation in April 2001 near Tokyo.[65] Extension of the life of landfills has also been developed in Japan by producing a system firing suitably processed waste paper and plastic, which melts incinerator fly ash, thus effecting a volume reduction. This and other advanced environmental equipment for processing ashes is described in the *Database on Japanese Advanced Environmental Equipment* published by the Japan Society of Industrial Machinery Manufacturers (JSIM) in 2001 to introduce the advanced environmental equipment made by leading Japanese manufacturers.[66]

Petroleum Coke

Petroleum coke is a by-product from the upgrading of heavy oil fractions to produce lighter hydrocarbons (i.e. gasoline), diesel and middle oil ranges (e.g. lubricating oils). While the coking process has many variants, all involve thermal cracking of the feed. There are two major types of petroleum coke: delayed coke, and fluid or Flexi-coke. Delayed coke is produced at about 415–450 °C, while fluid coking occurs at about 480–565 °C. Both methods produce a coke residue composed mainly of solid hydrocarbon polymers containing primarily semi-graphitic or amorphous carbon. Since fluid coke results from a higher temperature process, it contains significantly less volatile matter than delayed coke. The fluidisation process by which it is made also ensures that it is significantly finer.

Cokes produced in both processes often have high sulfur (up to 8 %) and high nitrogen (up to 4 %) content. In addition, the cokes can have high vanadium and nickel contents both concentrated from the parent feedstock. The ash content is typically less than 1 %, but can be as high as 5 %. The coking process also ensures that content of volatile matter is low, typically 4–6 % for fluid cokes and 7–13 % for delayed cokes.[67]

The high sulfur content and low reactivity have ensured that petroleum coke combustion is only possible when a fluidised bed system is employed. The issue of utilisation of ash from FBC of petroleum coke has been discussed by Anthony.[68] The low ash and high vanadium contents of petroleum coke seriously restrict the use of such ashes, which also contain high levels of unreacted calcium oxide (Table 1.20[69,70]). Some limited use of bottom ash for treating acidic waste is reported, but much is simply sent for disposal.[68]

Sewage Sludge

Sewage sludge triggers a waste disposal problem that will continue to increase as the population increases. The material is characterised by very high moisture and ash contents. In addition, high nitrogen content also causes potential NO_x and nitrous oxide emission problems. Other problems are its pathogenic nature and the presence of toxic heavy metals. These factors will, rightly or wrongly, have some effect on the reuse potential of this ash. Sewage sludge has been burned successfully using FBC techniques for many years. If the moisture content is relatively low, the fuel will burn better.

Sewage sludge ash, of similar size distribution to PFA, is produced during the combustion of dewatered sewage sludge in an incinerator. The specific size range and properties of the sludge ash depend to a great extent on the type of incineration system and the chemical additives introduced in the wastewater treatment process. At present, two major incineration systems, multiple hearth and fluidised bed, are employed in the USA. Approximately 80 % of the incinerators used in the USA are of multiple hearth type. Operating temperatures can vary, depending on the type of furnace, but can be expected to range from approximately 650 °C to 980 °C in the incinerator combustion zone. High operating temperatures > 900 °C can result in partial fusion of ash particles and the formation of clinker, which end up in the ash stream. Lime may also be added to reduce the slagging of sludge during incineration.[71] Since sludge is almost always dewatered prior to combustion, pre-treatment of the sludge to enhance the dewatering process may include the addition of ferrous salts, lime, organics and polymers. Ash produced at treatment plants that introduce ferrous salts or lime for sludge conditioning and dewatering contains significantly higher quantities of iron and calcium, respectively. The pH of sludge ash can vary from 6 to 12, but sludge ash is generally alkaline. Sludge ash from multiple hearth incinerators will usually consist primarily of fine material mixed with some larger sand-

Table 1.20   Petroleum coke composition.

	Delayed coke[70]	Fluid coke[69]	Flexi-coke[69]
Moisture (%)	2.93	4.04	2.6
Ash (%)	2.27	0.27	4.57
Volatile matter (%)	10.92	8.64	6.66
Fixed carbon (%)	83.88	87.05	86.16
Carbon (%, as C)	85.76	84.12	87.03
Hydrogen (%, as H)	3.55	3.4	0.77
Nitrogen (%, as N)	2.46	1.75	0.83
Sulfur (%, as S)	2.08	5.62	2.35
Oxygen (%, as O)	0.95	0.8	1.85
GCV (MJ kg^{-1})	33.97	30.2–35.0	30.2–32.6
Ash constituents %			
Sulfur (%, as SO_3)	32.98		
Calcium (%, as CaO)	26.36		
Silicon (%, as SiO_2)	15.94		
Vanadium (%, as V_2O_5)	7.12		
Iron (% as Fe_2O_3)	4.93		
Aluminium (%, as Al_2O_3)	4.14		
Sodium (%, as Na_2O)	1.68		
Magnesium (%, as MgO)	0.75		
Nickel (% as NiO)	0.58		
Undetermined	Balance		
	Oxidising		
Initial deformation (°C)	1152		
Softening point (°C)	1154		
Hemisphere point (°C)	1157		
Fluid point (°C)	1160		

like particles. The formation of larger particles is normally the result of higher operating temperatures and the formation of clinker. Fluidised bed furnaces produce only a very fine ash.

Sludge ash consists primarily of silica, iron and calcium, but the composition can vary significantly depending on the sludge conditioning operation. Sludge ash has been reported as having little measurable pozzolanic or cementitious activity.[72–74] Trace metal concentrations (e.g. lead, mercury, cadmium, zinc and copper) in sludge ash are typically higher than those in natural fillers or aggregate. This has resulted in some reluctance to use this material; however, recent investigations (leaching tests) suggest that these trace metal concentrations are not excessive and do not pose any measurable leaching problem.[72,73,75]

Incineration of sewage sludge (dewatered to $\sim$ 20 % solids) reduces the weight of feed sludge requiring disposal by approximately 85 %. Sludge ash has been previously recycled as a raw material in Portland cement concrete production, as aggregate in flowable fill, as mineral filler in asphalt paving mixes, and as a soil conditioner mixed with lime and sewage sludge.[72,74] Other potential uses that have been reported include a lightweight aggregate from elevated-temperature firing or sintering of ash or a mixture of ash and clay, and the use of ash in brick manufacturing[76] and as a sludge dewatering aid in wastewater treatment systems.[77] Applications that could potentially make use of sewage sludge ash in highway construction include the use of ash as part of a flowable fill for backfilling trenches or as a substitute aggregate material or mineral filler additive in hot mix asphalt.[72,73,78] Despite these uses, most of the sludge ash generated in the USA is presently land-filled.

A review of the combustion of sewage sludges has been published by Werther and Ogada.[79] A European perspective on the disposal and recycling routes for sewage disposal was completed in 2001 for the European Commission by Arthur Andersen and SEDE.[80] Sewage sludge compositional data is shown in Table 1.21.[40,81]

Tyres

Automotive tyres represent a growing waste problem; for example, in the USA over 270 million scrap tyres were produced in 2001, with an approximate weight of over 5 million tonnes. The European Commission recognises the problem and now considers scrap tyres as a 'priority waste stream'. A working group is developing legislation to reduce their environmental impact with harsh targets likely to be set.

Table 1.21   Sewage sludge analysis data.

	Dried[40]	Typical undigested dried[41]	Sludge[81] Raw basis	Sludge[81] Dry basis	Sludge[81] Dry, ash-free
Moisture (%)	3.0	nil	3.6	Nil	Nil
Ash (%)	45.1	27.1	45.8	47.5	Nil
Volatile matter (%)	49.5	63.1	44.2	45.9	87.4
Fixed carbon (%)	2.4	9.8	6.4	6.6	12.6
Carbon (%)	25.0		23.5	24.4	46.5
Hydrogen (%)	4.9		4.2	4.3	8.2
Nitrogen (%)	3.2		3.41	3.54	6.74
Sulfur (%)	1.1		0.91	0.94	1.79
Chlorine (%)	< 0.1		0.03	0.03	0.06
Oxygen (%)	17.7		18.3	19.0	36.2
GCV (MJ kg^{-1})		17.23 23.64[a]	9.6	10.0	19.0
NCV (MJ kg^{-1})	10.58				
Ash data %					
SiO_2		47.25			
Al_2O_3		15.69			
Fe_2O_3		9.37			
CaO		8.56			
MgO		3.31			
TiO_2		1.28			
Na_2O		2.23			
K_2O		1.77			
P_2O_5		9.03			
SO_3		1.02			

[a] Dry, ash-free.

The typical composition of tyres is as follows: rubber/elastomers, 45–47 %; carbon black, 20–22 %; metal wire, 16–25 %; zinc oxide, 1–2 %; sulfur, 1 %; and additives, 5–8 %. Car tyres may also contain 5–6 % textiles.[82]

Tyres have some intriguing properties, which make their reuse or disposal the subject of very careful consideration. For example, the stability and nonleachable nature of tyres might make them seem ideal candidates for landfill. However, they do not compact well and their shape allows water to collect and provide breeding conditions for mosquitoes and vermin. In addition, technological developments over the past years have resulted in higher quality tyres that have longer life spans. The drawback is that scrap tyres are now more difficult to recycle because of their complex chemistry. As a result, reclaiming the rubber from scrap tyres for reuse in new tyres is not favoured by some of the tyre manufacturers. An obvious alternative to land-filling would seem to be combustion, as tyres also have very high heat content ($\sim 27-39$ MJ kg^{-1}). However, they also contain steel wire and zinc, which can cause problems in heat recovery boilers and steam generators. In stoker-fired systems, the zinc can cause gas pass blockages and the steel can form low-melting slags, which also prevent full boiler operation. In the case of FBC systems, the partly oxidised steel wire can cause blockages in the system and the use of glass fibre reinforcement can also encourage bed agglomeration.[83] Furthermore, disposal of the FBC ash could be difficult, as enhanced zinc levels have been reported.[84] The best method for the disposal of used tyres would seem to be in cement kilns. In this high-temperature process, all the constituents of the tyres are incorporated into the cement clinker product. The steel wire is oxidised and all of the combustible matter is consumed, with the full heat content of the tyre being realised. This application has been tried in a number of countries and accounts for the disposal of some 45 % of all scrap tyres in Germany.

Pulp and Paper Wastes

The disposal of residual fibre and other mill wastes generated by the pulp and paper industry is becoming an increasingly serious concern, particularly with the current trend towards recycling paper products. At present, there is considerable interest in the application of FBC technology for burning de-inking and other wood paper pulp-derived sludges, both in incineration and energy-from-waste projects. It is reported[85] that paper pulp waste contains around 33 % ash, the chemical composition of which is 50–55 % silica, 28–35 % alumina and 5–10 % magnesia, with smaller amounts of calcium and titanium oxides. When mixed with clay and fired, this material was found to produce a porous ceramic composite material that could have application as an insulating material.[85]

Food Processing Waste

The wastes produced by the food and beverage industry form a complex category. For many food processing plants, a large fraction of the solid waste produced by the plant comes from the separation of the desired food constituents from undesired ones in the early stages of processing. Undesirable constituents include tramp material (soil and extraneous plant material), spoiled food stocks, fruit and vegetable trimmings, peels, pits, seeds and pulp. Where possible, these materials are converted into animal feed. Otherwise, they are collected for comminution and discharge to a sewer, incinerated or buried in landfill. In a few cases, the materials are composted or used as fuel.

The practice of land-filling is becoming less favourable, due to the generation of foul odours as communities expand and reside in close proximity to food processing plants. Leaching of undesirable constituents (salts, soluble organics) into soil and groundwater is also an important concern. Composting is another disposal option; however, like land-filling, odour and leaching of soluble constituents are limiting factors. Composted material is valued as a soil amendment or potting soil, but widespread use and marketability is constrained by shipping cost. Disposal of solid wastes to domestic sewers is becoming less favourable due to increased sewer rates and the reluctance of municipal sewage treatment plants to accept such waste streams with their high biological oxygen demand (BOD), and in some cases high salt content. Incineration or use as a fuel are options in certain cases, but are limited to those solid wastes that have a relatively low water content, and can be further dried with ease. The moisture content of suitable fuels is about 10 % or less. Such low moisture content solid wastes that are potentially suitable for incineration undergo limited processing, such as size reduction and minimal drying.[86]

The wide range of different residues and the relatively small quantities available mean that there are currently few applications for incineration and hence reuse of ash from such materials. However, pressures on alternative disposal methods suggest that future uses are likely to increase. One exception is the use of fruit processing residues as components in lightweight brick manufacture.[87]

Auto-shredder Residue (ASR)

Another important class of waste fuel, and likely to increase due to legislation, is auto-shredder residue (ASR). This fuel is derived from all of the nonrecyclable portions of the motor car and contains significant plastic wastes, rubber, metallic components, glass and so on. It is reported that 350 000 to 400 000 tonnes of ASR are disposed of to landfill every year.[88] However, processes have been developed to separate the foam (which is present as 10 % of the material) and other fractions from the ASR.[89] The high heat content of ASR, around 19 MJ kg^{-1}, indicates that this material would be a useful fuel in a thermal process. The high ash content of around 35 % also suggests application in cement kilns.[88]

Plastics

In 1992 almost 16 % of plastics waste from Western Europe was recovered for conversion into energy, with Sweden (56 %) and Switzerland (72 %) being most successful. Because of the high calorific value of plastics, their use as a fuel has long been regarded as an option for many developed countries. Thus, polyethylene and polypropylene contain an average of 40 MJ kg^{-1} of heat, while coal contains only 30 MJ kg^{-1} and wood only 15 MJ kg^{-1}. However, not all waste material containing plastics is as harmless as certain film and packaging waste; for example, ASR, which is expected to increase as automotive recycling is imposed by legislation, contains significant quantities of copper, zinc and lead. Its overall ash content is high, typically 30–60 %, and it also contains from 3–18 % iron and a heating value range of 9–20 MJ kg^{-1}.[90]

The particular properties of plastics mean that they are seldom burned alone. FBC technology does enable this material to be burned, but its quality is such that it is often more profitable to use it blended with coal (see Section 1.5.3, page 46 on coal and plastics). The successful use of plastic packaging waste as a supplementary fuel in cement kilns has been demonstrated, with no apparent differences in the emissions or quality of the cement clinker.[91]

1.6.4   Residues from MSW

MSW combustion for energy recovery generates residues at several points. Solids retained on furnace grates following combustion and solids passing through the grate (siftings) are generally referred to as bottom ash. Entrained particulates that are trapped and residues from scrubber systems, which are subsequently removed by fabric filters and/or ESPs, are normally referred to as air pollution control (APC) residue. In some cases, especially in Europe, ESPs are used to remove particulates before wet scrubbers. This stream may be considered as an APC residue or as a fly ash. Entrained particulates and condensed vaporised metals trapped in heat exchangers generate a small quantity of ash, referred to as heat recovery ash. This is either combined with the APC residue or the bottom ash. Approximately 80 % of the residues generated are bottom ash.[92]

The physical characteristics of bottom ash resemble an aggregate. It is a heterogeneous mixture of slag, metals, ceramics, glass, other noncombustibles and uncombusted organics. It is also a very porous, lightweight aggregate with high specific surface areas. Up to 20 % has a particle size of > 10 cm, consisting of metals, slags and construction-type material. The < 10 cm fraction is more uniform, with up to 10 % fine material. The major elements in bottom ash are oxygen, silicon, iron, calcium, aluminium, sodium, potassium and carbon.

The residues from dry/semi-dry APC systems are fine particulate mixtures of fly ash consisting of reaction products, primarily calcium chloride and unreacted lime used for acid gas emission controls. The fly ash is the coarsest of the APC residues, followed by ESP dust and fabric filter residues. APC residues by their nature are usually highly soluble in water (25 % and 85 % by weight). Major elements in APC residues are oxygen, silicon, calcium, aluminium, chlorine, sodium, potassium, sulfur and iron. Although many metals are present as their oxides, there are also significant amounts of metal chlorides, sulfates and carbonates. The APC residue contains significantly higher concentrations of cadmium, lead and zinc than bottom ash. The APC fraction also contains higher concentrations of soluble salts.[92]

There is a perception that ash disposal limits public acceptance of new MSW combustion facilities. The view is that heavy metals and organics may be leached from the residues after disposal. Work has confirmed that the frequently used Toxicity Characteristic Leaching Procedure (TCLP) showed that bottom ash was not toxic for barium, lead and mercury, in contrast to fly ash.[93] The subject of leachability remains a controversial issue however, and work continues to establish the nature of MSW residues and how they might best be disposed or utilised.

The potential problems associated with leachable components have resulted in the widespread disposal of MSW residues to lined landfill rather than finding uses for them, particularly in the USA. However, prior to disposal the residues are subject to a number of treatment options, including ferrous metal recovery, compaction, classification, solidification/stabilisation, chemical extraction and vitrification.[92]

MSW residues are not routinely utilised in the USA, but a number of European countries, including the Netherlands, Germany, Denmark, Sweden, France and Switzerland, and Japan are able to find uses for significant amounts of this material. Uses for bottom ash include road bases, sound barriers, and as aggregates in concrete and asphaltic concrete. APC residues are used as grout in coal mines and fine aggregate in asphaltic concrete.

The continuing use of processed bottom ash should be considered subject to certain controls, as it has value in certain applications. Problems with the composition and form of APC residues will limit their use in the future.

1.7 Other Fuel Ashes

1.7.1 Orimulsion® Ash

Orimulsion® is a fuel based on a very heavy bitumen found in vast amounts in the Orinoco basin of Venezuela. This material is converted into an emulsion with water by a process developed by BITOR in 1989. Approximately 70 % by weight of the bitumen is emulsified with 30 % water to produce a relatively low viscosity emulsion. It is designed to replace heavy fuel oil (HFO) and is similar to it in many respects. Being water-based, it cannot be preheated: unlike HFO, however, its lower viscosity means that it requires only modest preheating before atomisation. The incorporation of water droplets within the bitumen particles produces what is known as secondary atomisation, resulting in improved burnout and the formation of finely-divided ash particles.

Although the base material contains relatively high concentrations of vanadium, sulfur and other trace elements, such as nickel, emission control technologies exist to allow its safe combustion. It can be used as a direct replacement for fuel oil in electricity generation and has been burned successfully in Canada (from 1994 at Dalhousie), Denmark (from 1995 at Asnaes) and Italy (from 1998 at Brindisi and 1999 at Fiume Santo).[94]

The original Orimulsion® formulation contained a water-soluble magnesium additive and an ethoxylate-based surfactant. The magnesium was intended to combat corrosion and the effects of vanadium, and the surfactant to stabilise the emulsion. The surfactant, which was found to have some effect on wildlife, has been replaced by a more benign material and the magnesium has been eliminated. The improved product is known as Orimulsion®-400 (Table 1.22).[94]

After combustion, Orimulsion® ash is collected and processed at the power station into a dust-free product suitable for metal recovery in

Table 1.22   The composition of Orimulsion® 400.

	As fired[94]	Dry basis[94]
Moisture (%)	29.0	Nil
Ash (%)	< 0.1	< 0.1
Carbon (% 'as C')	60.0	84.5
Hydrogen (% 'as H')	7.7	10.8
Nitrogen (% 'as N')	0.50	0.75
Sulfur (% 'as S')	2.8	3.95
GCV (MJ kg^{-1})	30.0	42.3
NCV (MJ kg^{-1})	28.0	39.7
Density (kg m^{-3})	1010	No data
Bitumen droplet diameter (μm)	10	—
Viscosity at 30 °C (mPa.s)	200–350	—

specialised plants. Since October 1997, 6000 tonnes a year of granulated ash have been processed in a BITOR joint venture company in the UK to recover vanadium and nickel.[94]

1.7.2   Heavy Fuel Oil Ash

Heavy fuel oil normally contains less than 0.1 % by weight of ash. The main heavy metals of commercial interest are vanadium and nickel, the amounts depending on the origin of the crude oil.[95] Oil-firing produces much less ash per unit of energy than coal and it is very different in nature. The inorganic portion in HFO ash is typically around 0.05 % by weight, and when collected from a boiler it is often acidic and highly carbonaceous. For this reason, ESP collection is inappropriate and dust collectors or bag-filters are necessary. In certain applications, additives are incorporated into the fuel oil to alleviate the effects of acidity and the presence of low-melting, corrosive vanadium compounds. The additives have frequently been based on magnesium compounds such as oxide or hydroxide. Their presence obviously increases the flue gas dust burden but lowers the overall vanadium content of the ash by dilution. It is felt that the presence of such additives would not necessarily affect the extraction of valuable metals from ash.

   A number of investigations into the processing of ash from the combustion of fuel oil in power stations have recently been reported.[96–99]

1.7.3 Oil Shale

Oil shale is defined geologically as a sapropelite containing a variable amount of transformed organic matter (10–67 %) that on destructive distillation yields shale oil. A well-known misnomer, the fine-grained rock is neither shale nor does it contain oil. The appearance of a shale results from the foliated arrangement of the organic matter with the material matrix.

Oil shales are abundant, being found in over 30 countries, and direct combustion or refining to extract the oil and various chemicals are the major uses. The organic matter consists primarily of volatile matter (80–90 %) and the low calorific value, typically ≤ 11 MJ kg^{-1} NCV, has limited the exploitation of reserves in most countries. However, the relatively high heating value and availability in Estonia has resulted in over 90 % of the electricity being generated from oil shale. Consequently, most of the research relating to the use of oil shale has been carried out in Estonia, but little has been published in English.[100]

Dry Estonian oil shale consists of the organic fraction, carbonates and sandy-clay minerals. The major constituent of the carbonate fraction is typically calcite, whereas dolomite and siderite are present in lesser amounts. The sandy-clay materials consist of quartz, orthoclase, hydromuscovite and marcasite.

The presence of reactive calcium in the oil shale ash allows for the capture of sulfur dioxide emissions. However, the low calorific value and high ash content of this fuel results in large quantities of combustion residues.

Despite the technical problems, the use of oil shale as a source of energy has been increasing for more than two decades. In addition to Estonia, it has been used in the USA, states of the former USSR, Germany, China, Israel, Greece and Jordan.[101,102]

The by-product from oil shale combustion is oil shale ash (OSA). Its disposal may be costly and the quantities involved are large, so a number of initiatives have been made to find uses for OSA. However, the properties of this material vary widely, ranging from a high-silica, pozzolanic version to a substance with high calcium oxide content, which has cementitious properties of its own. Consequently, the properties of each OSA need to be established to enable suitable applications to be developed. The uses developed so far include concrete and pavement construction,[101] cementless building units[103] and fly ash binders.[104] In Estonia a number of uses have been, or are in the process of being, more fully investigated. These include flue gas cleaning, wastewater treatment, leather

Table 1.23 Analysis of inorganic fractions of oil shale (in %).

	Estonia[100]	Estonia[100]	Jordan[101]	Israel HCOSFA[102]	Israel OSFA[102]	Estonia Fine[105]	Estonia Finest[105]
SiO_2	26.5	24.9	35.4	16.3	18.0	20–28	30–35
Al_2O_3	7.3	6.9	3.8	5.6	7.1	6–8	10–12
Fe_2O_3	5.6	1.2	2.0	3.6	3.4	4–6	4–5
CaO	45.4	31.3	39.7	52.3	51.4	46–58	28–35
MgO	5.8	1.3	4.0	0.8	0.6	3–4	2–3
Na_2O	0.3	0.34	0.5	0.3	0.4	0.1	1–2
K_2O	3.7	2.65	0.6	0.4	0.6	1–2	4–6
SO_3	0.05	0.21	4.0	10.7	9.8	7–12	12–15
CO_2	No data	25.7	No data	No data	No data	2–3	1–2
FeS_2	No data	5.2	No data	No data	No data	No data	No data
Loss-on-ignition	No data	No data	7.3	5.1	6.0	No data	No data

treatment, filler for rubber products, road construction, pre-cast concrete production, mine filling and acidic soil treatment.[105] Table 1.23 shows details of typical oil shales and ash from different counties.[100–103,105]

1.8 Gasification

1.8.1 Process Details

In the integrated gasification combined cycle (IGCC), coal is gasified under pressure at high temperature (up to 2000 °C) under reducing conditions to produce a fuel gas or syngas containing mainly hydrogen and carbon monoxide. The gas is cooled and cleaned before being combusted in a gas turbine to produce power. Sensible heat from the fuel gas and residual heat from the exhaust gas of the gas turbine is used to raise steam for a steam turbine which is part of a combined cycle, generating more electricity. Typically, the gas turbine produces 65 % of the power while the steam turbine produces 35 %.

Combustion of coal under sub-stoichiometric conditions – that is, gasification of coal – generates reduced species such as sulfides in some of the by-products. These materials may be regarded as hazardous if the release of sulfides is above a specified level. To overcome this problem, reduced forms of sulfur can be oxidised and removed as calcium sulfate using a calcium-based sorbent. The most significant ash property is its fusion characteristics, as these must be low for all types of gasifiers.

There are several different gasification processes: fixed bed gasifiers (producing dry ash and granulated vitreous solids), entrained flow gasifiers (producing vitreous slag) and fluidised bed gasifiers, (producing dry or agglomerated ash, depending on the operating temperatures and fusion temperature of the ash).[30]

Fixed bed gasifiers, such as the Lurgi system, operate by passing air or oxygen and steam up through a bed of coal. During operation, further coal is added to the top of the bed and dry ash is removed from the grate at the bottom. Other fixed bed gasifiers such as the British Gas Lurgi (BGL) remove slag through a tap in the bottom.

In an entrained flow gasifier, such as the Shell or Texaco process, pulverised coal (or atomised fuel oil) flows co-currently with the oxidising medium, typically oxygen. The key characteristics of entrained flow gasifiers are their very high and uniform temperatures (usually more than 1000 °C) and the very short residence time of the fuel within the gasifier. For this reason, solids fed into the gasifier must be very finely divided and homogeneous, which in turn means that entrained flow gasifiers are not suitable for feedstocks such as biomass or wastes, which cannot be readily pulverised. The high temperatures in entrained flow gasifiers mean that the ash in coal melts and is removed as molten slag.

In a fluidised bed (FB) gasifier, such as the KRW process, the fluidising medium and oxidant is air. The key feature of the FB gasifier is that the fuel ash must not be allowed to become so hot that it softens and causes bed agglomeration. This may be achieved by maintaining the bed temperature usually below 1000 °C, and it is for this reason that air and not oxygen is used as the fluidising medium. This, in turn, means that FB gasifiers are best suited to relatively reactive fuels, such as biomass. Advantages of the FB gasifier include the ability to accept a wide range of solid feed, including household waste (suitable pre-treated) and biomass such as wood. It is also preferred for very high ash coals, particularly those in which the ash has a high melting point, because gasifier and fixed bed systems lose significant amounts of energy in melting the ash to form slag.

1.8.2   Solid Residues from Gasification Processes

The morphology and mineralogy of residues from coal gasification are both complex and subject to great variability caused by differences in the temperature of formation and slag or ash composition. If temperatures are high, the ash will be dominantly glassy (noncrystalline). If rapid cooling occurs, stable (equilibrium) crystalline assemblages are unable to form in the liquid before solidification, resulting in the formation of a vitreous slag. Slower cooling allows equilibrium conditions to be reached and crystallisation to occur. The bulk chemical composition of the coal mineral matter will determine the mineralogy of crystalline ash or devitrified slag.

Gasifier Slags

Gasifier slag is typically a glassy inert product consisting almost entirely of mineral matter from the coal. It is physically similar to bottom ash from coal combustion and blast furnace slag. The composition of the slag depends on the mineral composition of the feed coal and the gasification process. Some systems use fluxes, such as limestone, to maintain or control slag viscosity, and this is reflected in the slag composition.

Many of the slags produced by gasification are similar in appearance. For example, the British Gas/Lurgi slag is described[106] as an odourless, carbon-free, granular amorphous glassy frit. It has the appearance of coarse, black sand, and is hard but brittle. A range of shapes may be present, including plates, spheres, needles and irregular fragments. Up to 5 % free iron may also be present. Slag from the Shell process is usually dark grey to black in colour and composed of irregular, fractured, glassy granules and small spherical particles.[107] Texaco slag has a smooth, vitreous texture with some open porosity and undergoes conchoidal fracture. Coarse fragments are dull grey in colour, becoming bottle green on crushing. Most of the finer material is similar in colour to the coarser fragments, but a small fraction is composed of dark grey to black amorphous material.[108]

Ashes and Agglomerates

In the fluidised bed gasifier, two types of ash particles are present:

- Low melting point aluminosilicate agglomerates formed from clays, quartz and pyrite, or minerals containing calcium, magnesium, sodium and potassium as the carbon in a coal particle is gasified.
- Individual mineral particles and small amounts of pyrite or fluxing agents, which do not melt as the char is gasified. This includes the more refractory minerals from the coal.

Agglomerates grow by the combination of small molten drops and by the capture of free ash by molten material. Individual particles with between 15 and 80 % iron(III) oxide tend to melt and coalesce to form agglomerates, whereas particles outside this range tend to form free ash.[109] In coals with a low iron content, significant formation of ash agglomerates can still occur by the binding action of small amounts of iron aluminosilicate with unmelted mineral matter. The iron aluminosilicate binder arises from reaction between clays and the decomposition products from pyrite.

Gasification ashes from various lignites have also been studied.[110] These residues were typically agglomerated, partially melted, crystalline and glassy materials formed from the high-temperature reaction of the inorganic constituents of the source fuel. The lignite ashes were typically light grey in colour and had a consistency from fine powder to agglomerate.

Disposal of Gasification Residues

To determine the likely environmental consequences of dumping or utilising gasification residues, an understanding of the leaching characteristics of slag or ash is required. Knowledge of the morphology and composition of residues may aid prediction of leaching behaviour. However, empirical leaching tests are required to confirm the actual behaviour of residue components, particularly the more soluble constituents.

Leaching tests have shown that IGCC slags are relatively inert. Under current disposal regulations, they require no special processing before transport and disposal; however, future environmental regulations may make disposal more difficult.

In contrast, however, potential problems may arise with the disposal of IGCC residues from FB gasifiers with in-bed desulfurisation. This is due to the high contents of calcium oxide, sulfide and sulfate in these residues and the alkalinity of their leachate. Dust problems may arise during the handling of dry residues that contain large amounts of calcium oxide. In addition, uncontrolled hydration of such ashes can cause a violent exothermic reaction, followed by solidification and hardening.

The expected increases in cost, reductions in landfill availability and technical difficulty of disposing of such ashes has lead to consideration of their utilisation rather than their disposal. For example, the fused mineral ash from the Sasolburg gasification plant in South Africa is to be used in the construction of roads. It is claimed that it is cheaper and faster and easier to use than the material it replaces.[30] Unfortunately, although IGCC residues have the strength and resistance requirements to be used as unstabilised road bases, they are too fine in size and this limits their usefulness.

1.9 Summary

This chapter provides background information on the nature and composition of coal in a way that is aimed at providing an understanding of the scope and limitations of traditional ash by-products. It also serves as an introduction to some of the less commonly used materials that are now increasingly finding their way into the pool of by-products and for which new and imaginative uses are sought. To meet current electricity needs, the use of coal/renewable blends is expected to continue in the short to medium term. The effect on fly ash quality of generating power from such blends will continue to be studied as the need for more and more non-fossil fuel combustion systems increases. The sole use of non-fossil fuels in less conventional power plants will increase and utilisation of the residues from these more diverse materials will present a challenge to the ingenuity and innovative capabilities of both chemists and engineers in the future.

References

1   Anon., *An investigation of the pozzolanic nature of coal ashes*. Engineering News, 71(24): 1334–1335 (1914).
2   Davis, R.E., Kelly, J.W., Troxell, G.E. and Davis, H.E., *Proportions of mortars and concretes containing Portland – pozzolan cements*. ACI Journal, 33: 577–612 (1935).
3   Davis, R.E., Carlston, R.W., Kelly, J.W. and Davis, H.E., *Properties of cements and concretes containing fly ash*. ACI Journal, 33: 577–612 (1937).
4   Davis, R.E., Davis, H.E. and Kelly, J.W., *Weathering resistance of concretes containing fly ash cements*. Proc. American Concrete Institute, 37: 281–296 (1941).
5   Sear, L.K.A., *Properties and use of coal fly ash; a valuable industrial by-product*. Thomas Telford Ltd: London (2001).
6   Anon., *What are CCPs? Production & Utilisation*. European Coal Combustion Products Association (ECOBA) (2000).
7   Anon., *2001 Coal Combustion Product (CCP), Production and Use*. American Coal Ash Association (2001).
8   Barnes DI, *Personal communication – The impact of recent technical developments in energy generation on ash quality* (2002).
9   Anon., *Coal*, Environmental Literacy Council.
10 Anon., *Coal – Power for Progress*. World Coal Institute (2003).
11 Anon., *Coal Facts*, p. 2 (2000).
12 Speight, J.G., *The Chemistry and Technology of Coal*. Marcel Dekker Inc.: New York (1994).
13 Francis, W., *Coal: Its Formation and Composition*. Edward Arnold: London (1961).
14 ASTM, *D388 Classification of coals by rank*. American Society for Testing and Materials: Philadelphia, PA (1991).
15 Visser, W.A., (ed.), *Geological Nomenclature*. Royal Geological and Mining Society of the Netherlands (1980).
16 Thompson, A.W., *Personal communication – Use of high ash coals for steam generation* (1986).
17 Raask, E., *Mineral Impurities in Coal Combustion*. Hemisphere Publishing Corporation (1985).

18 Jones, M.L. and Benson, S.A., *An overview of fouling/slagging with Western coals.* In *Effects of Coal Quality in Power Plants.* Research Project 2256. Electric Power Research Institute, Coal Combustion Systems Division, Palo Alto, CA (1987).

19 ASTM, *C618-91 Fly ash and raw or calcined natural pozzolans for use as a mineral admixture in Portland cement concrete.* American Society for Testing and Materials: Philadelphia, PA. p. 3 (1991).

20 Bryers, R.W., *Fireside slagging, fouling, and high-temperature corrosion of heat-transfer surface due to impurities in steam-raising fuels.* Prog. Energy Combust. Sci., **22**: 29–120 (1996).

21 Meij, R., *Status Report on Health Issues Associated with Pulverised Fuel Ash and Fly Dust – Introduction and Summary Revision (version 2.1).* KEMA (2003).

22 Clarke, L.B. and Sloss, L.L., *Trace elements – emissions from coal combustion and gasification.* IEA Coal Research: London (1992).

23 Meij, R., van der Kooij, J., van der Sluys, J.L.G., Siepman, F.G.C. and van der Sloot, H.A., *The emission of fly ash and trace species from pulverised coal fired utility boilers.* In VIth World Congress on Air Quality. Paris, France (1983).

24 Meij, R., *Prediction of environmental quality of by-products of coal-fired plants; elemental composition and leaching.* In WASCON '97. Houthem St. Gerlach, The Netherlands: Elsevier, Amsterdam (1997).

25 Meij, R., Te Winkel, B.H. and Scholten, R.D.A., *Status Report on Health Issues Associated with Pulverised Fuel Ash and Fly Dust, Section 2, Environmental impact associated with airborne pulverised fuel ash.* KEMA (2000).

26 Meij, R., *Composition and particle size of and exposure to coal fly ash.* J. Aerosol Sci., **31** (Supplement 1): S676–S677 (2000).

27 Meij, R., Te Winkel, B.H. and Overbeek, J.H.M., *Particle size of suspended coal fly ash.* J. Aerosol Sci., **32** (Supplement 1): S595–S368 (2001).

28 Anon., *Cyclones.* In Steam – its generation and use, S.C. Stultz, and J.B. Kitto (eds). The Babcock & Wilcox Company, pp. 14-1 to 14-11 (1992).

29 US Department of Energy, *Clean Coal Technology Demonstration Program* (2000).

30 Sloss, L.L., *Trends in the use of coal ash.* IEA Coal Research: London, p. 64 (1999).

31 Thompson, A.W., *Personal Communication – CFBC ash streams* (1985).

32 Cobb, J.T., Clifford, B.V., Neufeld, R.D., Pritts, J.W., Beeghly, J.H. and Bender, C.F., *Laboratory evaluation and commercial demonstration of hazardous waste treatment using clean coal by-products.* In 1997 International ash utilisation symposium: pushing the envelope. Lexington, KY, University of Kentucky Center for Applied Energy Research/FUEL/US DOE/FETC (1997).

33 Winschel, R.A., Wu, M.M. and Burke, F.P., *Synthetic aggregates from coal-fired fluidized-bed combustion residues.* In Coal Science, J.A. Pajares and J.M.D. Tascon (eds). Elsevier Science B.V. (1995).

34 Sami, M., Annamalai, K. and Wooldridge, M., *Co-firing of coal and biomass fuel blends.* Prog. Energy Combust.Sci., **27**: 171–214 (2001).

35 Final Report of Project DIS-0506-95-UK, *European Commission DG XVII, 'European Co-combustion of Coal, Biomass and Wastes',* 74 pp. (2000).

36 Beekes, M.L., Gast, C.H., Korevaar, C.H., Willeboers, W. and Penninks, F.W.M., *Co-combustion of biomass in pulverised coal-fired boilers in the Netherlands.* In 17th World Energy Congress. Houston, Texas (1998).

37 Annamalai, K., Sweeten, J., Chen, C.J. and Thien, B., *Evaluate Selected Aspects – Manure/Coal Blends in Boiler Burners.* Western Regional Biomass Energy Program funded by US DoE.

38 Ohlsson, O., *Results of Combustion and Emissions Testing when Co-firing Blends of Binder-enhanced Densified Refuse Derived Fuel Pellets in a 440 MW$_e$ Cyclone-fired Combustor*, vol. 1: *Test Methodology and Results*. Subcontract report no. DE94000283, Argonne National Laboratory, Argonne, IL (1994).

39 Makansi, J., *Traditional control processes handle new pollutants*. Power. **131**(10): 11–18 (1987).

40 Richers, U., Scheurer, W., Seifert, H. and Hein, K.R.G., *Present Status and Perspectives of Co-combustion in German Power Plants*. Forchungszentrum Karlruhe GmbH: Karlsruhe, Germany, pp. 1–16 (2002).

41 Ireland, S.N., McGrellis, B. and Harper, N., *On the technical and economic issues involved in the co-firing of coal and waste in a conventional pf-fired power station*. Fuel, **83**(7–8): 905–915 (2004).

42 Mark, F.E. and Rodriguez, M.J., *Energy Recovery of Greenhouse PE Film: Co-combustion in a Coal Fired Power Plant*. Association of Plastics Manufacturers in Europe: Brussels, pp. 1–16 (1999).

43 Schoen, L.A.A., Beekes, M.L., van Tubergen, J. and Korevaar, C.H., *Mechanical Separation of Mixed Plastics from Household Waste and Energy Recovery in a Pulverised Coal-fire Power Station*. Association of Plastics Manufacturers in Europe: Brussels, pp. 1–19 (2000).

44 Mark, F.E. and Vehlow, J., *Co-combustion of End of Life Plastics in MSW Combustors*. Association of Plastics Manufacturers in Europe: Brussels, pp. 1–21 (1999).

45 Nielsen, C., *Utilisation of straw and similar agricultural residues*. Biomass and Bioenergy, **9**(1–5): 315–323 (1995).

46 Jensen, P.A., Stenholm, M. and Hald, P., *Deposition investigation in straw-fired boilers*. Energy and Fuels, **11**(5): 1048–1055 (1997).

47 Sander, B., *Properties of Danish biofuels and the requirements for power production*. Biomass and Bioenergy, **12**(3): 177–183 (1997).

48 Olanders, B. and Steenari, B.-M., *Characterization of ashes from wood and straw*. Biomass and Bioenergy, **8**(2): 105–115 (1995).

49 Hein, K.R.G. and Bemtgen, J.M., *EU clean coal technology – co-combustion of coal and biomass*. Fuel Processing Technology, **54**: 159–169 (1998).

50 Jenkins, B.M., Bakker, R.R. and Wei, J.B., *On the properties of washed straw*. Biomass and Bioenergy, **10**(4): 177–200 (1996).

51 Thompson, A.W., *Personal communication – Fibrowatt* (2003).

52 Thompson, A.W., *Personal communication – Energy Power Resources Ltd* (2003).

53 Miles, T.R. Jr, Baxter, L.L., Bryers, R.W., Jenkins, B.M. and Oden, L.L., *Alkali Deposits Found in Biomass Power Plants – A Preliminary Investigation of their Extent and Nature*. National Renewable Energy Laboratory: Golden, CO (1995).

54 Stultz, S.C. and Kitto, J.B., (eds), *Steam – Its Generation and Use*. Babcock & Wilcox (1992).

55 Singer, J.G. (ed.), *Combustion – Fossil Power Systems*. Rand McNally (1981).

56 Thompson, A.W., *Personal communication – The use of bagasse ash in mortar* (2003).

57 Armesto, L., Bahillo, A., Cabanillas, A., Veijonen, K., Otero, J., Plumed, A. and Salvador, L., *Co-combustion of coal and olive oil industry residues in fluidised bed*. Fuel, **82**: 993–1000 (2003).

58 Masghouni, M. and Hassairi, M., *Energy applications of olive oil industry by-products: I The exhaust foot cake*. Biomass and Bioenergy, **18**: 257–262 (2000).

59   Thompson, A.W., *Personal communication – The combustion of olive oil waste to generate electricity* (2002).
60   Lundborg, A., *Biomass utilisation in Sweden*. In Ash and Particulate Emissions from Biomass Combustion. I. Obernberger (ed.)., Institute of Chemical Engineering, Technical University of Graz, Austria (1998).
61   Zevenhoven, M., *The Utilisation of Biomass Ash*. Abo Akademi: Abo, Finland. pp. 1–56 (2001).
62   Ruth, L.A., *Energy from municipal solid waste: a comparison with coal combustion technology*. Prog. Energy Combust.Sci., 24: 545–564 (1998).
63   Nugteren, H.W., *Personal communication – Chlorine content of MSW from Germany* (2003).
64   Anon., *Refuse-derived fuel*. Cardiff University, Engineering Department, Waste Research Station.
65   Maekawa, H., Aoyama, S. and Kagamida, M., *Environment technologies in the cement industry – toward zero emissions*. GCL: Environmental Special Issue 2002, pp. 23–26 (2002).
66   Anon., *Database on Japanese Advanced Environmental Equipment*. Japan Society of Industrial Machinery Manufacturers (2001).
67   Hobson, G.D. and Pohl, W., *Modern Petroleum Technology*, 4th edn. Applied Science Publishers Ltd: Barking/Institute of Petroleum (1973).
68   Anthony, E.J., *Fluidized bed combustion of alternative solid fuels; status, successes and problems of the technology*. Prog. Energy Combust. Sci., 21: 239–286 (1995).
69   Stephan, B., Yee, B.H. and Rosenquist, W.A., *Petroleum coke as a viable alternative fuel*. In POWER-GEN International '96. Orlando, Fla (1996).
70   Placer, F., Olivo, C.A. and Grace, D., *Petroleum coke testing in a coal-fired fluidized bed boiler plant: operating experience*. In Power-Gen '93. Dallas, TX. (1993).
71   Anon., *Utilization of ash from incineration of municipal sewage sludge – draft literature review*. Wastewater Technology Centre. Environment Canada (1992).
72   Anon., *Sewage Sludge Ash Use in Bituminous Paving*. Metropolitan Waste Control Commission, Minnesota Department of Transportation, Minnesota Pollution Control Agency (1990).
73   Khanbiluardi, R.M., *Ash Use from Suffolk County Wastewater Treatment Plant*. Sewer District No. 3, City University of New York (1994).
74   Anon., *Analysis of Sludge Ash for Use in Asphalt, Concrete, Fertilizer and Other Products*. Metropolitan Council of Twin Cities Area, Publication No. 12-82-103 (1982).
75   Anon., *Sewage Sludge Ash Use in Bituminous Paving, Report on Additional Testing*. Braun Intertec Environmental (1991).
76   Anderson, M., *Encouraging prospects for recycling incinerated sewage sludge ash (ISSA) into clay-based building products*. J. Chem. Tech. Biotechnol, 77: 352–360 (2002).
77   Micale, F.J.A., *A Mechanism for Ash-assisted Sludge Dewatering*. USEPA (EPA-600/2-76-297) (1976).
78   Wegman, D.E. and Young, D.S., *Testing and evaluating sewage sludge ash in asphalt paving mixtures*. In 67th Annual Transportation Research Board Meeting. Washington, DC (1988).
79   Werther, J. and Ogada, T., *Sewage sludge combustion*. Prog. Energy Combust.Sci., 25: 55–116 (1999).

80 B/2, E.D.E.-., *Disposal and Recycling Routes for Sewage Sludge Part 3 – Scientific and technical report.* European Commission (2001).

81 Spliethoff, H., Scheurer, W. and Hein, K.R.G., *Effect of co-combustion of sewage sludge and biomass on emissions and heavy metals behaviour.* Trans. I. Chem. E, 78 (Part B): 33–39 (2000).

82 Anon., *Technical guidelines on hazardous wastes: identification and management of used tyres.* Basel convention on the control of trans-boundary movements of hazardous wastes and their disposal. Secretariat: Geneva, Switzerland, pp. 2–10 (1999).

83 Rasmussen, G.P., *Fluidized bed systems for steam generation from scrap tires.* In 7th International Conference on Fluidized Bed Combustion, Philadelphia, PA (1982).

84 McGowin, C.R. and Howe, W.C., *A growing business opportunity: Alternate fuel co-firing in fluidized bed boilers.* In Applications of Fluidized Bed Combustion for Power Generation, EPRI (1993).

85 Dasgupta, S. and Das, S.K., *Paper pulp waste – A new source of raw material for the synthesis of a porous ceramic composite.* Bull. Mater. Sci, 25(5): 381–385 (2002).

86 Phillips, R.J. & Associates, *Extraction, Separation and/or Reuse of Solid Wastes from Food Processing* (1996).

87 Moedinger, F. and Anderson, M., *A responsible approach to the recycling of by-product wastes into commercial clayware.* In Sustainable Waste Management (Proceedings of the International Symposium on Advances in Waste Management). Thomas Telford: London (2003).

88 Lanoir, D., Trouve, G., Delfosse, L., Froelich, D. and Kassamaly, A., *Physical and chemical characterization of automotive shredder residues.* Waste Management & Research, 15: 267–276 (1997).

89 Argonne National Laboratory, Office of Technology Transfer, *Recovering Foam from Scrapped Autos*; at http://www.anl.gov/techtransfer/Available_Technologies/ Environmental_Research/autoshredder.html

90 Mark, F.E., Fisher, M.M. and Smith, K.A., *Energy Recovery from Automotive Shredder Residue.* Association of Plastics Manufacturers in Europe, Brussels. pp. 1–21 (1998).

91 Mark, F.E. and Caluori, A., *Fuel Substitution for Cement Kilns through Source Separated Plastics Packaging Waste – the Effects on Metal Content in Clinker and on Emissions.* Association of Plastics Manufacturers in Europe: Brussels, pp. 1–15 (1998).

92 Wiles, C.C., *Municipal solid waste combustion ash: State-of-the-knowledge.* J. Hazardous Mat., 47: 325–344 (1996).

93 Kosson, D.S., van der Sloot, H., Holmes, T.T. and Wiles, C.C., *A comparison of solidification/stabilization processes for treatment of municipal solid waste residues, Part II – Leaching properties.* In Municipal Waste Combustion – 2nd Annual International Conference, Pittsburgh, PA (1991).

94 Anon., *Orimulsion in Europe.* Bitor Energy News (2000).

95 Cunningham, A.T.S. and Datschefski, G., *The effects of heavy fuel oil composition on particulate emissions from power station boilers.* In The Heavy End of the Barrel – Future Trends in Oil Firing. Portsmouth, Hampshire, UK (1981).

96 Alemany, L.J., Larrubia, M.A. and Blasco, J.M., *A new use of boiler ash: recovery of vanadium as a catalytic VPO system.* Fuel, 77(15): 1735–1740 (1998).

97   Vitolo, S., Seggiani, M. and Falachi, F., *Recovery of vanadium from a previously burned heavy oil fly ash*. Hydrometallurgy, **62**(3): 145–150 (2001).
98   Amer, A.M., *Processing of Egyptian boiler ash for extraction of vanadium and nickel*. Waste Management, **22**: 515–520 (2002).
99   Abdel-latif, M.A., *Recovery of vanadium and nickel from petroleum fly ash*. Minerals Engineering, **15** (Suppl. 1): 953–961 (2002).
100   Aunela-Tapola, L.A., Frandsen, F.J. and Hasanen, E.K., *Trace metal emissions from Estonian oil shale fired power plant*. Fuel Processing Technology, **57**: 1–24 (1998).
101   Smadi, M.M. and Haddad, R.H., *The use of oil shale as in Portland cement concrete*. Cement & Concrete Composites, **25**: 43–50 (2003).
102   Yoffe, O., Nathan, Y., Wolfarth, A., Cohen, S. and Shoval, S., *The chemistry and mineralogy of the Negev oil shale ashes*. Fuel, **81**: 1101–1117 (2002).
103   Freidin, C. and Motzafi-Haller, W., *Cementless building units based on oil shale and coal fly ash binder*. Construction and Building Materials, **13**: 363–369 (1999).
104   Freidin, C., *Hydration and strength development of binder based on high-calcium oil shale fly ash*. Cement & Concrete Research, **28**(6): 829–839 (1998).
105   Hanni, R., *Energy and valuable material by-product from firing Estonian oil shale*. Waste Management, **16**(1–3): 97–99 (1996).
106   Thompson, B.H. and Vierrath, H.E., *The BGL gasifier – experience and application*. In Sixth Annual International Pittsburgh Coal Conference, Pittsburgh, PA (1989).
107   Perry, R.T., Salter, J.A., Baker, D.C., Potter, M.W., Thompson, C.M. and Eklund, A.G., *Environmental characterisation of the Shell coal gasification process: III Solid by-products*. In Seventh Annual International Pittsburgh Coal Conference, Pittsburgh, PA (1990).
108   Choudhry, V. and Hadley, S.R., *Synthetic Lightweight Aggregate from Cool Water Slag: Bench Scale Conformation Test*. EPRI (1990).
109   Carty, R.H., Mason, D.M., Kline, S.D. and Babu, S.P., *Formation and growth of coal ash agglomerates*. Amer. Chem. Soc. Div. of Fuel Chemistry Preprints, **33**(2): 34–41 (1988).
110   McCarthy, G.J., Stevenson, R.J. and Hassett, D.J., *Characterisation, Extraction and Reuse of Coal Gasification Solid Wastes. Vol. 1 – Characterisation of Coal Gasification Wastes (Final Report)*. University of North Dakota, Mining and Mineral Resources Research Institute: Grand Forks, ND (1988).

2

Established Uses of Combustion Residues

Michael Anderson[1] (2.1–2.3, 2.7–2.9) and Rod Jones[2] and Michael McCarthy[3] (2.4–2.6)
[1] University of Staffordshire, UK
[2] University of Dundee, Scotland
[3] University of Dundee, Scotland

List of Abbreviations and Acronyms

AAC	Autoclaved Aerated Concrete
CKD	Cement Kiln Dust
BEA	British Electricity Authority
BRE	Building Research Establishment
CEGB	Central Electricity Generating Board
ESP	Electrostatic Precipitator
FAMB	Fly Ash Bound Mixtures
FBA	Furnace Bottom Ash
FGD	Flue Gas Desulfurisation
ISSA	Incinerated Sewage Sludge Ash
LOI	Loss On Ignition
M/C	Micro-encapsulation/Containerisation
MOL	Milk-Of-Lime
MSWF	Municipal Solid Waste Fly Ash
PC	Portland Cement
PFA	Pulverised Fuel Ash
RH	Relative Humidity

Combustion Residues: Current, Novel and Renewable Applications   Edited by Michael Cox,
Henk Nugteren and Mária Janssen-Jurkovičová   © 2008 John Wiley & Sons, Ltd

S/S   Solidification/Stabilisation
TRL   Transport Research Laboratory
UKQAA   United Kingdom Quality Ash Association
w/w   Weight for Weight

2.1 Introduction

This chapter is devoted to providing an overview of the existing disposal practices and uses of long-standing and newer 'bulk' combustion residues. Inevitably, the majority of the text is devoted to Pulverised Fuel Ash (PFA). This is because since the invention of pulverised coal firing in the United States at the very beginning of the 20th century, this technology has been adopted as the mainstay electricity generating source in over 50 separate countries. As a result, by 1992 an estimated 459 Mt of PFA was being produced annually worldwide.[1] In contrast, more recently there has been a widespread movement to use incineration to reduce the growing volumes of sewage sludge and municipal solid waste. However, this is yielding types of ashes which, although insignificant in comparative tonnage to PFA, are far less benign. Consequently, there is an increasing need to develop and promote alternative and more environmentally acceptable solutions to reduce the current reliance on landfill.

The chapter begins with a summary of the bulk disposal procedures currently being widely applied to all three combustion residues, plus reference to possible recycling opportunities for the two newer ash types. The well-established commercial uses for PFA are then examined, beginning with its use in agriculture, followed by an overview of its versatile role in concrete, grout, fill and highway construction. A more recent and expanding use in toxic waste amelioration is then reviewed. The chapter concludes with a summary of the various commercial building products in which PFA has already been successfully incorporated.

2.2 Disposal Approaches for Three Prominent Combustion Residues

In many countries, land-filling has invariably been the long-accepted option for the disposal of the majority of wastes including combustion residues. It is relatively simple to carry out and, provided that sufficient space exists within economic haulage distance of the source, it has generally remained the cheapest means available. This procedure is reviewed below within the context of long-standing practices, but in the case of the newer combustion residues, mention is also made of legislative pressures that will have an increasing influence on their continued disposal by this route in the future.

2.2.1 Casebook Examples of the Land-filling of Bulk PFA

At a modern coal-fired power station, PFA, after being captured by the fine particulate filter system, can be handled or disposed of in several ways (Figure 2.1), depending broadly on whether it has an immediate market, or needs to be stored in anticipation of future arising markets.

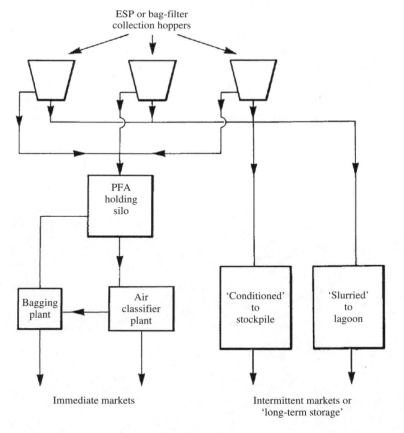

Figure 2.1 A schematic diagram showing PFA disposal options at a power station.

PFA for which an immediate use cannot be found is commonly managed in one of two separate ways. Firstly, it may be drawn from its initial dry accumulation/holding point and, following conditioning with water to typically 16–18 % (dry weight basis), trucked to suitable disposal locations. Secondly, it may be sluiced out of the holding location with water as a slurry and then piped directly to adjacent lagoons (Figure 2.2), where it is allowed to settle, with the supernatant water continually drawn off. When these lagoons are eventually filled, they are left to dry out and the PFA subsequently removed by excavator and stockpiled, thereafter being commonly transported to more permanent disposal placements. Compared with other combustion residues, PFA can be considered as broadly innocuous – although if allowed to blow about it can be justifiably claimed to be an irritant. Until 2007, PFA was free of waste regulation controls in the United Kingdom, and individual power stations generally left to decide their own 'most economic' disposal strategy. With many early power stations, ownership of large tracts of land surrounding them allowed disposal to be kept within the perimeter fence. But in the 1960–70 period, a number of much larger (2000 MW) power stations were commissioned. This greatly increased ash output and required more innovative solutions to be developed, and some examples of these are cited below to illustrate the scale and versatility of such undertakings.[2]

Figure 2.2 Slurried PFA discharging into a power station disposal lagoon.

Filling Old Brick Workings

Brickmaking since the early 1900s around the Peterborough area of Bedfordshire (UK) has produced deep quarries that now cover an area in excess of 400 hectares, into which, between 1970 and 1990, around 50 Mt of PFA from three separate large power stations was deposited. After emplacement and consolidation, 15–30 cm of topsoil was overlain: over time, slow weathering of the PFA has contributed to the soil fertility and the land has now been returned to productive agricultural and amenity use.

Restoring a Coal Mining Subsidence Area

In North Yorkshire (UK), the disposal procedure at two power stations has been to pump the PFA slurried with water (at the rate of 35 000 tonnes per week) to a common disposal point. On arrival, the PFA is dewatered in a vacuum filtration plant to produce a 'cake' that is transported by truck to marginal land prone to subsidence due to earlier coal mining. Here, it is placed, rolled and sculptured by large earth-moving equipment into a relief profile of undulating 'artificial' hills, to merge with the surrounding topography. As the operation continues, these hills are subsequently grassed and planted with trees to complete the re-habitation, which now covers over 300 hectares of land and has elevated the original level of the topography by approximately 50 metres.

Building an Artificial Topographic Landform

The Barlow land reclamation scheme utilises surplus PFA produced at Drax power station in North Yorkshire (UK), which at 4000 MW capacity is the EU's largest coal-fired electricity generating plant. Here, the dry PFA is pumped from the 3500 tonne holding silo, conditioned with water to 20 % (dry weight basis) and then transported by belt-conveyor in this damp form to the nearby disposal site where it is emplaced and sculptured to form a continuous elevated landform that blends in with the topography of the adjoining countryside and that is under constant restoration to pasture-land.

Disposal and recycling procedures of this kind take many different forms in different countries,[3-11] but all have a common goal in establishing appropriate long-term solutions to the unrelenting problem of maintaining a satisfactory and cost-effective disposal/recycling strategy for dealing with this major volume by-product.

2.2.2 Bulk PFA Disposal with Reference to the Aqueous Environment

Irrespective of whether PFA is disposed of into holes in the ground or landscaped above ground level, consideration of the possible interaction with groundwater invariably needs to be addressed. If rainfall percolates through a thickness of stockpiled PFA, it will pick up soluble material during its downward journey and this will ultimately reach the underlying bedrock. It will also displace any original water already held in the body of the deposit. Where the strata below contains or leads to potable, a potential aquifer then before giving permission for a scheme to proceed, the relevant licensing authority will need to be satisfied that there will be no threat to the water quality of the affected catchment area. Where ash is disposed of in mounds, there are also other surface-water issues to be addressed, namely run-off from the slopes of the mound and possible percolation from the mass of the deposit into low-flow drainage systems at the edges of the site. The volumes of such water may be large in relation to the natural water in the pre-existing drainage system and the potential dilution effect may be low. It has consequently been common practice to monitor such drainage situations and, where necessary, to collect it for return and reuse as conditioning water, washing down plant equipment and dust suppression. Other precautionary measures – such as installing geo-textile underlinings and/or topsoil covering of these mounds – are also widely practiced. Studies to enable a more detailed understanding and regular monitoring of the hydrological interaction taking place at such disposal sites have been carried out as a precautionary measure in many countries over a considerable period of time.[12–18]

2.2.3 Land-filling of Incinerated Sewage Sludge Ash (ISSA)

In the UK, incinerated sewage sludge ash (ISSA) has until recently been classified as a 'special waste', but under a recent EU Landfill Directive it has been reclassified as a 'hazardous waste' and can now only be consigned to designated hazardous waste sites. The common handling procedure for ISSA, which today is most commonly produced by fluidised-bed incineration, is to condition it with water to typically 16–20 % (dry weight basis) and transport it in road haulage skips or trucks to appropriate disposal sites. Yet in a few instances, due to the excellent liquid absorption property of this by-product, it has been re-routed to licensed waste transfer-stations and mixed with various other toxic liquid effluents, to enhance their handling and disposal properties. References specifically directed to the disposal of ISSA have not been widely available in the past. However, an international movement towards an increasing use of incineration for this waste and growing environmental concerns surrounding the safety and cost of land-filling the resulting ash has led to an increased level of interest and exploration of alternative options.[19–21]

2.2.4 Land-filling of Municipal Solid Waste Fly Ash (MSWF)

This fine particulate ash is strongly deliquescent and, together with its high heavy metals' content, causes widespread difficulty in handling and disposal. Procedures vary according to local circumstances. Some authorities or companies operating incinerators in the UK take the precaution of encapsulating the ash within impervious bags prior to its disposal. (Figure 2.3). Others merely condition it with water to typically

Figure 2.3 Sealed double-thickness polypropylene bags of MSWF awaiting transport to a hazardous waste landfill site.

10–14 % (dry weight basis) before transporting it in road haulage skips or sheeted trucks to appropriate landfill sites. Here, arrangements are often in place to cover it with a layer of the coarser (and less reactive) hearth/bottom ash arising from the same incineration process. However, the newly introduced EU Landfill Directive now brings more invasive control over such wastes by restricting their future disposal to specially designated hazardous waste sites. Over the past decade, incineration of municipal solid waste has become the preferred option in many EU countries and advanced economies. Consequently, there has been considerable international research to assess the environmental impact on its disposal to landfill, as well as efforts to identify alternative more environmentally friendly alternatives.[22–25] As available landfill space in many places continues to decreases and legislative requirements relating to this disposal practice become more demanding, the costs involved will continue to rise substantially. However, in many countries the land-filling of this ash still remains the most cost-effective option for the present.

2.3   PFA and its Agricultural Applications

The spreading of PFA onto agricultural land has been carried out since at least the 1940s. Initially in the UK, the practice was to strip off the soil from a designated disposal area to a depth of up to 30 cm, depending on the available soil depth. The ash was then laid down and the removed soil replaced on top. This procedure subsequently allowed a full range of crops to be grown. At that time it was assumed that nothing would grow on ash alone, but because of the problems and expense involved in the above practice, an extensive research programme was initiated.[26] Central to this work was the recognition that PFA contained most of the essential nutrients required for plant growth and in this context it was re-appraised as a 'special' kind of soil, rather than merely as a support for a covering layer of natural soil. The initial aim of the investigation was to identify what substances present in the ash were likely to interfere with plant growth. The main problem that emerged was with boron, which was confirmed to be present in PFA in a highly soluble form. As such, it presented a major plant toxin, for while the boron 'available' to plants in normal soil was typically between 1 and 2 ppm, within PFA this was increased by a factor of up to 15 times. Another property of the ash recognised to cause difficulties was the presence of a relatively large amount of soluble salts: their alkaline behaviour when released was capable of raising the soil pH to 8.5 or higher, compared with a more normal range of 6.0–7.0. The next stage of the study was to identify which plants would tolerate such adverse conditions and a programme of pot and field planting experiments was carried out with a wide variety of plants which, after trialling, were then able to be grouped as 'tolerant' and 'intolerant'. The results[27] showed that the clover family was the most tolerant. White sweet clover thrived under the most toxic conditions, followed by white wild clover, red clover and lucerne. Grasses (particularly the rye varieties) were also proven to be tolerant. Among the arable crops, those of the beet and cabbage families thrived best. Moreover, sugar beet, fodder beet and mangles actually benefited from the presence of boron, provided that it was not too high. Rye was the only cereal crop that was really found to be tolerant of the ash in field trials that took place at the time (Figure 2.4). The most appropriate form of ash for this agricultural application was found to be that reclaimed from disposal lagoons, as most of the associated soluble salts had been leached out and removed.

Figure 2.4 The harvesting of rye grown on PFA in the late 1950s in Flintshire, UK.

Another well-established use of PFA in agriculture is as a soil conditioner. In specific instances – such as where clay bands are impeding downward drainage – it has been found possible to intermix these with PFA to increase water percolation and improve the overall soil texture, thereby promoting enhanced root growth. Acidic soils can also be effectively neutralised with the addition of high-lime PFAs.

A further agricultural application, developed and established in Japan, is a unique fertiliser containing PFA, marketed as 'Extra Green Compound Fertiliser'.[28] Its manufacture involves nitrogen- and potassium-containing compounds plus dried seaweed, which are mixed with approximately 30 % (dry weight basis) PFA and added to a phosphate solution, which is then caked and granulated. The resulting fertiliser is then dried to yield the final marketed product.

More recently, investigations have been carried out in a number of different countries to create artificial soils incorporating PFA. To generate the essential carbon : nitrogen ratio for soil bacteria to thrive, composting has been the favoured route, in which the PFA is mixed with organic by-products such as sewage sludge, pulped food waste and paper-making sludge.[29-31] However, increasing concerns, particularly with the ultimate destination of the minute quantities of heavy metals in the ash and their possibility of entering the food chain, have hindered a wider commercial exploitation of this technology.

2.4   Uses of Fly Ash in Concrete

2.4.1   A Short History of the Use of Fly Ash in Concrete

As noted in Chapter 1, fly ash has a long and venerable history of use as a component of concrete. The classic work of Davies and co-workers[32-34] provided the fundamental underpinning of modern specifications and use, and identified the need to assess fly ash quality through fineness, based on 45 μm sieve retention, and loss-on-ignition (LOI). Indeed, these properties continue to be adopted in most specifications worldwide. The USA carried out further developments in the use of fly ash and specifications began to be introduced in many countries (Table 2.1).

The increased use of fly ash in concrete mirrored the expansion of major infrastructure, particularly dams, around the world between the 1940s and 1960s.[35-42] In the UK, the first British Standard for the material was introduced in 1965 (first edition of BS 3892[43]). However, this covered fly ash as a fine aggregate (i.e. inert) and not as a cement. Around this time, the power industry in the UK initiated several comprehensive research projects to examine fly ash properties and their variability.[44] It should be noted that during this period coal firing was carried out under so-called 'base load conditions' – that is, at high temperature – as emission-reducing technology was yet to be introduced. This research

Table 2.1 Historical development of the requirements of the major characteristics of fly ash in various early specifications.

Some early specifications for fly ash	LOI (max. %)	Fineness		Water requirement (max. %)	Pozzolanic activity index (min. %)
		Specific surface (min. m² kg⁻¹)	45 μm sieve retention (max. %)		
Davis et al. (1937)[33]	7.0	250	12.0	103	Min. 5 N mm⁻² with lime
US Bureau of Reclamation (1951)	5.0	300	12.0	103	—
Nebraska State (1952)	10.0	—	—	—	—
ASTM C350 (1954)	12.0	280	—	—	100
BS 3892 (1965)[43] Zone A	7.0	125–275	—	—	—
Zone B		275–425			
Zone C		>425			
ASTM C618 (1968)	12.0	650	—	105	85
Tennessee Valley Class I	6.0	650	<12.0	—	85
Authority Class II	6.0	500	12.0–22.0	—	75
G-30 (1968)[42] Class III	6.0	—	22.0–32.0	—	75
ASTM C618 (1969)	12.0	—	12.0	105	75
ASTM C618 (1971)	12.0	—	34.0	105	75
BS3892, Part 1 (1982)	7.0	—	12.5	<95	85

began to establish the influences of combustion conditions and collection systems on the physical and chemical properties of fly ash. In addition, it was recognised that to use the material as a cement component in concrete construction, tight control of its properties would be required to ensure acceptability in a market used to working solely with Portland cements.

An Agrément Certificate was obtained for fly ash in 1975[45] and BS 3892 underwent revision in 1982,[46] permitting fly ash use as a cement component in structural concrete (Part 1 only). Material to BS 3892, Part 1 was required to have a fineness limit of 12.0 % by mass retained on a 45 μm sieve and a LOI of less than 7.0 %. This standard was further revised in 1997,[47] but with the same general property requirements applying. The fineness range was broadened (up to 40 % by mass retained on a 45 μm sieve) with the publication of the European fly ash standard, BS EN 450, in 1995.[48] This was recently revised in 2005 in two parts[49,50] covering definitions, specifications and conformity criteria, and conformity evaluation. This introduced several changes to the way in which the material is categorised and also coverage of co-combustion fly ash.

During the 1980s, with the acceptance of fly ash as a cement component, research and publications on the use of the material in the UK,[51-67] the USA and other countries with significant coal-fired electricity production, particularly Australia, Canada and South Africa, increased substantially.

2.4.2 Fly Ash Types

Fly ash can be divided into two specific types in relation to its use as a cement component in concrete:

(i) *Low-lime fly ash* (Class F to ASTM C618[68]) is produced from bituminous coal sources and anthracites. The resulting ash is pozzolanic and can be used with Portland cement (PC) or another 'activator' to give cementitious products. This is by far the most common fly ash type.

(ii) *High-lime fly ash* (Class C to ASTM C618[68]) results from brown coal/ lignite combustion. This fly ash exhibits cementitious properties, as there is free lime as well as pozzolanic phases. This ash is common to particular regions of Canada and the USA, where lignite is the dominant coal source.

In Europe, EN 197-1[69] covers fly ash use in cement and EN 206[70] its use in concrete, where further information on cement combinations and their application may be found.

In this section, the properties and effects of fly ash in concrete are confined to low-lime fly ash. The reader is also directed to a number of textbooks for detailed information on fly ash use in concrete[71-75] and the following provides a summary of the key issues raised in these works and elsewhere.

2.4.3 Key Fly Ash Characteristics with Regard to its Use in Concrete

Loss-on-ignition (LOI)

The LOI is used as an indicator of the unburned coal residue in fly ash and reflects the combustion conditions under which the material was produced. The LOI can influence the colour, water demand, fineness and reactivity of the material, with performance in concrete generally becoming poorer as this increases.

Limits are, therefore, normally set for LOI in national standards. The recently introduced BS EN 450-1[49] gives LOI in three categories: (i) Category A, not greater than 5.0 %; (ii) Category B, between 2.0 and 7.0 %; and (iii) Category C, between 4.0 and 9.0 %. Unburned residual coal can also significantly influence air-entrainment in concrete, and indeed this has restricted its use in highway structures in temperate climates.

Sulfate Content

The presence of sulfate in fly ash can cause expansion when used in concrete due to ettringite formation and, as a result, its content is limited in standards. BS EN 450-1[49] sets the limit for sulfate content, expressed as sulfuric anhydride (SO_3), at 3.0 % by mass.

Alkali Content

The alkalis arising from PC and fly ash can react with certain silicate aggregates, leading to the formation of an expansive gel, which in time can cause cracking and degradation of concrete. A limit on total alkali content as Na_2O (equivalent) of 5.0 % by mass is given in BS EN 450-1[49] and it is also noted that fly ash from pulverised coal only is deemed to satisfy this. It is generally agreed that fly ash inhibits alkali–aggregate reaction, but the reader is referred to specialist reports[76] on this process for more detailed advice.

Water Requirement

Water requirement is generally considered to be a good indicator of fly ash 'quality'; that is, ashes that reduce water requirement compared to PC may be expected to perform well in concrete, as a cement component. The water requirement (applicable to fineness Category S to BS EN 450-1[49]) is measured using a standard flow test and mortar, with 30 % fly ash in cement. Category S fly ash (with fineness less than 12.0 % retained on a 45 μm sieve) should have a water requirement of not greater than 95 % of its Portland cement mortar reference. It is argued that the test is not sensitive to changes in fly ash quality, particularly when its inherent variability is considered.

Activity Index

BS EN 450-1[49] uses an activity index test to provide a measure of fly ash reactivity. This adopts a fixed water : cement ratio and 25 % fly ash by mass as a cement component in the mortar, with mixes therefore having potentially variable flow. In this test, fly ash mortar strength is expected to be at least 75 % of the PC reference mix at 28 days and 85 % by 90 days.

Fineness

Fineness is probably the single most important characteristic of fly ash in relation to its use in concrete, with the basic rule being 'the finer the better'. Fineness is widely specified as a limit on the mass of material retained on a 45 μm sieve. Although a full particle size distribution is probably more indicative of quality, the simplicity of the 45 μm sieve test means that a full particle size distribution is rarely specified.

BS EN 450-1[49] defines two categories of fineness: (i) Category N, the fineness should not exceed 40.0% by mass retained on a 45 μm sieve and should not vary by more than ± 10.0 percentage points from the declared value; and (ii) Category S, the fineness should not exceed 12.0 % by mass. The fineness variation (± 10.0 percentage points) does not apply in this case. Details on the effect of fly ash fineness on concrete performance can be found in Chapter 4.2.2, page 209.

Co-combustion

Fly ash produced from the firing of coal with co-combustion material (e.g. wood-based products, paper sludge, municipal sewage sludge
etc.) can be used as a cement component in concrete. Limits on co-
combustion materials in the fuel, the ash and their calorific contribution
are given in BS EN 450-1,[49] where means of establishing co-combustion
fly ash suitability and environmental compatibility are also considered.
Further information on the influence of co-combustion on fly ash properties and its use in concrete can be found in Chapter 4.2.4, page 212.

2.4.4 Fresh Concrete Properties

The spherical nature of fly ash particles is considered to produce 'a ball
bearing-like' effect, allowing PC particles to move past each other more
easily, and is analogous to the effect of a plasticiser dispersing these, and
improving concrete workability. In addition, fly ash particle surfaces
adsorb mix water due to their hydrophilic nature, resulting in a 'water-
entrainment' effect, and wider distribution of water in fresh concrete.
These influences also reduce bleeding and aid the development of a more
cohesive and stable mix.

There are likely to be other factors involved, which affect fresh
concrete. Indeed, fly ash may have a retarding effect on the early hydration of C_3S (tricalcium silicate, Ca_3SiO_5) and C_3A (tricalcium aluminate, $Ca_3Al_2O_6$), thus increasing workability retention compared to PC
concrete, although setting times are generally similar.

There are, therefore, several factors associated with fly ash that influence concrete workability, as summarised in Table 2.2, and some fly
ashes should allow a reduction in water demand. This varies from ash to

Table 2.2 The influence of fly ash on fresh and hydrating concrete.[71]

Influence	Effect
Physical	Reduced cement flocs; improved uniformity of the microstructure; improved distribution of mix water; additional sites for Portland cement hydration; possible reduction in microcracking by 'blunting' crack growth due to the presence of hard, spherical fly ash particles
Chemical	Preferential encouragement of Portland cement silicate phase hydration
Pozzolanic	Increased amount of cementitious material

ash and also depends on the way in which the concrete constituents are proportioned. The generalised use of plasticisers, however, both masks the influence of fly ash on fresh properties and enhances its effectiveness, particularly at lower water : cement ratio.

The use of fly ash also improves other characteristics of plastic concrete, such as cohesiveness, pumpability and finish, although surface colour will tend to be darker than PC concrete, due to the presence of residual carbon particles. Occasionally, the carbon particles can cause 'staining' on concrete surfaces, since with their low density they can 'float' out of the mix.

2.4.5 Hydration and Pozzolanic Reactions

Fly ash has a complex influence on the setting and hydration of cement and strength development of concrete. The dispersing effect of fly ash on PC particles opens an increased area of the material to hydration, potentially enhancing the microstructure. Further benefits, through the wider distribution of water in fresh concrete and reduced bleeding, enhancing aggregate/paste interfaces, may be achieved.

Fly ash can also enhance the hydration of the PC silicate phases, forming additional gypsum, thus retarding the hydration of C_3A and promoting C_3S dissolution through adsorption of Ca^{2+} ions on fly ash particle surfaces. The fine nature of fly ash also provides increased sites for the precipitation of C–S–H (calcium silicate hydrate).

The pozzolanic reaction between the glass portion of the fly ash particles and the $Ca(OH)_2$ by-product of PC hydration is given elsewhere, where more detailed reviews of C_3S – fly ash and C_3A – fly ash interactions are also presented.[73]

One important effect of fly ash use in concrete is reduced problems due to heat of hydration; that is, with increased fly ash levels in the mix, there is a lower rate of heat evolution and peak temperature reached. Historically, this effect was one of the main drivers for fly ash use in large dam construction.

Clearly, fly ash has several effects on the developing concrete structure. While the timing and magnitude of these influences is both complex and difficult to quantify individually, their overall effect on concrete strength development is well established and certainly enhances strength development from 28 days, in the presence of moisture.

2.4.6   Engineering Properties of Concrete

It is important to recognise the way in which research work comparing the behaviour and properties of PC and fly ash concrete has been and continues to be reported. In many cases, a direct mass replacement of PC is used and/or an equal water : cement ratio, giving concrete of different standard strength, and with fly ash concrete usually slightly lower. As a result, many of the properties of fly ash concrete have been reported as being lower or inferior. Particular care must be exercised when interpreting data presented in this way, as concrete will mainly be specified with a particular strength requirement, irrespective of the cement type used. Therefore, arguably, the only real basis of comparison between different concrete mixes is by equal strength; that is, how the concrete is normally used in practice. On this basis, fly ash concrete will normally provide significantly superior performance when compared to PC concrete. It should also be noted that to achieve comparable strength, the water : cement ratio of fly ash concrete normally requires to be reduced, either by increasing the cement content or, more effectively, by reducing the free water content. The following discussion assumes that concrete mixes are of equal strength.

Compressive and Tensile Strength

For engineers, the compressive strength of concrete is one of its key structural requirements and, despite its limitations, 28 day cube or cylinder strength continues to be the primary method of assessment. However, there are often further requirements for early strength development to allow for example formwork to be stripped. Typical data for age–strength relationships are given in Table 2.3.[71] This indicates that fly ash concrete behaviour is broadly similar to PC concrete up to 28 days, although there can be some lag up to seven days, and increases compared to that of an equivalent PC concrete thereafter, provided that the constituent proportions are correctly designed.[51–53] It should be recognised that these advantages can only be realised when appropriate curing procedures are carried out.

The effects of accelerated curing (at high temperature) on fly ash reactions are used effectively in precast concrete production. The impact on strength development of the high temperatures achieved in mass concrete or large sections is also noteworthy, even though fly ash reduces the peak temperature reached. For further information, reference should be made to work published elsewhere.[56,58]

Table 2.3 Typical strength age factors for standard cured (20 °C) PC and fly ash concretes.[71]

Design 28 day strength (N mm^{-2})^a	Strength factor					
	3 days	7 days	28 days^b	3 months	6 months	1 year
PC concrete						
20	0.39	0.63	1.00	1.26	1.40	1.48
30	0.42	0.65	1.00	1.24	1.35	1.44
40	0.46	0.68	1.00	1.22	1.32	1.42
50	0.48	0.70	1.00	1.19	1.29	1.38
60	0.50	0.72	1.00	1.16	1.24	1.32
Fly ash concrete						
20	0.40	0.60	1.00	1.50	1.91	2.40
30	0.41	0.63	1.00	1.43	1.72	2.25
40	0.44	0.66	1.00	1.36	1.60	1.76
50	0.46	0.69	1.00	1.30	1.46	1.58
60	0.49	0.71	1.00	1.24	1.36	1.44

^{a} Using 100 mm cubes.
^{b} Concrete proportioned to give equal strength at 28 days standard curing.

The relationships between tensile and compressive strength for PC concrete have been found to generally apply to fly ash concrete.

Modulus of Elasticity (E-value)

Given the same strength class, it can be expected that there will be similar stiffness for fly ash and PC concretes, and that the conventionally used equations to estimate this parameter in design codes from standard cube or cylinder strength are equally applicable. Many investigations have demonstrated that, using a wide range of fly ash qualities, when concrete is proportioned on an equivalent strength basis, fly ash has no significant effect on the E-value.

Creep and Drying Shrinkage

The use of fly ash in concrete can reduce creep, although only where the pozzolanic reactions are able to be supported over an extended time; that is, where there is significant long-term strength increase. It is generally agreed that creep deformation of fly ash concrete follows the same trend as that of Portland cement concrete with respect to compressive strength. Although structural design codes make no allowance for these, it is likely that a lower creep coefficient will reduce the long-term deflection of structural elements made with fly ash concrete.

It is normally assumed that the inclusion of fly ash in concrete mixes does not significantly change the rate or ultimate level of shrinkage occurring and that shrinkage in concrete with or without fly ash has a similar magnitude.

2.4.7 Concrete Durability

Permeation Properties

The permeation properties – that is, absorption, permeability and diffusion – are the principal mechanisms by which aggressive liquids, ions and gases pass into concrete. These are indicative of the concrete microstructure and durability performance. The use of fly ash generally improves these properties, making concrete more resistant to environmental deterioration. The pozzolanic reactions convert free lime in concrete to cementitious gel, blocking pores, enhancing permeation and reducing the level of portlandite that can be leached.[71] The only negative issue arising from this is that it results in a lower pore fluid alkalinity. While this reduces the risk of alkali–aggregate reaction, it may increase the rate of carbonation.

Carbonation

As noted above, the pozzolanic reactions result in reduced pore fluid alkalinity and this effect is not wholly offset by the improvement in concrete permeation properties. Overall, the rate of carbonation may be slightly higher than Portland cement concrete, depending on the mix proportions. There are no reported instances of these effects causing any problems in correctly specified and compacted concrete in real structures. In turn, concrete specifications for durability do not distinguish between PC and fly ash concrete, although care has to be exercised at higher fly ash contents; say, > 40 % by mass. Rates of reinforcement corrosion in carbonated concrete are generally similar between fly ash and PC concretes, where the main controlling factor is the environment conditions, in particular relative humidity.[60,61]

Chloride Ingress

Chloride from the environment, either de-icing salts or in a coastal exposure, can penetrate concrete and when present in sufficient quantities at the site of reinforcement lead to corrosion, which can threaten the serviceability of structures. Work investigating the use of fly ash in concrete[63-65] indicates that the material is effective in reducing rates of chloride transmission (by absorption and diffusion). The main benefits of fly ash in this respect are due to[66] (i) its high alumina content and ability to bind chloride, (ii) the large number of fine particles to adsorb chloride and (iii) the decreased interconnected porosity and pathways into concrete. Furthermore, it appears that it is the quantity of fly ash[63] rather than its quality that is the critical factor influencing resistance to chloride ingress. Other work[67] suggests that at a given level of chloride in concrete at the depth of reinforcement, less corrosion may occur in concrete containing fly ash.

Sulfate Attack

The use of fly ash improves the sulfate resistance of concrete, and can be used in all but the most extreme exposure conditions, where additional concrete protection is, in any case, required.[77]

Alkali–Aggregate Reaction

Fly ash is widely used in concrete to minimise the risk of alkali–aggregate reaction, although this remains a complex area, where specialist advice should always be sought.[76]

Frost Resistance

As noted earlier, the presence of residual carbon can cause variability in the entrainment of air in concrete with fly ash. The quantity of admixture required to give a specific air content will also generally be greater for fly ash concrete. Thus, it is necessary to exercise greater control with air-entrained fly ash concrete. The latest generations of admixtures have improved this situation, and variability in the entrained air content is less with chemical systems specifically designed for use with fly ash.

In terms of actual frost resistance, once air is entrained to the required level, there is little difference in performance compared to that of PC concrete.

2.4.8 Summary Points

Fly ash is now well established as a cement component (addition) for structural concrete and it enhances nearly all of its properties. However, fly ash is changing, mainly due to low-emission coal combustion methods and this is leading to an increase in coarseness and loss-on-ignition. Fundamentally, this has not changed what fly ash is or does in concrete, but users will have to adjust mixes to compensate for these effects. It is possible that producers will use more processing of fly ash to enhance its properties.

Whatever the long-term future is for coal-fired power stations, it is clear that for the foreseeable future this source of power will remain a significant contributor to the world's energy needs. Conversely, the use of fly ash remains stubbornly around 50 %, as it has done for the past 30 years, and this is not acceptable either for the conservation of material resources or the environment. Many countries have used taxation or material levies to facilitate the growth of a 'recycling' culture, but this remains a highly controversial way of bringing about changes to the bulk materials market. There is no technical reason why 100 % fly ash should not be used, as is the case for furnace bottom ash in many countries. One aspect that may produce a rethink on the use of fly ash is climate change and the effect of CO_2 emissions during Portland cement production. Although concrete is a relatively green material, it still remains the case that approximately a tonne of CO_2 is produced for every tonne of Portland cement. The more widespread use of fly ash, even at a 10 % level, could greatly help the overall situation. As the new mega-economies of China and India develop further, so will the need for more concrete for infrastructure, and fly ash must play an important role in minimising the environmental impact of this.

The major challenge for researchers, producers and users, as well as regulators, is to improve the utilisation of fly ash, including all grades – that is, dry run-of-station, conditioned, stockpile and lagoon – as well as the fine proportion. This will mean that processing will become more widespread, and that new markets must be identified and developed for all of the resulting processed grades, that make best use of their particular characteristics.

2.5   Application of Fly Ash in Grouts

Grouting is the technology concerned with the introduction of a filler material (grout) into the voids of or between other materials (man-made or natural) in order to fill or strengthen them, or to achieve structural action. Applications range from the bulk filling of mines to improving the physical properties of the ground, to providing the bond between tendons and sheaths in pre-stressed concrete construction.[78–80]

The main requirements of grouts are that: (i) following production, they have appropriate fluid properties to be transported to their required position within the construction, normally by pumping, which may be over a range of distances; and (ii) thereafter, they develop a hardened structure, with specific properties for the particular application. These are, therefore, the main factors considered in formulating and specifying grout.

There are a variety of grout types that may be used in grouting operations and a range of techniques that may be adopted to achieve particular effects. Specialist information on these can be obtained from relevant textbooks and literature covering the wider subject area.[75,78–83] The aim of this section is to provide a summary considering the use of fly ash and its impact on the main characteristics of cementitious grouts.

2.5.1   Materials and Grout Compositions

The use of fly ash as a component of cementitious grouts is a relatively mature technology, which dates back to early applications of the material in construction. The characteristics of fly ash for use in grout in the UK are described in BS 3892, Part 3,[84] which was published in 1997. This applies to grouts for various applications, except those in ducts for pre-stressing tendons, which are covered in BS EN 447[80] (and where there are restrictions on cement types that may be used).

The main property requirements of fly ash in BS 3892, Part 3 include fineness, which should be $\leq 60\%$ retained on a $45\,\mu m$ sieve, LOI $\leq 14.0\%$ and SO_3 content $\leq 2.5\%$ (all by mass). The material may be used in conditioned form, where this is agreed between supplier and purchaser. Thus, a wider range of fly ash qualities may be used in cementitious grouts than in concrete. The standard also describes tests to measure fluidity and strength activity to enable these properties of fly ash grout to be evaluated in relation to particular applications.

While fly ash can be used alone with water, in low-strength, bulk-filling applications,[85] it is normally combined with other components, most often Portland cement (PC). Given the differences in physical properties and reaction characteristics of these materials, it is possible by adjusting their relative quantities in the grout to control the fluid, setting and hardened properties. Indeed, the main parameters used in proportioning fly ash grouts are the water : solids ratio (similar to the water : cement ratio in concrete) and the fly ash : PC ratio.

Other grout compositions with fly ash include combinations with lime, cement and clay, cement and sand, or in specialist grouts.[85,86] These are used to achieve specific effects, for example, the use of lime can improve pumpability, while clay can reduce bleeding.[85] A range of chemical admixtures can also be used with fly ash grouts to similarly modify properties.[78,86]

2.5.2 Handling and Placement

The main issues of concern associated with recently produced grout during construction operations are that the material has: (i) appropriate fluid characteristics, including stability, so that it can be transported to its location in the construction; and (ii) particle characteristics suitable for the specific void system.

Fluidity

Cement-based grouts after mixing usually exhibit the behaviour of a Bingham fluid; that is, they have both viscosity and cohesion,[87] which has implications for placement of the material. In practice, this aspect of grout is frequently evaluated using methods including flow cones or a flow channel,[78,84] which measure the time or distance for fixed volumes of grout to flow through a small-diameter outlet, or move along a channel.

Results obtained using a flow channel (the Colcrete™ flow channel) for fly ash/PC grouts (with a water : solids ratio of 0.4 but without admixtures) up to a ratio of 10 : 1 indicate that flow increased with fly ash content and was influenced by fly ash PC from different sources.[88] Other work[86] suggests that decreases in fluidity may occur with increasing fly ash level in grout (proportioned by mass), as thickening occurs.

Given the reaction characteristics of fly ash and the levels commonly used in grouts, the handling time of the material is likely to be extended. Work examining variations in fluidity with time indicates that in fly ash : PC grout with a ratio of 3 : 1 or more, handling could be carried out satisfactorily for around 4 hours.[86,89]

Bleeding

Bleeding is defined as the autogenous flow of water within, or emergence from, newly placed grout.[87] The process occurs as the quantity of water for transporting grout is greater than that required for reactions with the cement components.[78] As in concrete, excessive bleeding is undesirable, since it can negatively influence the properties of the hardened material.[75] In addition, where there is bleeding during grouting, an assessment of the environmental implications associated with fly ash leachate may be necessary in certain applications.[90] This represents an area that has been the subject of a recent investigation and where further information can be obtained.[91]

The main factors influencing bleeding of grout have been described as,[75] (i) the relative quantities of solids/fluid, (ii) particle densities of the solids/fluid, (iii) the fineness of the particles present and (iv) the setting characteristics of the grout. Given its characteristics fly ash may be expected to give benefits compared to PC with respect to bleeding.[85] An evaluation of the impact of fly ash in grout on this process[86] suggests that the rate of bleeding and total bleeding tend to increase with fly ash content. It was also noted that re-absorption of bleed water into grout can occur with time, and this may be greater at lower fly ash levels (due to increasing water combination by hydration at higher Portland cement contents).

2.5.3 Compressive Strength

There are a range of applications in which grout may be used and requirements for strength can vary from a few N mm^{-2} to levels around that of structural concrete. As noted above, the proportioning of fly ash grouts is mainly based on the water : solids and fly ash : PC ratios, and hence these can facilitate control of both the fluid properties and strength.[92] The typical ranges for these, as noted in the literature, are between about 0.35 and 0.60 and 1 : 4 to 20 : 1 (by mass), respectively.

As might be expected, strength normally reduces with increasing water : solids and fly ash : PC ratios. The impact of both parameters on strength tends to reduce as the fly ash : PC ratio increases.[92] It was also noted in the same work that finer fly ashes do not necessarily give increased strength in grout, as has been reported in concrete, and this may relate to particle packing with coarser material.

Another benefit associated with fly ash in grout is the long-term strength development achievable beyond 28 days.[88] This depends on grout compositions, since at high fly ash contents, limited lime availability may influence the degree of pozzolanic reactions occurring. Given the long-term strength potential, it has been suggested that specification of grout strength at later ages may be more appropriate in certain applications than at 28 days.[75,92]

2.5.4 Durability Performance

Given their geotechnical applications, one of the main issues of concern with regard to durability of fly ash grout is where sulfate is present in the ground. The literature covering fly ash in concrete indicates that in this type of exposure, the material is beneficial in limiting sulfate attack and its use is recommended in relevant guidance.[77] The effects of fly ash are mainly due to the dilution of PC (and hence sulfate-reactive components) and reduction in $Ca(OH)_2$ through pozzolanic reactions. In concrete, benefits may also be achieved through enhanced permeation characteristics, which influence sulfate transportation rates.[71]

Work carried out measuring expansion of grout exposed to sulfate solutions (both sodium and magnesium sulfate) indicates that improvements are observed when fly ash is used in grout; in particular, at high levels.[92,93–95] This is considered to be due to the dilution effects highlighted above for concrete, the nature of the sulfate product formation and the potential of the fly ash grout structure to accommodate expansive products. On the basis of these studies, classifications to define performance and/or give guidance on suitable compositions for sulfate-bearing conditions have been suggested.[92,93]

2.6 Use of Fly Ash in Fill and Pavement Construction

One of the main large-volume applications of fly ash is in the construction of general and structural fills, and the material is also finding increasing use in pavement construction. The following provides a summary of fly ash use in these applications. It should be noted that the United Kingdom Quality Ash Association (UKQAA),[96–100] provides an extensive list of design information and data for engineers and specifiers, as well as details about the construction technology.

2.6.1   Applications in Fill

Conditioned or wet-stored (stockpile or lagoon) fly ash is normally used in fill applications, an example of which is shown in Figure 2.5. A gradual understanding of the behaviour of fly ash as fill has been developed over the years. Indeed, in early work[101] examining water addition to fly ash, it was found that (at optimum moisture content) beyond its initial properties after compaction, the California Bearing Ratio increased by two to three times by 28 days. It was suggested that a weak pozzolanic reaction occurred between the added water and the small amount of 'free' lime believed to be present in the fly ash. Other work[102] suggested a link between strength development and reactions of fly ash free lime, calcium sulfate and other water-soluble contents. Precipitation of hydration products in compacted fly ash have been observed,[103] appearing to give a

Figure 2.5 A roadway embankment constructed of compacted fly ash (Ash Marketing Division, National Power Plc).

bonding matrix, which was considered to be responsible for the enhanced strength. Similar effects have been noted in more recent work[104] examining conditioned fly ash (stored for different periods of time) for use as a cement component in concrete. Reference has also been made[105] in the compaction of fly ash to the development of cohesion due to suction forces occurring during the process, which may be affected by moisture conditions in the material.

Details of fly ash for use in earthwork applications, in the U.K., are covered in the *Manual of Contract Documents for Highway Works*[106] and its guidance notes,[107] (Which include general and structural fill, reinforced earth and use with cement to form capping) and describe fly ash as a cohesive material. These distinguish between conditioned, stockpile and lagoon fly ash and their use in general (stockpile and lagoon, Class 2E) and structural (conditioned, Class 7B) fill, and describe various material property requirements, test methods and limits for acceptability in these applications. The optimum moisture content of fly ash as fill and associated maximum dry density, relevant to compaction, typically lie between 15 and 35 % and 1.1 and 1.6 $Mg\,m^{-3}$, respectively, by standard test.[105] Compaction requirements for fly ash in these applications are given as end-product specifications (95 % of maximum dry density),[106] but it is noted in the literature that material variability and moisture content control may be issues with regard to the achievement of this. This type of approach to specifying fly ash in general fill has also been examined in a recent TRL report,[108] where it is suggested that a method specification, commonly used for natural fill materials, could be adopted in this case.

Issues relating to the nature of fly ash influencing the details of fill construction are considered in references 106 and 107 and in the *Design Manual for Roads and Bridges*.[109] Case studies in the literature[110] also demonstrate how fly ash has been used effectively in practical situations. Similarly, details of the construction process and other issues associated with fly ash that may influence performance in fill, such as environmental aspects and dealing with the risk of frost heave, can also be found in UKQAA technical information.[96]

2.6.2 Fly Ash Bound Mixtures (FABM) in Pavements

Fly ash bound mixtures (FABM) used in various components of pavements (eg road-bases, capping layers) comprise fly ash combined with either lime (quick or hydrated), lime/gypsum or Portland cement (and relatively low water contents), although they may also contain other constituents, such as granular fillers. The properties of fly ash for use in these mixtures are described in a European Standard (EN: 4227–4),[111] which covers both siliceous and calcareous fly ash, and provides limits on physical properties, chemical composition and requirements for the material's reactivity. Fly ash (siliceous) in this application can be coarser and of higher LOI than that used in concrete and either dry or conditioned.

The use of fly ash bound mixtures in pavement construction in the U.K. is covered in the *Manual of Contract Documents for Highway Works, 800 Series*.[112] and a European Standard (EN: 14227–3).[113] These documents give mixtures for various material combinations with fly ash and their performance requirements. The mechanisms involved during the reaction process of fly ash with lime (similar to those noted in fill) have been described as:[114] (i) combination of lime with soluble sulfite/sulfate compounds in fly ash to form gypsum, which results in an initial set after 24–72 hours; and (ii) a slower reaction involving the lime and fly ash alumino-silicates, which begins immediately and may continue for several months. These hardening characteristics with lime offer advantages in relation to the construction period and to the mixture properties finally achieved.[97]

More detailed coverage of mix design, and use in different pavement layers, construction techniques and the performance of these mixtures can be found in UKQAA publications.[97–99] Case studies describing the use of fly ash bound mixtures in UK pavements and the technical issues associated with these can also be found in the literature.[115]

2.7 PFA as an Ameliorator of Liquid and Solid Toxic Wastes

2.7.1 Stabilisation/Solidification (s/s) Technology

Stabilisation/solidification (s/s) systems for the remedial treatment of a wide variety of liquid/solid toxic wastes have been used in the UK for approximately 20 years. However, elsewhere, such as in the USA, Japan and some other EU member states, these technologies have been established for over 30 years.[116] Stabilisation and solidification technologies are different, but complementary, and in some instances overlap. They have been broadly defined in the following way:[117]

- *Stabilisation.* This procedure involves adding appropriate chemicals to a contaminated material to produce more chemically stable constituents; for example, by precipitating soluble metal ions out of solution. It may not result in an improvement in the physical characteristics of the waste – for example, a treated liquid may still remain a relatively fluid sludge – but the treatment will have reduced the toxicity and mobility of the hazardous constituents within it.
- *Solidification.* This procedure involves adding appropriate reagents to a contaminated material to decrease its fluidity/friability and thereby reduce access by external mobilising agents such as wind and water. It does not necessarily require a chemical reaction between the contaminant and the solidification agent, although such reactions may take place, depending on the nature of the reagent.

Cement is recognised to be the most popular treatment additive for a wide variety of situations requiring the effective stabilisation of aggressive wastes, variously accompanied by other appropriate enhancing additives and/or to promote solidification. Yet the use of PFA is also well established in a supporting role, due to its widespread availability and relative cheapness. Most frequently, it is used in combination with cement, CKD or lime. The properties of PFA, which makes it ideal in many such applications, can be broadly summarised as follows:

(1) the alkalinity arising from its surface coating of soluble salts;
(2) the possession of pozzolanic properties;
(3) its generally inert (dominantly aluminosilicate) composition;
(4) a high specific surface area per unit mass;
(5) good flow characteristics and ease of handling in mixing operations and emplacement; and
(6) a low permeability, due to its good particle packing efficiency once emplaced and compacted.

An example of s/s technology using PFA where both mechanisms work together can be cited from the Hanford disposal operation in the USA, which involved the land-filling of mixed low-level radioactive liquids.[118] Here, a mixture of cement, PFA and clay was first dry-blended before mixing with a dilute phosphate/sulfate solution derived from decontamination activities at a nearby nuclear reactor. The resulting grout-slurry was then pumped through a pipeline into specially built reinforced concrete disposal vaults (Figure 2.6). After hardening, the rigid matrix formed, created a physical barrier to the release of the

Figure 2.6 PFA/OPC/clay grout incorporating low-level radioactive wastes for disposal and burial in purpose-built concrete vaults (Reproduced by courtesy of *Concrete International*, American Concrete Institute, 'Concrete – design & construction').

toxic constituents through solidification. At the same time, the cementitious material in the grout maintained the high alkalinity of the waste (pH 13) that contained multivalent cations and brought about a reduced mobility for these species. Moreover, some metal ions were reported to be completely stabilised by their incorporation within the crystal structures of the minerals formed during the overall process.

A common example of the application of solidification technology using PFA that has been widely adopted follows a procedure where the ash is blended with an appropriate level of cement, plus the targeted toxic waste, then mixed into a slurry form and cast into rigid moulds. After hardening, the concrete blocks (monoliths) produced immobilise the toxic species by locking them up within an insoluble cementitious amalgam. This approach has been found appropriate for the containment of many hazardous wastes, such as low-level radioactive salts, mercury, arsenic, cadmium, pesticides and high-concentration phenols. The concrete monoliths are then commonly emplaced in appropriate landfill sites, with minimal risk of the release of the encapsulated contaminants. The added attraction of using PFA in this role is that its presence brings enhanced workability during the slurry mix preparation, a low water:cement ratio requirement and long-term strength gain once in place.

Nevertheless, it is widely recognised that each s/s project under consideration must be viewed as unique and that, consequently, the specific procedure selected will always be dependent upon the unique properties and characteristics of the particular waste under consideration.[119]

2.7.2   Related Technologies

Micro-encapsulation/containerisation (m/c) is a closely related technology to s/s, but relies on a physical enrobing process and has the advantage of being able to use PFA alone. Here, rather than chemical or rheological reactions, the waste being dealt with is targeted to be evenly dispersed and locked within an entrapping PFA matrix. This approach has been successfully used for stabilising intermediate level hazardous wastes such as ore-refining tailings in South Africa.[120] In particular, applying it using high-lime PFA capitalises both on the ash's high surface area absorption capacity as well as its pozzolanic nature. The results reported are said to offer a good economy in scale by successfully tackling and disposing of fairly large quantities of reasonably hazardous materials at a modest cost, while at the same time affording acceptable environmental protection. Nevertheless, within the context of this general application, it is always necessary to consider the impact of such treatment in relation to any possible deleterious effect that the selected PFA itself might introduce through the activation and release of toxic metallic species that may have become associated with it at its time of formation, and which would otherwise have been absent from that particular waste-stream undergoing remedial treatment.

A further derivation of the broad m/c approach also recorded from the USA is the 'Poz-O-Tec' process, which was developed in the early 1970s with the initial objective of stabilising FGD scrubber sludges. The sludge was first dewatered and then mixed with PFA on a typical 1:3 (dry) w/w basis. Subsequently, a small quantity of lime was added to activate the mix and further mixing undertaken. The resulting product was self-hardening and suitable for landfill-capping.[121] This same basic technology has subsequently been successfully expanded to deal with a wide variety of other toxic wastes.

Using PFA as a 'entrapping barrier' to prevent the migration of toxic wastes also provides a valuable function in the construction and operation of landfill sites. Notably in the USA, it has been widely used to line large containment structures, particularly with the employment of high-lime PFAs, which have been shown to be notably successful due to their physical and chemical absorption capacity, enabling them to form an effective impermeable barrier against migrating contaminant leachates.[122,123]

2.8   Commercial Building Products Incorporating PFA

Over the long period in which PFA has been widely available, considerable international research has been carried out to develop it as a feedstock for a variety of building materials, as it offers a potential end-use for substantial quantities of this material. The following summary describes the more successful developments that have subsequently entered the marketplace.

2.8.1   Autoclaved PFA Concrete Blocks

This type of PFA-based building product is known variously as 'autoclaved aerated concrete' (AAC) or simply as 'gas concrete'. Its original development goes back to the late 19th century and initially used silica sand and lime as the basic ingredients. Full-scale block production was established in Sweden during the late 1920s. Subsequently, a modified technology in which PFA was used to substitute for the siliceous component was developed and is now in widespread manufacture in the UK,[124] the former Soviet Union, China and India. Basically, this derived product is composed of cement, lime, aluminium powder and PFA (usually about 55–60 % by dry weight). Following the typical process technology shown diagrammatically in Figure 2.7, the main ingredients are discharged in the correct proportions from weigh-bins and pre-mixed. They are then transferred into a slurry-mixer, with hot water and a small amount of aluminium powder added. An immediate reaction begins as the aluminium reacts with the lime to form minute hydrogen bubbles. As this reaction starts, the mix is poured into large metal moulds (typically 4–5 m^3), to half-fill them. The continuing evolution of hydrogen causes the mix to rise and finally fill the moulds, yielding a body with the consistency of mousse, during which time the hydrogen subsequently diffuses, to be replaced by air. After approximately 2 hours the product has hardened to the consistency of cheese, the mould boxes

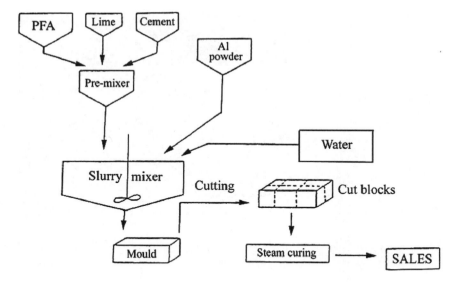

Figure 2.7   A schematic view of the autoclaved aerated concrete (AAC) process for building blocks.

are stripped off and the block cut into smaller-sized building blocks. Packs of these blocks are then transported to autoclaves, where they are cured by high-temperature steam (typically 200 °C at 10 bars pressure) for an average time of 3–4 hours, after which the product is ready for sale. The micro-cellular structure (Figure 2.8) gives these blocks excellent insulation properties: they are also lightweight, of moderate strength and used extensively by the building and construction industry.

2.8.2   Sintered Lightweight Aggregate from PFA

The UK Building Research Establishment, in association with Laing Research and Development Ltd,[125] first developed a method of producing lightweight aggregate from PFA by a sintering process in the mid-1950s. Using this technology, a commercial product (Lytag)[126] was manufactured at a number of factories in the UK during the 1960–70 period. Similar factories have also since been commissioned elsewhere, notably in the Netherlands and Poland. Although minor differences in plant design are apparent, the basic Lytag process can be used to illustrate the broad technology (Figure 2.9). Initially, PFA is conveyed pneumatically from the power station into a storage silo, from which

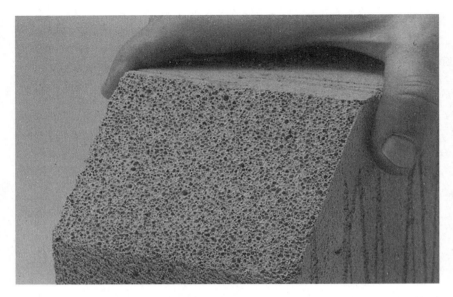

Figure 2.8   The highly porous structure of an autoclaved aerated concrete block product.

it is discharged by feeder and carried to rotary mixers, where water is added to moisten it. Depending on the carbon content of the feed, it may also be necessary at this stage to add pulverised coal to maintain an optimum level of 4 % carbon in the mix. It is then fed onto rotating inclined pans, where the rotation of the pans causes the PFA particles to be held together by capillary forces derived from the water content, typically 15–20 % (wet weight basis). The green pellets thus formed are fed onto a sinter-strand, where the surface is ignited under a hood by burners (firing downwards), using either oil or gas. Thereafter, using a downward draft, the heat is drawn gradually through the bed and sintering takes place throughout its entire thickness. A temperature in the region of 1200–1250 °C is needed for the sintering operation, with most of the energy required to raise the entire bed to this temperature supplied by the carbon fuel in the pellets. The finished product is hard, mainly spherical to sub-rounded granules with brown outer shells and honeycombed black cores, typically possessing a 40 % void ratio. After passing through the sintering zone, the fired pellets are cooled by a downward draft of air. The product then passes into a crusher that separates the fused clusters of pellets, which are then fed onto a vibrating screen and split into a number of grades required by industry (4–14 mm, 6–14 mm, 4–8 mm and 0–4 mm).

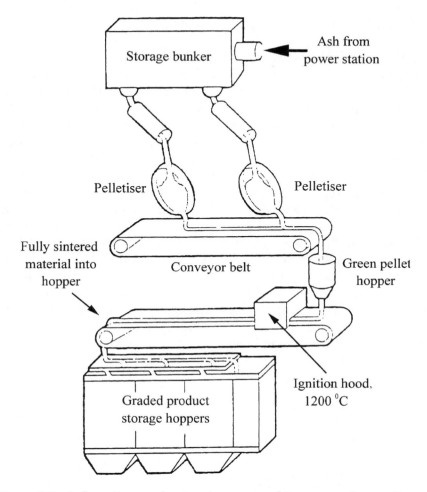

Figure 2.9 A flow diagram showing the process of Lytag manufacture (Reproduced by courtesy of Lytag Ltd).

Large quantities of Lytag have been used in the concrete building block industry for its combination of high strength, low density and good thermal insulation properties.

Although it is more expensive tonne for tonne when compared to FBA, which is used in the same role, it still gives a commercially viable product. Many other uses have since been developed, such as its use in the creation of lightweight structural concrete, vehicle speed arrestors, land drainage, pipe beddings and as a horticultural (moisture-retaining) medium. By 1999, over 15 Mt of PFA had been used in the manufacture of this product in the UK since it originally entered the market.

2.8.3 Humid Cured Aggregate from PFA

Unlike Lytag, which is the product of a high-temperature firing procedure, this is a different technology that uses the natural pozzolanic properties of PFA to produce a granulated 'cold-bonded' aggregate that does not require firing. It was developed in the UK in the early 1980s, where it was known as 'Granulite',[126] but at approximately the same time it was also being developed in the Netherlands under the trade name 'Aardelite'.[127] The Granulite manufacturing steps are shown schematically in Figure 2.10. The operation commences with freshly collected PFA being delivered to the processing plant in its still warm condition, whereupon it is conveyed pneumatically into two holding silos. Each silo serves a separate process-line equipped with a weigh-hopper, mixer, table feeder and pelletiser. The active ingredient is milk-of-lime (MOL), which is produced on site by slaking quicklime (CaO).

The PFA is discharged from the silos via a weighing-hopper into the mixer. MOL is then delivered from a small weigh-hopper into the mixer, where the two ingredients are blended on a batch system. On mixing, these two ingredients interact to promote a chemical reaction involving the aluminosilicate PFA and calcium ions in the MOL. The reaction

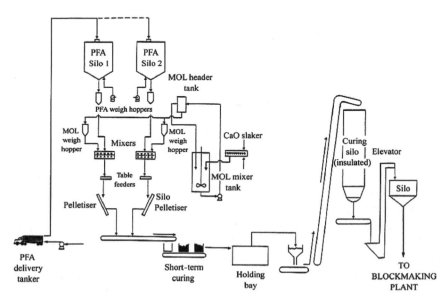

Figure 2.10 The flow-line of the Granulite humid-cured aggregate process (Energy Support Unit (ETSU), Harwell Laboratory, UK).

rate is relatively slow, with pozzolanic bonds only beginning to form as the batch is discharged on a table-feeder, from which it is continuously delivered to the pelletisers. These inclined rotating pans transform the mix into rounded to sub-rounded pellets, typically between 3 and 8 mm in diameter, the size being controlled by the speed of rotation of the pan. The pellets are formed and compacted by the constant rolling movement and the addition of water via spray-jets located around the circumference of the pan. The formed pellets are constantly discharged over the lower rim of the pan and conveyed into the curing part of the process.

The hardening reaction begins with a short-term curing phase in insulated enclosures. After about 24 hours, the pellets are sufficiently strong to be conveyed into the insulated curing silo for further hardening, with an overall curing period of approximately 48 hours. The hardened pellets are then conveyed to an adjacent factory to be used in the manufacture of lightweight concrete blocks for the construction industry. The energy requirement for this process has been shown to be approximately 5 % of that required by the Lytag sintering route.

Although problems due to PFA variability caused the UK Granulite factory to close down after a relatively short time, the Aardelite process proved more commercially successful, and has since been widely used in the manufacture of lightweight concrete and in road construction. This process has since been replicated at commercial plants in Florida (USA), India and Spain. During the first year of operation in the Netherlands, 70 kt of PFA was used in production, which was mainly incorporated as a lightweight aggregate in concrete blocks manufactured in Belgium and the UK.

2.9 The Use of PFA in Ceramic Products

The use of combustion residues in ceramic products dates back to a very early period where coal ash was frequently mixed with the clay used to make bricks and tiles, to prevent them being damaged by cracking during drying. Later, in Victorian times, London's dustbin refuse, which contained a high proportion of semi-burned coal from domestic fires, was collected and sieved. The fine ashes were then mixed with the brickmaking clay, while the remaining coarser unburned coal was used to help fire the kilns.[128]

Modern interest in the possible applications of combustion residues for use in ceramic products is closely associated with the first appearance of PFA in the early part of the last century. While the potential of using other more recent combustion residues in this function will be examined in Chapter 4, to date, PFA still remains indisputably the largest volume combustion residue for potential ceramic application. Yet its use in this industrial sector has never developed to any sizable extent. The reasons for this will become clear in the following review of early research into its application in this role. However, first, to emphasise this potential, it is important to highlight the close affinity that PFA shares with more familiar ceramic raw materials, which first attracted the interest of this particular manufacturing sector.

2.9.1 The Ceramic Properties of PFA

As noted in Section 1.2.3, in addition to the combustible organic material, coal also contains inorganic minerals, including clays that remained substantially unaltered during the coalification process. During combustion in a power station boiler at 1500–1600 °C, the encapsulated clay minerals are converted into molten spheres that are carried out of the boiler by the exhaust gas stream into a cooler region (about 700–800 °C), where they solidify into glassy solids typically with a mullite $(3Al_2O_3.2SiO_2)$ composition. It is this preliminary 'heat-treatment', turning raw clays into a glassy-phase solid, which makes PFA such a potentially attractive ceramic raw material.

This advantageous property can be illustrated with reference to the role played by traditional raw clays used to manufacture bricks. At an early stage in the manufacturing process, water is added to the brick-making clay to achieve plasticity, thus allowing the shaping process to take place. The bricks then require drying and during this operation the presence of the clay minerals promotes a hardening of the product, allowing it to be subsequently handled without damage during transfer to the kiln for firing. During this final stage, as the heat in the kiln is raised, the clay minerals commonly melt into a liquid phase as the highest temperature is approached. This developing mobile liquid comes into intimate contact with other minerals present in the bricks that have remained stable during the firing process, and on cooling it becomes the 'glue' that binds the whole matrix of mixed minerals and rock debris together, giving the finished products their strength and durability.

In contrast, in the power station boilers the clay minerals present in the coal have already been subjected to a temperature frequently much higher than the typical brick kiln, which has converted them to a fully fused glassy state. Consequently, unlike freshly quarried brickmaking clay, it requires far less heat for the PFA particles, once formed into brick shapes, to be re-heated and fused together to form durable ceramic products. As commonly 60–70 % of the total process-energy needed for the manufacture of bricks is required during the firing stage, there are potentially considerable cost-savings available by introducing this 'pre-fired' material as a direct replacement for raw clay. It has been this wide recognition, coupled with the need to find a further satisfactory high-volume disposal solution for PFA, which over many years has stimulated international interest in using it as a ceramic raw material. The most important historical research and development undertaken in its promotion in this role, the results achieved, and the technical and commercial difficulties encountered are examined below.

2.9.2 A Short History of the Use of PFA in Brickmaking

The first power station using pulverised fuel in the UK was commissioned in London in 1919, and by the mid-1920s numerous other plants using this firing technique were becoming established throughout the country. At the same time, a similar pattern of expanding replication was spreading throughout mainland Europe.

A UK literature survey[129] records that within a few years of commissioning such power stations, a number of brick manufacturers were conducting independent trials with PFA. In most instances, few precise records of their progress and conclusions have survived. This is understandable, as such investigations were the result of informal initiatives between brickworks and local power station management. The outcomes would have been judged on the spot, with little need to make the findings known beyond those that were directly involved. However, from the late 1940s, as the volume of PFA being produced continued to increase rapidly, more systematic investigations were carried out by various research organisations that provided detailed reports on their progress and results. These investigations can be considered under two separate categories: (a) the development of experimental bricks made totally of PFA; and (b) bricks containing variable replacement levels of PFA for normal clay. The results and conclusions arising from both approaches are summarised below.

Bricks Made Totally from Coal Ash Residues (PFA/FBA)

During the mid-1950s, with the electricity-generating industry in the UK under state ownership, the laboratories of the (then) British Electricity Authority (BEA) began an investigation to develop a commercially acceptable 100 % PFA brick. The impetus for this was purely economic, as it was recognised that such a product would not require clay and could therefore be manufactured at power stations as a new business enterprise. During this development work, 100 % PFA bricks (with various bonding agents) were formed by semi-dry pressing. This research was subsequently developed by the N.W. Region of the Central Electricity Generating Board (CEGB).[130] After overcoming a number of fabrication difficulties, full-scale making and firing trials were carried out in collaboration with a commercial kiln manufacturer. The physical properties of the experimental bricks produced were satisfactory and a structure composed of the fired product was built on the roof of a local power station, where it remained for over 30 years, unaffected by weather conditions, until the station closed (Figure 2.11).

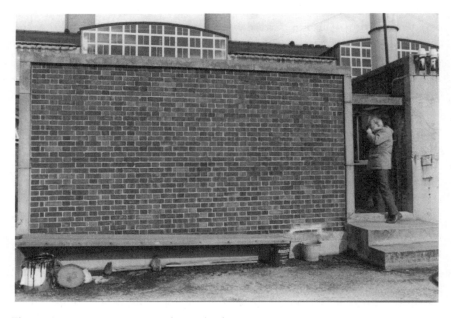

Figure 2.11 An experimental panel of 100 % PFA bricks at Northfleet power station in Kent, UK, built to study their durability.

However, difficulties were experienced in full-scale production and, in particular, the problem of overcoming the obstacle of the very narrow firing tolerance of the bricks, which resulted in a high level of rejects (due to both over- and under-firing), led to the conclusion that the process technology was unlikely to be commercially competitive against normal clay brick products, and the project was discontinued in 1961.

In the USA, a similar project with the same objective was carried out over the period 1965–72 by the United States Department of the Interior, at the Coal Research Bureau of West Virginia University.[131] Early realisation of the problem of the narrow firing tolerance of 100 % PFA bricks ultimately led to the development of a semi-dry pressed product containing approximately 80 % PFA and 20 % of milled furnace bottom ash (FBA), also produced abundantly at power stations. This mixture was bonded with ∼ 3 % (dry) w/w sodium silicate (water-glass), yielding a brick containing a total of 97 % power station ash residue. The inclusion of the coarser FBA helped by creating a more 'open' structure, which overcame an earlier problem of laminations caused by air being trapped within the mix during the pressing operation, Moreover, the FBA also had a thermally stabilising influence on the bricks during their subsequent firing, thereby avoiding the serious problem of the very narrow firing tolerance encountered in the earlier BEA investigation. The product surpassed all ASTM requirements established for clay bricks in the USA. A pilot plant was then set up to produce sufficiently large numbers of bricks to acquire all the technical and economic information pertinent to the proposed establishment of a full-scale factory. Yet, although the experimental product was shown to be substantially lighter than conventional clay bricks and was also comparable in manufacturing cost, the technology was not taken up commercially. One of the main reasons was that in appearance these bricks were very bland compared with existing clay bricks and consequently considered likely to be unpopular if offered on the commercial market.

Clay Bricks Containing Variable Levels of PFA

There are many technical papers describing the use of PFA as a partial replacement for clay in ceramic products. Most frequently, these refer to bench-scale undertakings, although in some cases limited factory trials are also reported. However, the most comprehensive investigation recorded concerns a programme carried out in the UK by the laboratories of the Building Research Station (now BRE Ltd)[132] over the period

1952–58 and which can be used to exemplify the general findings that have been broadly recorded in other studies.

Initially, a series of factory trials in collaboration with commercial brick manufacturers were carried out that involved incorporating different replacement levels of PFA into various clay feedstocks. These trails were of relatively short duration, as their timing coincided with an acute national shortage of bricks, and consequently the decision was soon made to concentrate solely on the development of high PFA bricks as a new market product to help cope with this shortfall. Supported by government funding, the contract given to the Building Research Station was to create prototype brickworks in which PFA would be a major component of the fired product. All the common commercial brickmaking methods were investigated, and particularly good results were achieved using both extrusion and soft-mud forming techniques,* which yielded satisfactory bricks containing up to 70 % PFA (dry) w/w. On testing, they successfully met the British Standard for clay bricks and samples were subsequently built into heavy-weathering piers. These piers were designed to replicate the harshest of exposure conditions and the majority have survived intact for over 50 years (Figure 2.12). In operational terms, the 'concept' factory was considered capable of effective competition against traditional clay brick manufacturing operations most notably in terms of lower firing costs due to the high level of PFA incorporated in the new product. Unfortunately, final commercialisation of this novel technology was never accomplished. This was partly due to political objections by the established ceramic industry, which feared a new competitive product appearing on the market. But also a number of technical difficulties were identified during the development programme and considered likely to cause continuing problems had the planned brickworks gone into full-scale production. Foremost of these was the variability in quality of the PFA available for use. Variability of any 'additive' raw material used in brickmaking is always a matter of concern to manufacturers and great importance is always placed on 'consistency', as any unpredictable fluctuations carries the potential risk of causing high levels of reject bricks and consequential loss of income. At the time of the investigation, serious fluctuations were being encountered with the supply of PFA that would have caused insurmountable production difficulties at full production scale.

---

* See details in Chapter 4.3.4.

Figure 2.12 Experimental piers of clay bricks containing up to 70 % w/w substitution of PFA erected in the 1950s at the Building Research Station, near Watford in Hertfordshire, UK.

Today, there is very much more severe environmental legislation surrounding the burning of fossil fuels, which particularly affects power stations. In addition, there is a more diversified electricity supply market, where alternative energy sources to coal are also competing, As a result, PFA variability has shown to be even more pronounced and remains just as obstructive to its successful large-scale use in ceramic products. The essence of this continuing problem of inconsistent quality and the ways in which it can be minimised to create a new wider market for the future are discussed in Chapter 4.

References

1 Sloss, L.L., Smith, I.M. and Adam, D.M.B., *Pulverised Coal Ash – Requirements for Utilisation.* IEA Coal Research, London (IEACR/88), pp. 12–13 (1996).
2 Bissell, T.B. and Wallis, D.M., *The handling and disposal of ash from CEGB power stations.* In AshTech 84: 2nd International Conference on Ash Technology and Marketing, Barbican Centre London, 16–21 September, pp. 229–238 (1984).
3 Brown, J. and Snell, P.A., *The use of ash in land reclamation.* In First International Ash Marketing and Technology Conference, London, 22–27 October, pp. 89–104 (1978).
4 Bahor, M.P., *The siting and design of ash disposal facilities in the United States.* In Proceedings of the Second International Conference on Ash Technology and Marketing, Barbican Centre, London, 16–21 September, pp. 641–645 (1984).
5 Kneissl, P.J., *Ash utilisation and disposal in Austria.* In Proceedings of 'Ash – a Valuable Resource', Second International Symposium Eskom Conference Centre, Pretoria, Republic of South Africa, 2–6 February, 1: Ci, pp. 1–16 (1987).
6 Joshi, R.C. and Achari, G., *Fly ash research and utilisation in Canada.* In Energy and Environment: Transitions in Eastern Europe, Prague, Czechoslovakia, 20–23, April, pp. 419–431 (1992).
7 Jo, Y.M., Guang, D. and Raper, J.A., *Characterisation of Australian and New Zealand fly ashes for utilisation,* 6th Australian Coal Science Conference, Australian Institute of Energy, 17–19 October, pp. 237–246 (1994).
8 Michael, M.D. and Fedorsky, C.A., *Rehabilitation of ash disposal sites in Eskom.* In Proceedings of 'Ash – a Valuable Resource', Second International Symposium, Eskom Conference Centre, Pretoria, South Africa, 21–23 February, 2: 461–481 (1994).
9 Chandramouli, R., Motza, G., Dagaonkar, A.V. and Mahadew, D., *Practical solution for an environmental problem at Trombay power station India.* In Power Gen Europe 95, Amsterdam, the Netherlands, 16–18 May, pp. 603–626 (1995).
10 Kamada, H. and Tezuka, M., *Present activity and prospective on effective utilisation of coal ash in Japan.* In 11th International Symposium on Use and management of Coal Combustion By-products, Orlando, Fla, 15–19, January, pp. 61.1–61.14 (1995).

11 Edens, T.F., *Recovery and utilisation of pond ash*. In 1999 International Ash Utilization Symposium, Center for Applied Energy Research, University of Kentucky, Paper 96 (1999).

12 Meij, R. and Winkel, H., *Health aspects of coal fly ash*. In 2001 International Ash Utilization Symposium, Center for Applied Energy Research, University of Kentucky, Paper 21 (2001).

13 Behel, D., *TVA research on coal combustion by-products: uses and environmental impacts*. In 2001 International Ash Utilisation Symposium, Center for Applied Energy Research, University of Kentucky, Paper 96 (2001).

14 Riepe, W., *The leaching behaviour of ashes under real atmospheric conditions*. In Proceedings of 'Ash – a Valuable Resource', International Symposium, Eskom Conference Centre, Pretoria, South Africa, 2–6 February, **3**: 12i (1987).

15 Harpley, D.P., Smith, A.C.S. and Geldenhuis, S.J.J., *Predicting contaminant migration from in-pit ash disposal and implications for water quality monitoring*. In Proceedings of 'Ash – a Valuable Resource', Second International Symposium, Eskom Conference Centre, Pretoria, South Africa, 2–6 February, **3**: 12ii (1987).

16 Gutierrez, B., Pazos, C. and Coca, J., *Characterisation and leaching of coal fly ash*. Waste Management and Research, **11**: 279–286 (1993).

17 Erbe, M.W., Keating, R.W. and Hodges, W.K., *Evaluation of water quality conditions associated with the use of coal combustion products for highway embankments*. In 1999 International Ash Utilization Symposium, Center for Applied Energy Research, University of Kentucky, Paper 31 (1999).

18 Nugteren, H.W., Janssen-Jurkovičová, M. and Scarlett, B., *Improvement of environmental quality of coal fly ash by applying forced leaching*. In 1999 International Ash Utilization Symposium, Center for Applied Energy Research, University of Kentucky, Paper 32 (1999).

19 UK Water Industry Research, *Sewage Sludge Incineration*. Report 01/SL/07/1, UKWIR, London (2001).

20 Pan, S.C. and Tseng, D.M., *Sewage sludge ash characteristics and its potential applications*. Water Science and Technology, **44**(10): 261–267 (2001).

21 Johnson, C.A., *Characterisation and leachability of sewage sludge ash*. In Proceedings of the International Symposium of Sustainable Waste Management, University of Dundee, 9–11 September, Thomas Telford: London, pp. 353–362 (2003).

22 Bagchi, A., *Characterisation of MSW incinerator ash*. J. Environ. Engineering, **115**(2): 447–452 (1989).

23 Hamemik, J.D. and Trantz, G.C., *Physical and chemical properties of municipal solid waste fly ash*. ACI Materials Journal, **88**(3): 294–301 (1991).

24 Pluss, A. and Ferrell, R.E., *Characterisation of lead and other heavy metals in fly ash from municipal waste incinerators*. Hazardous Waste & Hazardous Materials, **8**(4): 275–292 (1991).

25 Kirby, C.S. and Rimstidt, J.D., *Interaction of municipal solid waste ash with water*. Environ. Sci. Technol. **28**(3): 443–451 (1994).

26 Barber, E.G., *Ash and Agriculture*. Central Electricity Board publication 3/63 G271 (1963).

27 Townsend, W.N. and Gillham, E.W.F., *Pulverised fuel ash as a medium for plant growth*. In 15th Symposium of the British Ecological Society, London (1973).

28 Kamada, H. and Tezuka, M., *Present activity and prospective on effective utilisation of coal ash in Japan*. In 11th International Symposium on Use and

Management of Coal Combustion By-products, Orlando, Fla, American Coal Ash Association/Electric Power Research Institute, 2, 50:1–50:15 (1995).

29 Beaver, T., *Adding coal ash to the composting mix*. BioCycle, 36(3): 88–89 (1995).

30 Renolds, K., Kruger, R. and Rethman, N., *The manufacture and evaluation of artificial soil (SLASH) prepared from fly ash and sewage sludge*. In 1999 International Ash Utilization Symposium, Center for Applied Energy Research, University of Kentucky, 1 (1999).

31 Mittra, B.N., Karmakar, S., Swain, D.K. and Gosh, B.C., *Fly ash – a potential source of soil amendment and a component of integrated plant nutriment supply*. In 2003 International Ash Utilization Symposium, Center for Applied Energy Research, University of Kentucky, 28 (2003).

32 Davis, R.E., Kelly, J.W., Troxell, G.E., and Davis, H.E., *Proportions of mortars and concretes containing Portland–pozzolan cements*. ACI Journal, 32: 80–114 (1935).

33 Davis, R.E., Carlston, R.W., Kelly, J.W., and Davis, H.E., *Properties of cements and concretes containing fly ash*. ACI Journal, 33: 577–612 (1937).

34 Davis, R.E., Davies, H.E., and Kelly, J.W., *Weathering resistance of concretes containing fly ash cements*. In Proceedings, American Concrete Institute, 37: 281–296 (1941).

35 Anon., *Nebraska paves with fly ash concrete*. Roads and Streets, 95 (March): 70–72 and 133 (1952).

36 Anon., *Fly ash use as concrete additive questioned by building inspectors*. Engineering News – Record, 152, 22 (March): 22 (1954).

37 Fulton, A.A. and Marshall, W.T., *The use of fly ash and similar materials in concrete*. Proc. Inst. Civil Eng. Part 1, 5: 714–730 (1956).

38 Brink, R.H. and Halstead, W.J., *Studies relating to the testing of fly ash for use in concrete*. Proceedings of the American Society for Testing and Materials, 56: 161–1214 (1956).

39 Allen, A.C., *Features of Lednock Dam, including the use of fly ash*. Paper No. 6326. Proc. Inst. Civil Eng., 13 (August): 179–196 (1958).

40 Howell, L.H., *Report on Pulverised Fuel Ash as a Partial Replacement for Cement in Normal Works Concrete*. CEGB, East Midlands Division (1958).

41 Richardson, L. and Bailey, J.C., *Design, Construction and Testing of Pulverised Fuel Ash Concrete Structures at Newman Spinney Power Station, Parts I, II, III*. CEGB Research and Development Report (1965).

42 Tennessee Valley Authority, *Fly ash for use as an admixture in concrete*. TVA G-30 (1968).

43 BSI, *BS 3892, Pulverised fuel ash for use in concrete*. British Standards Institution: London (1965).

44 Thorne, D.J. and Watt, J.D., *Investigation of the composition, pozzolanic properties, and formation of pulverized-fuel ashes*. The British Coal Utilization Research Association, Information Circular No. 265, Document No.C/5371 (1965).

45 Anon., *Pozzolan – a selected fly ash for use in concrete*. The Agrément Board Certificate No. 75/283 (1975).

46 BSI, *BS 3892, Part 1: Pulverised fuel ash for use as a cementitious component in structural concrete*. British Standards Institution: London (1982).

47 BSI, *BS 3892, Part 1: Specification for pulverised-fuel ash for use with Portland cement*. British Standards Institution: London (1997).

48 BSI, *BS EN 450, Fly ash for concrete – definitions, requirements, and quality control*. British Standards Institution: London (1995).

49 BSI, *BS EN 450-1, Fly ash for concrete – definitions, requirements and conformity criteria*. British Standards Institution: London (2005).

50 BSI, *BS EN 450-2, Fly ash for concrete – conformity evaluation*. British Standards Institution: London (2005).

51 Dhir, R.K., Munday, J.G.L., and Ong, L.T., *Strength variability of OPC/ fly ash concrete*. Concrete, **15**(6), (a): 33–37 (1981).

52 Dhir, R.K., Apte, A.G. and Munday, J.G.L., *Effect of in-source variability of pulverized fuel ash upon the strength of OPC/fly ash concrete*. Magazine of Concrete Research, **33**(117), (b): 119–207 (1981).

53 Dhir, R.K., Munday, J.G.L. and Ong, L.T., *Investigation of the engineering properties of OPC/pulverized fuel ash concrete: strength development and maturity*. Proc. Inst. Civil Eng., Part 2, **77**(a): 239–254 (1984).

54 Dhir, R.K., Ho, N.Y. and Munday, J.G.L., *Pulverized-fuel ash in structural precast concrete*. Concrete, **19**(6), (b): 32–36 (1985).

55 Dhir, R.K., Jones, M.R., Munday, J.G.L. and Hubbard, F.H., *Physical characterization of UK pulverized fuel ash for use in concrete*. Magazine of Concrete Research, **37**(131): 75–87 (1985).

56 Dhir, R.K., Ong, L.T. and Munday, J.G.L., *Fly ash in structural precast concrete; full scale factory trails*. Concrete, **20**(1): 40–42 (1986).

57 Dhir, R.K., Munday, J.G.L. and Ong, L.T., *Investigations of the engineering properties of OPC/pulverized-fuel ash concrete: deformation properties*. The Structural Engineer, Part B, R&D Quarterly, **64B**(2): 36–42 (1986).

58 Dhir, R.K. and Jones, M.R., *PFA concrete: Influence of simulated in-situ curing on elasto-plastic load response*. Magazine of Concrete Research, **45**(163): 139–146 (1993).

59 Dhir, R.K. and Jones, M.R., *Development of fly ash use in precast concrete manhole units*. Cement and Concrete Research, **22**(1): 35–46 (1992).

60 Dhir, R.K., Jones, M.R. and Munday, J.G.L., *A practical approach to studying carbonation of concrete*. Concrete, **19**(10): 32–34 (1985).

61 Dhir, R.K., Jones, M.R. and McCarthy, M.J., *PFA concrete: Carbonation induced reinforcement corrosion rates*. Proc. Inst. Civil Eng., **94** (August): 335–342 (1992).

62 Dhir, R.K. and Byars, E.A., *Permeation properties of cover to steel reinforcement*. Cement and Concrete Research, **23**(3): 554–566 (1993).

63 Dhir, R.K., Jones, M.R. and Seneviratne, A.G.M., *Diffusion of chloride ions in concretes: Influence of fly ash quality*, Cement and Concrete Research, **21**(6): 1092–1102 (1991).

64 Dhir, R.K., Jones, M.R. and Elghaly, E.A., *Fly ash concrete: Exposure temperature effects on chloride diffusion*. Cement and Concrete Research, **23**(5): 1105–1114 (1993).

65 Dhir, R.K. and Byars, E.A., *PFA Concrete: chloride diffusion rates*. Magazine of Concrete Research, **45**(162): 1–9 (1993).

66 Jones, M.R., McCarthy, M.J. and Dhir, R.K., *Chloride resistant concrete*. In Proceedings of the International Conference Concrete 2000, R.K. Dhir and M.R. Jones (eds), E & FN Spon: London, pp. 1429–1444 (1993).

67 Dhir, R.K., Jones, M.R. and McCarthy, M.J., *PFA Concrete: chloride-induced reinforcement corrosion*. Magazine of Concrete Research, **46**(169): 269–278 (1994).

68 ASTM, *C618-05 Standard specification for coal fly ash and raw or calcined natural pozzolan for use in concrete.* American Society for Testing and Materials (2005).

69 BSI BS EN 197-1, *Composition, specifications and conformity criteria for common cements,* British Standards Institution: London (2000).

70 BSI, *BS EN 206-1, Concrete: Specification, performance, production and conformity,* British Standards Institution: London (2000).

71 Swamy, R.N., *Cement replacement materials.* In Concrete Technology and Design. Glasgow: Surrey University Press, volume 3 (1986).

72 Helmuth, R., *Fly Ash in Cement and Concrete.* Portland Cement Association: Skokie, IL (1987).

73 Joshi, R.C. and Lohtia, R.P., *Fly Ash in Concrete: Production, Properties and Uses.* Advances in Concrete Technology, Gordon & Breach: Amsterdam, 2 (1997).

74 Wesche, K., *Fly Ash in Concrete: Properties and Performance.* RILEM Technical Committee 67-FAB: Use of Fly Ash in Building. E & FN Spon: London, RILEM Report, 7 (1991).

75 Sear, L.K.A., *Properties and Use of Coal Fly Ash: a Valuable Industrial By-Product.* Thomas Telford: London (2001).

76 *Alkali – Silica Reaction – Minimising the Risk of Damage to Concrete.* The Concrete Society, Third Edition, Technical Report 30 (1999).

77 Building Research Establishment, *Concrete in Aggressive Ground.* Special Digest 1 Third Edition, (2005).

78 Houlsby, A.C., *Construction and Design of Cement Grouting in Rock Foundations.* John Wiley & Sons, Inc.: New York (1990).

79 CIRIA, *Grouting for Ground Engineering.* Construction Industry Research and Information Association Report C514 (2000).

80 BSI, *BS EN 447, Grout for prestressed tendons – grouting procedures.* British Standards Institution: London (1997).

81 Bowen, R., *Grouting in Engineering Practice.* Applied Science Publishers: London (1981).

82 Domone, P.L.J. and Jefferis, S.A., *Structural Grouts.* Blackie Academic and Professional: London (1994).

83 Henn, R.W., *Practical Guide to Grouting Underground Structures.* Thomas Telford: London (1996).

84 BSI, *BS 3892-3, Pulverized-fuel ash: Specification for pulverized fuel ash for use in cementitious grouts.* British Standards Institution: London (1997).

85 Anon., *Pulverized-fuel ash for grouting.* Technical data sheet 3, United Kingdom Quality Ash Association (2002).

86 Edmeades, R.M. and Mangabhai, R.J., *PFA grouts: selection of materials and flow properties.* In Proceedings of a National Seminar: The Use of PFA in Construction, Dundee, February, pp. 75–88 (1992).

87 BSI, *BS EN 12715, Execution of special geotechnical work – grouting.* British Standards Institution: London (2000).

88 Dhir, R.K., *Sulfate resistance of PFA grout.* Research Report, University of Dundee (1994).

89 Anon., *Engineering with ash – grout.* National Ash Technical Brochure (1993).

90 Anon., *Frequently Asked Questions.* United Kingdom Quality Ash Association (2003).

91 Matthews, J.D., Quillin, K.C., Watts, K.S. and Tedd, P., *Laboratory and field data for PFA grouts.* BRE Report 220192, Garston, Watford (2006).

92 Hughes, D.C., *Fly ash grouts: performance of hardened grouts.* In Proceedings of a National Seminar: The use of PFA in Construction, Dundee, February, pp. 88–103 (1992).

93 McCarthy, M.J., Dhir, R.K. and Jones, M.R., *Benchmarking PFA grouts for magnesium sulphate exposures,* Materials and Structures, **31**(103): 335–342 (1998).

94 Hughes, D.C., *Sulphate resistance of OPC, OPC/fly ash and SRPC pastes: pore structure and permeability.* Cement and Concrete Research, **15**: 1003–1012 (1985).

95 Hughes, D.C., *Sulphate resistance of fly ash/OPC grouts.* In Proceedings of the International Conference on Blended Cements in Construction, R.N. Swamy (ed.). Elsevier Applied Science: London, pp. 336–350 (1991).

96 Anon., *Pulverised fuel ash for fill applications.* United Kingdom Quality Ash Association. *Technical Datasheet No. 2.* May 2007.

97 Anon., *Fly ash in highways construction – An overview of the use of FABM and ESC.* United Kingdom Quality Ash Association. *Technical Datasheet No. 6.* May 2007.

98 Anon., *Fly ash in highways construction – detailed descriptions of FABM and ESC.* United Kingdom Quality Ash Association. *Technical Datasheet No. 6.1,* May 2007.

99 Anon., *Fly ash in highways construction – laboratory mix design.* United Kingdom Quality Ash Association. *Technical Datasheet No. 6.2,* May 2007.

100 Anon., *Fly ash in highways construction – structural design.* United Kingdom Quality Ash Association. *Technical Datasheet No. 6.3,* May 2007.

101 Raymond, S., *Pulverized fuel ash as embankment material.* Proceedings of the ICE, **19**: 515–536 (1961).

102 Sutherland, H.B. and Finlay, T.W., *A Laboratory Investigation of the Age Hardening Characteristics of Pulverized Fuel Ash,* volume 1. University of Glasgow (1964).

103 Raask, E., *Pulverized fuel ash constituents and surface characteristics in concrete applications.* In J.G. Cabrera and A.R. Cusens (eds). The use of PFA in concrete: International symposium, vol. 1. University of Leeds, pp. 5–16 (1982).

104 McCarthy, M.J., Tittle, P.A.J. and Dhir, R.K., *Characterisation of conditioned PFA for use as a cement component in concrete.* Magazine of Concrete Research, **51**(3): 191–206 (1999).

105 Clarke, B.G., *Structural fill.* In Proceedings of a National Seminar: The use of PFA in Construction, Dundee, February, pp. 21–32 (1992).

106 Anon., *Manual of Contract Documents for Highway Works. Volume 1, Specification for Highway Works. Series 600, Earthworks.* The Stationery Office: London.

107 Anon., *Manual of Contract Documents for Highway Works. Volume 2, Notes for Guidance on the Specification for Highway Works. Series NG 600, Earthworks,* The Stationery Office: London.

108 Winter, M.G. and Clarke, B.G., *Specification of pulverized-fuel ash for use as general fill.* Transport Research Laboratory Report 519, TRL (2001).

109 Highways Agency, *Design Manual for Roads and Bridges. Earthworks: design and preparation of contract documents.* Volume 4, Section 1, Part 1, HA 44 (1991).

110 Cotton, R.D., *Construction of embankments.* In Proceedings of a National Seminar: The use of PFA in Construction, Dundee, February, pp. 45–56 (1992).

111 BSI, *BS EN 14227-4, Hydraulically bound mixtures – specifications. Part 4: fly ash for hydraulically bound mixtures.* British Standards Institution: London (2004).

112   Anon., *Manual of Contract Documents for Highway Works, Volume 1 Specification for Highway Works. Series 800, Road Pavements – Unbound Cement and Other Hydraulically Bound Mixtures.* The Stationery Office: London (2004).

113   BSI, *BS EN 14227-3, Hydraulically bound mixtures – specifications. Part 3: fly ash bound mixtures.* British Standards Institution: London (2004).

114   Mullin, P E., *Landfill procedures for coal ash surplus: permanent disposal sites or storage for utilisation.* In American Coal Ash Association, Symposium 8, Paper 45, pp. 1–12 (1987).

115   Waste Resources Action Plan (WRAP). *Fly ash bound material* (online), last accessed October 2007. Available from the WRAP (Aggregain) website (2007).

116   Carey, P.J., Barnard, L., Hills, C.D. and Bone, B., *A review of solidification/stabilisation technology and its industrial application in the UK.* In Sustainable Waste Management, R.K. Dhir, M.D. Newlands and T.D. Dyer (eds). Thomas Telford: London, pp. 391–400 (2003).

117   Construction Industry Research and Information Association (CIRIA), *Remedial Treatment for Contaminated Land, Vol VII, Ex Situ Remedial Methods for Soils, Sludges and Sediments.* CIRIA Special Publication 107 (1995).

118   Benny, H.L., *Hanford site grout treatment facility.* Concrete International Design & Construction, **12**(7): 14–18 (1990).

119   Roy, A., Eaton, H.C., Cartlidge, F.K. and Tilebaum, M.E., *Solidification/stabilisation of a heavy metal sludge by a Portland cement/fly ash binding mixture.* Hazardous Waste & Hazardous Materials, 8(1): 33–41 (1991).

120   Boswell, J.E.S., *The utilisation of coal ash in hazardous waste management.* In Proceedings of 'Ash – a Valuable Resource', Second International Symposium, Eskom Conference Centre, Pretoria, Republic of South Africa, **1**: 121–127 (1994).

121   Smith, C.L., *Case histories in full-scale utilisation of fly ash-fixated FGD sludge.* In Proc. 9th International Ash Use Symposium, Florida, Electric Power Research Institute, **1**: 34, 1–13 (1991).

122   Edil, T.B., Sandstrom, L.K., and Berthouex, P.M., *Interaction of inorganic leachate with compacted pozzolanic fly ash.* J. Geotech. Eng., ASCE, **118**(9): 1410–1430 (1992).

123   Kusterer, T., Kula, J.R. and Kramer, M.K., *The potential use of coal ash in the construction and operation of a municipal solid waste landfill.* In 11th International Symposium on Use and Management of Coal Combustion By-products, Orlando, Fla, American Coal Ash Association/Electric Power Research Institute, **2**: 69, 1–14 (1995).

124   Carroll, R.A. and Payne, J.C., *Autoclaved cementitious products using pulverised fuel ash.* In Proceedings of 3rd RILEM International Symposium on Autoclaved Aerated Concrete, Zurich, Switzerland, 14–16 October (1992).

125   Harrison, W.H. and Munday, R.S., *An investigation into the production of sintered PFA aggregate.* Building Research Establishment Report CP 2/75, 32 pp. (1975).

126   Anon., *Utilisation of pulverised fuel ash in the manufacture of lightweight aggregate building blocks.* Energy Support Unit (ETSU), Harwell Laboratory, Didcot, Oxfordshire, *Final Report ED/253/236* (1989).

127   Boas, A. and Spannjer, J.J., *The manufacture and the use of artificial aggregates from fly ash produced according to the Dutch cold bonded 'Aardelite' process.* In 'Ash Tech 84', Central Electricity Generating Board, pp. 577–582 (1984).

128   Perks, R.-H., *George Bargebrick Esquire.* Meresborough Books: Rainham, Kent, pp. 3–14 (1981).

129   Anderson, M. and Jackson, G., *The history of pulverized fuel ash in brickmaking in Britain.* Trans. J. Inst. Ceram., **86**(4): 99–135 (1987).

130   Crimmin, W.R.C., Gill, G.M. and Jones, G.T., *The production of 100 % P.F. Ash bricks.* Central Electricity Generating Board, Internal Report, N.W. Region (1963).

131   Anon., *Production of fly ash-based structural materials.* Final Report, No. 69 (14-01-0001-488), Office of Coal Research, Department of the Interior, Washington, DC (1973).

132   Building Research Station, *Brickmaking with pulverized fuel ash – investigation on behalf of the British Electricity Authority.* Reports RSI 1679, 1707 and 1827 (1953–1958).

3

Limitations of Combustion Ashes: 'From Threat to Profit'

Henk Nugteren
Delft University of Technology, The Netherlands

Abbreviations and Acronyms

ADI	Acceptable Daily Intake
AFNOR	Association Française de Normalisation
ASTM	American Society for Testing of Materials
BCR	European Community Bureau of Reference (now Standards, Measurements and Testing Programme)
BMD	Building Materials Decree (Dutch)
BS	British Standard
CCSEM	Computer Controlled Scanning Electron Microscope (analyses)
CEM	Cement
CEN TC	Comité Européen de Normalisation – Technical Committee
CERCHAR	Centre d'Etudes et Recherches des Charbonnage de France
CNSLD	Chronic Non-Specific Lung Diseases
CUR	Civieltechnisch Centrum Uitvoering Research en Regelgeving (Dutch)
DIN	Deutsche Industriele Norm (German)
DNA	Desoxyribonucleïc Acid
EC	European Commission
EN	European Norm

Combustion Residues: Current, Novel and Renewable Applications   Edited by Michael Cox,
Henk Nugteren and Mária Janssen-Jurkovičová   © 2008 John Wiley & Sons, Ltd

EPA	Environmental Protection Agency (USA)
EPRI	Electric Power Research Institute
ESP	Electrostatic Precipitator
EU	European Union
FGD	Flue Gas Desulfurisation
Hacac	2,4-pentanedione
H_4EDTA	ethylenediaminetetracetic acid
H_2pnaa	bis(pentan-2,4-dionato)propan-1,2-diimine
Hprps	tetra-iso-propyldithioimidophosphinate
ISO	International Organisation for Standardisation
I-TEQ	International Toxic Equivalent
JST	Japanese Standard Tests
KEMA	Dutch Research Institute for the Energy Sector
KEMA-DAM®	Dust Assessment Methodology (KEMA)
L/S	Liquid : Solid ratio
LOI	Loss on Ignition
MINTEQA2	Computer code for mineral equilibria
MSWI	Municipal Solid Waste Incinerator
MWI	Municipal Waste Incinerator
NEL	No Effect Level
NEN	Netherlands Normalisation Institute
NO_x	Nitrogen oxides
PAH	Polycyclic aromatic hydrocarbons
PC	Portland Cement
PFA	Pulverised Fuel Ash
PM_4	Particulate Material $< 4\,\mu m$
PM_{10}	Particulate Material $< 10\,\mu m$
PM_{50}	Particulate Material $< 50\,\mu m$
PMF	Progressive Massive Fibrosis
PMI	Progress Materials Inc. (USA)
prEN	Preliminary (draft) European Norm (standard)
PVC	Polyvinyl chloride
RIVM	Rijks Instituut voor Volksgezondheid and Milieuhygiëne (Dutch Institute for Public Health and Environment)
RIZA	Rijksinstituut voor Integraal Zoetwaterbeheer en Afvalwater-behandeling (Directorate-General for Public Works and Water management) (Dutch)
SCR	Selective Catalytic Reduction
SEM	Scanning Electron Microscopy

SERVO	Selective Extraction and Recovery of metals using Volatile Organic extractants
SFE	Supercritical Fluid Extraction
SNCR	Selective Non-Catalytic Reduction
STI	Separation Technologies Inc. (USA)
TCLP	Toxicity Characteristic Leaching Procedure (USA)
TG-DTA	Thermogravimetric and Differential Thermographic Analysis
TLV	Threshold Limited Value
TSPM	Total Suspended Particulate Matter
XAFS	X-Ray Absorption Fine Structure Spectrometry
XRD	X-Ray Diffraction

3.1 Introduction

The quality of combustion residues is a rather broad term. It only has a meaning when it is used in a certain context. One may talk about the quality of a fly ash in relation to certain applications. For different applications, the technical requirements may be different, so that it is difficult to talk about the *quality* of ash. Thus the technical quality of a combustion residue may meet the requirements of one particular product, whereas it does not for another one. However, this has an advantage as well, as for each quality of ash a suitable application may be available. Thus, variation in quality should be regarded as a challenge and should lead to the development of suitable new applications for different ashes. This will be particularly true for co-combustion residues that are fairly new to the market. It will result in a larger quantity of ashes being reutilised and in market diversification, both of which are considered as positive contributions towards sustainable development.

There are two important aspects of quality, technical and environmental, which lead to the distinction of these two different kinds of quality:

- The technical quality of an ash refers to how the ash performs technically in a given application, and is mainly reflected by the chemical and physical properties of the ash. The technical quality is generally considered good if the ash performs equal to or better than other commercially available materials used for the same application.
- Environmental quality refers to the impact that ash has on the natural environment. Information on the environmental quality is obtained

by standardised (leaching) tests that are believed to be valid for prediction of the future behaviour of a combustion residue under natural
conditions. The results of such tests are compared to environmental
requirements set down in directives. The type of application allowed
for the ash is determined by the environmental quality. But the latter
also depends on the form in which the ash is exposed to the natural
conditions. Thus if the ash is landfilled or used as a granular material,
its behaviour under direct leaching becomes the key property to be
considered. But if it is used in a bound form, leaching will generally
be reduced and diffusion leaching of the bound products becomes the
main exposure route. The interaction between ash and the natural
environment, and thus the environmental quality, depends largely on
the concentrations of the trace elements and their chemical speciation,
although particle size distribution and major element concentrations
(Ca and S) do also have an influence.

This chapter deals mainly with available technologies that can influence and improve the technical and environmental quality of ashes.
Although technical and environmental qualities are treated separately,
such technologies may influence both at the same time. For example,
the removal of certain compounds by physical separation (e.g. unburned
carbon, cenospheres, magnetic fraction) is treated here under maintenance of technical quality, whereas it may also have effects on the
environmental quality, since some heavy metals may have a preferential
affinity with such compounds. Conversely, the wet chemical extraction
of heavy metals with the aim of improving environmental quality will
probably remove free lime as well, and thereby at the same time change
the technical quality. Yet another complication arises when dealing with
the separation or extraction of compounds contained in the ash. The
separated or extracted fraction then becomes a new product, and therefore such technology should strictly be treated in Chapter 4, the best
example being the recovery of cenospheres. Although cenospheres are
harvested because of their favourable properties – that is, large hollow
particles with a low effective particle density – the removal of cenospheres does change the grade of the remaining ash residue considerably and therefore can also be regarded as producing a better ash for
certain applications. The same applies to some extent to methods for
the removal of unburned carbon – in other words, lowering the LOI
and producing a carbon-rich fraction for the production of activated
carbon products – and the extraction of some elements (e.g. Ga, Ge).

Environmental impact as expressed by leaching intends to reflect long-term effects. Workers in a coal-fired power plant and the public living immediately in its vicinity are more directly exposed to ash and these (health) effects are treated in a separate section.

3.2 Technical Quality

3.2.1 Quality Requirements

In general, the PFA available from power stations is not homogenous. Although in modern power stations the main influencing factors, such as fuel blending and grinding and boiler loadings, are computer controlled and optimised, there is a range of operating variables that continually affect the PFA quality such that the variation of inhomogeneity can be kept within certain allowed ranges. Variations result from changes within the operating schedules arising from different running conditions of the boilers. Thus, during the time taken for a particular boiler to reach its optimum operating temperature from cold, the PFA produced will frequently contain a higher than normal proportion of unburned coal and coarser, poorly formed ash particles. Furthermore, once achieved, further adjustments to the loading level of the boilers regularly have to take place to meet the changing demand for electricity. These adjustments are made in various ways. For example, the number of grinding mills supplying pulverised coal to an individual boiler may be reduced, resulting in a change in the particle sizing of the ash. Under extremely low-level operating conditions, oil may be introduced into the boiler to stabilise the flame, in which case there may be some oil ash contamination of the bulk material. Today, the coexistence of several alternative electricity-generating options – that is, coal, gas, and nuclear – contributes towards further variability, owing to the policy of rapidly switching between supply sources to achieve maximum economy. The outcome has been the gradual emergence of PFA of increasing variability. The maximum efficiency base-load firing conditions under which power station boilers are designed to run are no longer the way that many are now operated. As noted later, in Chapter 4.3.2, other factors such as changes in the source of coal, combination of coal with other fuels, fitting of 'low-NO_x burners' and regular schedules of maintenance contribute to increasing variability of the ash. Consequently, the PFA in the receiving hoppers accumulates as an irregular deposit of variable quality.

So far, the only application for which specifications are well developed and highly specific is in the field of the cement and concrete industry. The long history of use of ash in the production of cement and as a substitute for sand in concrete has resulted in the development of many international standards.

The first British Standard for the use of PFA in concrete (BS 3892, first edition) was published in 1965,[1] although PFA was used in concrete long before this. The main property considered in this standard was the fineness, and three classes were distinguished. However, variability in the fineness of the fly ash when taken directly from a power station turned out to be so high that it became a major factor in concrete production, as it had a direct influence on water demand and concrete strength. Moreover, variations in loss of ignition (LOI) had an impact on colour and air entrainment. The fine fraction of ash is the most valuable for its pozzolanic activity, whereas the coarse fraction only acts as a filler replacing sand. The physical classification of ash results in a finer fraction that not only performs better as a pozzolan but also, with the significantly reduced variability, the material becomes more consistent. This led to a diversification of BS 3892 into Part 1 for classified fly ash counting fully towards the cement content of the mix,[2] Part 2 for PFA additions with less restrictive specifications but considered as an inert filler,[3] and Part 3 for use in cementitious grouts.[4] Thus, according to BS 3892, Part 1, between 15 % and 35 % of fly ash is permitted in Portland PFA cement (BS 6588) and between 35 % and 50 % in pozzolanic PFA cement (BS 6610).

With European harmonisation, BS EN 450 was introduced in 1995 as a new standard for fly ash,[5] whereas the Netherlands already had its own standard, NEN EN 450. Since physical classification of ash for concrete is not common practice on the European continent, the requirements on fineness are less strict in the EN 450 standard compared to BS3892, Part 1. Therefore, EN 450 fly ash may only be counted partially towards the cement content of the mix, which is specified in EN 206.[6] An overview of the use of fly ash in cement and concrete is given in Chapter 2.1, and in Chapter 4.2 the newest developments in this area are given. A thorough discussion on the British and European Standards is provided by Sear.[7] Regular updates of these standards are being made: for EN450, the last update was issued in 2005.[5]

In the Netherlands, the standards were set by the Civieltechnisch Centrum Uitvoering Research en Regelgeving (CUR). This institute coordinates research and publishes overview reports.[8,9] Standards on coal fly ash in cement and concrete were regularly updated,[10] and finally are all being replaced by the European EN 450 standard.

In the USA, the use of coal fly ash in concrete is regularised by the ASTM standard C618-03.[11] In addition to these most important features of fineness and loss on ignition, other properties considered in the standards for fly ash in cement and concrete include particle density, soundness, maximum allowable concentrations of SO_3, Cl and CaO, moisture content, water requirement, activity index and strength factor (the last two indicating the development of strength with time). The requirements as laid down in BS3892 Parts 1 and 2 and in EN 450 are summarised in Table 3.1.

Table 3.1 A summary of the requirements for the BS 3892 and EN 450 standards for pulverised fuel ash (after Sear[7]). Reproduced by permission of Thomas Telford Ltd.

Attribute	Requirements		
	PFA BS 3892, Part 1	PFA BS 3892, Part 2	BS EN 450, fly ash
Particle density	≥ 2000 kg m^{-3}	N/A	$\pm$ 150 kg m^{-3} of declared value
Fineness	$\leq$ 12 % retained on 45 μm sieve	$\leq$ 60 % retained on 45 μm sieve^a	$\leq$ 40 % retained on 45 μm sieve Must be within $\pm$ 10 % of declared value
Soundness	$\leq$ 10 mm 30 % fly ash + 70 % PC (BS 12 42.5)	N/R	$\leq$ 10 mm 50 % fly ash + 50 % CEM I 42.5
Sulfur: maximum present as SO_3	$\leq$ 2.0 %^{a}	$\leq$ 2.5 %^{a}	$\leq$ 3.0 %
Chloride	$\leq$ 0.10 %	N/A	$\leq$ 0.10 %
Calcium oxide	Expressed as total CaO $\leq$ 10 %	N/R	Expressed as free CaO $\leq$ 1.0 %, or $\leq$ 2.5 % if soundness satisfactory
Loss on ignition	$\leq$ 7.0 %^{a}	$\leq$ 12.0 %^{a}	$\leq$ 7.0 %^{b}
Moisture content	$\leq$ 0.5 %	$\leq$ 0.5 % unless conditioned ash used	Must be dry

Table 3.1   (Continued)

Attribute	Requirements		
	PFA BS 3892, Part 1	PFA BS 3892, Part 2	BS EN 450, fly ash
Water requirement	≤ 95 % of PC 30 % fly ash + 70 % PC (BS 12 42.5)	N/A	N/A
Activity index: ref. EN 450–EN 196-1	N/A	N/A	≥ 75 % at 28 days ≥ 85 % at 90 days 25 % fly ash + 75 % CEM I 42.5
Strength factor: ref. BS 3892, Part 1, Annex F	≥ 0.80 at 28 days	N/A	N/A

[a] Absolute limits. Other values are autocontrol limits.
[b] Permitted on a national basis only.

So far, there are no firm standards for applications other than cement and concrete. However, the requirements and even the properties to be considered may be different from those contained in the standards mentioned above. Thus, in order to apply ash successfully in novel applications and products, appropriate standards must be developed. In Chapter 4, it is suggested how such new standards could probably be developed, and it is indicated that a lot of testing and experimenting will be required to achieve this. If an ash does not conform to an appropriate standard for a certain application, or shows too large a variation in some properties, technologies for maintaining a consistent quality are available. As shown above, such technologies are normally based on physical separation techniques such as classification, in contrast to the more chemical approaches for improving environmental quality. Typical technologies are reviewed in the following section.

3.2.2   Quality Consistency

If the conditions (feed, combustion technology) are such that it may be expected that specifications of coal combustion residues will not comply with the technical and environmental requirements for any intended application, several options are available for favourable adjustments that may be made prior to, during or after the combustion process.

Pre-combustion   Coal preparation and coal cleaning, although essentially an operation to achieve the optimum balance between recovery of the energy content and the rejection of extraneous mineral matter, also includes options to improve ash quality. In general, coal preparation plants contain several circuits for grinding and removal of extraneous materials and the direct result is to decrease the ash content and increasing its homogeneity. The sulfur content in coal is an important environmental issue and therefore many coal cleaning plants include a separation section to remove sulfide minerals. This may be done either by density separation techniques, or by making use of surface properties of the minerals with flotation techniques, or even by chemical techniques using strong bases.[12] Although the main minerals removed are iron sulfides (e.g. pyrite, marcasite, pyrrhotite), base metal sulfides (e.g. chalcopyrite, sphalerite, galena) are also removed, together with elements that exchange with the metals and sulfur in such minerals. Thus the cleaned coal and also the ash will not only be lower in sulfur and iron, but also in Cu, Zn, Cd, Pb and Ni and in As, Sb and Se.

Another interesting possibility is to influence the elemental composition of the ash by blending different coals or fuels. If the composition of the constituent fuels is known, the KEMA TRACE MODEL® (see Chapter 1) provides an excellent tool to predict the composition of the resulting ash. This approach has been practised in coal-fired power stations in the Netherlands, but after liberalisation of the energy market it became too expensive. Apart from the elemental composition of the ash, this model also gives an indication of the leachability of the ash.

Syn-combustion   The addition of lime and limestone to coal influences the nature of the solid residues resulting from combustion, as the available sulfur is retained in the solid form instead of being removed with the flue gases as SO_2. This system is only practised for low-grade coals in fluidised bed combustion and thus generally raises the sulfur content in the ash. Lignite ashes therefore generally show a high gypsum concentration.

By on-line measurement of the LOI, the content of unburned coal can be monitored and controlled. If the LOI becomes too high, the combustion temperature may be raised to achieve an improved burnout. However, increasing the temperature will also generally result in increased NO_x emission, which has to be monitored as well. As a result, an optimum has to be found between these two counterbalancing requirements, which in practice is achieved by computer-controlled plant operation assuring the optimum conditions at any time.

Post-combustion Upgrading the ash after it is generated, although at first sight an end-of-pipe solution, is attractive because it has no impact on the combustion process, and it may allow flexibility in power generation and still produce a good ash quality. Many technologies for upgrading of ash have been and are being developed. These technologies focus on reducing the LOI and ammonia content, improving the particle size distribution, reducing the variability of ash properties, and the conversion of free lime and anhydrite into lime and gypsum. The driving force for the development of such technologies is the improvement of ash quality for use in the cement industry. Such upgrading technologies that have been implemented are mainly focused on the reduction of the LOI by removal of unburned carbon. Commercially operating facilities for unburned carbon removal and recovery are mostly located in the USA, whereas in Europe some sieving and blending techniques are used.

Possible technologies for improvement of technical quality are reviewed below.

Physical Classification

Most fly ashes as produced by the power stations do not comply with the BS 3892 Part 1 requirements. In the past, selection of suitable batches was a way of producing high-quality ashes, but the current procedure in the UK is to classify them. The requirement of < 12 % retained on a 45 μm sieve has its background in the performance of the ash in the concrete matrix. The finer ash has a higher specific surface area and therefore shows an increased pozzolanic activity. It fills in the smallest pores better and gives a denser product. Both are reasons for a gain in strength of the final product. Further, finer ash reduces the water demand. Sear[7] adds that the UK practice has learned that classified ash shows a reduced variability and therefore improves the consistency of the particle size distribution. Classification is normally carried out with common mineral processing equipment such as cyclones, normally with two or more placed in series. The use of cyclones for classification is a proven technology in mineral processing, and the cut-off size and cyclone efficiency may be influenced by setting the cyclone parameters.[13] However, the air-classifiers used for PFA upgrading in the UK are all 'forced vortex' type units, rather than the traditional in-line cyclones widely used for many years by the mineral processing industry. Although these use the same cyclone principle, they are designed to create their own centrifugal and air frictional drag-force field to achieve a more precise particle size cut. Typical UK fly ash will have 25 % retained on the 45 μm sieve, but after standard classification this is reduced to about 8 %, whereas approximately 17 % of the ash is discarded as the coarse fraction.[7] The coarse fraction may be sold for other purposes or ground and fed back into the classifier.

LOI Reduction

Carbon Burnout Progress Materials Inc. (PMI, USA) developed a carbon burnout process in co-operation with EPRI.[14] This process is now implemented on several sites to reduce LOI. The principle of this process is the burning of the unburned matter in a bubbling fluidised bed. The fly ash is fluidised by air at a temperature of 650–815 °C. The waste heat of the processed air is recovered by means of a heat exchanger. For an energy-neutral operation, a minimum LOI of the processed fly ash of about 5–6 wt% is required. Fly ash particles leaving the bed with the flue gas are captured by a cyclone. It is essential that the velocity of the flue gas is low so that the amount of particles in the flue gas can be kept low. Since 1999, a full-scale plant has been in operation at Wateree, Southeast Columbia (USA), with a capacity of 180 000 tonnes per year. In this plant, the LOI of the input fly ash is 6.5–18 wt%, and this is reduced to about 2.5 wt% for the output fly ash. The upgraded fly ash does not agglomerate and shows no loss of pozzolanity.

Carbon burnout by microwave radiation is under investigation in Canada.[15] The Microwave Carbon Burnout (MCB) process reduces the LOI to any prescribed limit and simultaneously eliminates adsorbed ammonia products. No complication with handling As and Hg pollution was experienced, as As remains in the solids and Hg is revaporised and incorporated in the gas stream, and recondenses onto the fly ash upon cooling.

Electrostatic Separation Carbon removal by electrostatic separation technologies has been applied for some years. A distinction must be made between mechanical transport and pneumatic transport tribo-electrostatic methods.

Separation Technologies Inc. (STI, USA) have developed a process based on the mechanical transport electrostatic separation of fine powders.[16,17] The STI process consists of a belt separator made up of two oppositely charged horizontal plates, about 8 mm apart. Between these two plates an open-mesh transport belt, consisting of a perforated plastic conveyor belt, moves with high velocity in both directions.

The fly ash is transported on this belt and, because of the high velocity, the ash particles enter a turbulent regime. Contact with the belt and between the particles causes the particles to become tribo-charged, with the unburned organic matter being positively charged and the inorganic particles negatively charged. As a result, the organic particles move to the negatively charged plate and the inorganic ones to the positively charged plate. A small movement in the direction of a plate causes the particle to be picked up by the transport belt and to be moved to the left or right (Figure 3.1). In this way, fly ash is separated into organic- and inorganic-rich fractions. The whole process of charging and separation takes less than one second. The humidity of both the air and fly ash are important factors for the performance of these electrostatic separation units. The separators are appropriately compact, with a 40 tonnes per hour unit occupying the size of a 30 foot container.

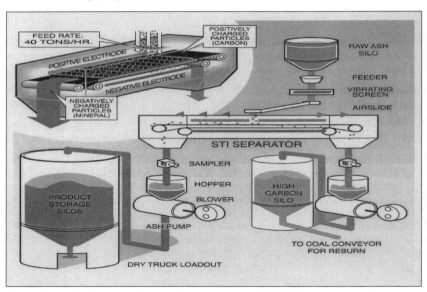

Figure 3.1 An electrostatic separator for the removal of carbon (Reproduced by permission of the University of Kentucky Center for Applied Energy Research and the American Coal Ash Association).[17]

In 1994, the first commercial full scale STI separator was installed at the New England Power Salem Harbor Station (USA), and in 2001 five commercial-scale separators with capacities up to 450 000 tonnes per year were in operation. They generate a uniform quality fly ash of approximately 0.5 % LOI from highly variable feed ashes ranging from 4 to 25 % LOI.

Instead of mechanical transport on a conveyor belt, pneumatic transport in air has been proposed as an alternative. This system was presented at the 2001 Kentucky Symposium[18,19] and is currently less commercialised compared to that using mechanical transport, but because of the faster transport conditions such a system would be even more compact and able to run at very high throughputs. An initial 1 tonne per hour demonstration plant was built in 2001 and appears to operate satisfactorily. Efficiency results are given only in relative numbers, the best performance being 30 % reduction in LOI content of the parent ash, with 74 % of the ash recovered in the cleaned product.

Silo-blending A clever method of classification by means of large screens combined with re-mixing, known as silo-blending, is used in the Netherlands to reduce LOI to conform to local requirements. If operated in an optimum way, other parameters – such as, for example, the $SiO_2 : Al_2O_3$ ratio – may be adjusted as well.

Flotation Flotation is a separation technology based on the difference of surface properties of the materials to be separated. Undoubtedly this technology is the most important and versatile technique for mineral processing and has been in long-term use in the mineral industry.[13] In the context of fly ash, the organic carbon particles are more hydrophobic, whereas the inorganic particles are more hydrophilic. The hydrophobic carbon particles will adhere to rising air bubbles in a flotation cell and move to the surface where they are skimmed off, whereas the other ash particles will sink to the bottom of the cell and removed as a slurry. The hydrophobic and hydrophilic properties may be further artificially enhanced by adding to the slurry flotation reagents, such as collectors to render the mineral surface hydrophobic or depressants to render the surface hydrophilic.

Unburned carbon can be removed successfully from fly ash using froth flotation.[20] A more integrated approach is represented by the FUEL-FLOAT™ method, in which a novel flotation reagent system is used[21] together with a patented unique processing configuration that integrates hydraulic classification with flotation and centrifugal classification.[22] This system separates ash into narrow ranges, recovering coarse carbon via gravity concentration, coarse lightweight aggregate via hydraulic classification and fine carbon via froth flotation.[23]

At the Michigan Technological University Institute of Materials Processing, a similar integrated process for fly ash beneficiation was developed in which froth flotation is used to remove carbon, after the cenospheres have been first removed from the slurry by gravitational separation.[24,25] The process also provides options for magnetic separation of Fe-rich particles; however, no classification step is included. An additional advantage of the wet process is the simultaneous removal from the ash of some $CaSO_4$ (basanite) that has retarding effects on cement in concrete mixtures. Plants ranging in capacity from 5 to 100 tonnes per hour can be built and total costs (operating costs plus capital costs) are estimated at between $6 and $12 per tonne. These costs are reported to include dewatering and drying. The cleaned ash product has a carbon content below 1 % combined with ash recovery of more than 90 %, with the production of a carbon concentrate product at an 80 % purity level.

Other Techniques Other, so far nonimplemented technologies for reducing the LOI include separation in fluidised acoustic beds and separation by a jet mill. Supercritical water oxidation was demonstrated by Hamley et al.[26] The advantage of this method would be the reliability of carbon removal and the utilisation of the heat content of carbon.

Reduction of Ammonia

Coal-burning power stations increasingly inject ammonia or ammonia-based reagents into the flue gas stream to enhance electrostatic precipitator performance and to remove nitrogen oxides (NO_x), using selective catalytic reduction (SCR) or selective noncatalytic reduction (SNCR) technologies to meet NO_x emission regulations. When ammonia contaminated fly ash (200–2500 ppm) is used in cement-based concrete applications, ammonia gas is emitted from the fresh concrete mix into the air, exposing concrete workers to this potential risk.

Removal of ammonia by alkaline water extraction creates a high pH to enforce the process.[27] This method produces an increased alkali content and, moreover, is not considered very economical due to the high drying costs involved. However, wetting fly ash with a few per cent water in combination with careful dosing of the alkali content (by lime addition) is claimed to render an economical process in which 100 % of the fly ash is recovered and the removed ammonia is allowed to be reused in the plant.[28,29] The process is owned by Separation Technologies Inc. and can be integrated with their carbon removal process. However, the energy costs are not favourable, as the thermal processing requires temperatures in the range of 300–500 °C to desorb or to decompose ammonium salts.[30] An option would probably be the use of microwave heating.

Ammonia is adsorbed in the ash mainly as ammonium salts, and another possibility may be to remove the ammonia by oxidation through reaction with hypochlorite.[31]

Lime and Anhydrite Conversion

CERCHAR in France has developed a process for conversion of free lime and anhydrite into calcite and gypsum, respectively, for application on fluidised bed combustion systems. The process is based on forced hydration of these components by mixing the ash with limited amounts of water and a short retention time at 15 bar.[32,33]

Magnetic Separation (Fe Removal)

In principle, Fe minerals (magnetite and hematite) can be removed using the magnetic separation equipment usually found in mineral dressing operations.[34]

Because MWI residues have a more appropriate structure and metal distribution, magnetic separators, and also heavy media and eddy current separators, are more commonly found at MWIs for the removal of iron, steel and nonferrous metals.

Separation of Cenospheres

Hollow spheres feature as common particles in coal ash, although their abundance is limited to a few weight per cent. They are normally at the coarser end of the size distribution, with an average particle size of approximately $100 \mu m$ (range $50–200 \mu m$), a particle density of only $400–600 \, kg \, m^{-3}$ and a still lower bulk density of $250–350 \, kg \, m^{-3}$. The shell consists of glass that is relatively rich in SiO_2 compared to the bulk ash and that contains fewer impurities. Therefore, cenospheres display interesting properties such as high particle strength, low water absorption, and good thermal and electrical resistance. As a filler, cenospheres provide up to four times the bulking capacity of normal industrial minerals. As a powder, the flow characteristics are excellent and the perfect spheres improve the rheology of fillers when mixed with them.[7] According to Sear, cenospheres can be used 'almost anywhere that traditional fillers can be used'. This would be true as long as their particle size is acceptable, because most other properties are favourable; however, the size is rather large compared to what is normally required for many industrial fillers (see Chapter 4.8).

Both size and density are properties that can be used for the separation of cenospheres.[35] Since their density is lower than water, in cases where ash is disposed of in lagoons, a layer concentrated in cenospheres can be skimmed off. Most integrated ash beneficiation processes now include a separation step for cenospheres. Such processes were mentioned above in the section on carbon removal.[23,24] Thus Ghafoori et al.[36] used variation in particle size with a dry sieving process prior to eventually using wet processes for cleaning. The ash was dry sieved at $75\,\mu m$ and $53\,\mu m$ to yield three fractions. The fine fraction has a reduced carbon content and is readily used for cement-concrete applications. The middle fraction is enriched in carbon and is further processed in a dry fluidised bed separator to remove carbon, whereas the underflow is combined with the fine fraction. In the coarse fraction, all the cenospheres are recovered.

Pneumatic transport, tribo-electric separation, as described in the section on carbon removal, is also believed to have potential for the simultaneous concentration of cenospheres.[37] Adjustment of the voltage of the electrostatic separator made it possible to produce a concentrate with a density of $1.6\,kg\,cm^{-3}$ from a feed density of $2.46\,kg\,cm^{-3}$.

Grinding and Micronisation

Micronisation of fly ash into an ultra-fine product ($< 1\,\mu m$) has been investigated by a number of researchers.[38] Although milling costs are high ($\in$200 per tonne), the products have a high added value and may compete with expensive raw materials such as silica fume, because of their ability to improve concrete strength considerably.

3.3   Environmental Quality

3.3.1   Introduction

Solid particulate matter is the unavoidable by-product of combustion. Because the solids exist as a fine powder and easily form aerosols, they have been regarded by the public as a nuisance since the widespread use of combustion practices. As early as 1257, Eleanor, Queen of England

and wife of Henry III, was obliged to leave the town of Nottingham due to the 'troublesome smoke from the coal being used for heating and cooking'.[39] Later, with the development of analytical chemistry, it was recognised that such airborne material contains variable amounts of many heavy metals and hazardous compounds. Therefore, the first environmental measures were concerned with controlling the release of such materials to the open air and avoiding exposure to workers and the public. Rapid development in filtration techniques and electrostatic precipitation of fine powders meant that by the beginning of the 21st century modern power plants and incinerators hardly emitted any particulate matter into the air. However, science makes progress and with the identification of new carcinogenic compounds and recognition of the health effects of ultrafine particles, rules and regulations are regularly being modified at regular intervals becoming stricter to protect the population. These aspects are reviewed in Section 3.4.

Once the residues in the form of particulate matter are safely collected from the combustion process, they start posing their own problems. Regardless of whether they will be disposed of in landfill, stored or reused in various applications, elements and compounds may be released from them by leaching. The extent to which this may happen depends of course on the form in which the residues are contained – that is, as loose granular material or bound in a composite matrix – and on the external conditions to which the residues are exposed; for example, dry/wet, pH–Eh conditions, temperature, weathering, abrasion and so on. A further complication is that long-term and short-term behaviour are equally important, as are the environmental impacts resulting from these. Prediction of leaching behaviour and emission levels from such materials is therefore a difficult task. The next question is then: what is acceptable to be received by the soil or water? The answer to this question must be translated into an allowable emission level from the material. Such levels must be predicted by using laboratory leaching and extraction tests that are representative of the actual leaching under field conditions. Many countries have set their own environmental standards and developed their own laboratory standard leaching procedures. Worldwide, a variety of leaching protocols have been developed, some of which are used for regulatory control purposes.[40] This is a major cause for confusion, since different testing methods lead to different results.

The EU is in the process of harmonising the tests used by the member states into European standard tests. The CEN TC292 committee and the Thematic Network for the Harmonization of Leaching/Extraction Tests looked into the subject and published a report in which leaching of different materials was reviewed and recommendations made for standardisation and harmonisation.[41] In 2002, the European standard EN 12457 (four parts) was issued.[42-45]

3.3.2 Standards To Be Met

Leaching data are collected from different leaching tests as described in the next section. The results must be compared to certain threshold values, which are highly variable from location to location and situation to situation, as well as in time. Further, they may also depend on the specific type of application in which the material is used.

Within the UK, ashes may be used in civil construction and other applications when permission is given by the Environmental Agency and/or Drinking Water Inspectorate. This is done on a contract-by-contract basis. An environmental risk analysis forms the basis of such acceptance and this implies that for each application a complete re-analysis would be required. The leachate quality threshold as published by the Drinking Water Inspectorate is applied to leachates from materials in contact with drinking water. Normally, a dilution factor of 10 is allowed for leachates in respect to this threshold, as being more representative of the true environmental risk.[7]

The primarily policy in the Netherlands is to produce residues that can be used as (building) products or as raw materials for making (building) products, rather than to produce waste. This means that long-term disposal of coal-firing residues is currently impossible. KEMA has conducted research and experiments with a wide range of potential applications. Today, the combined efforts of the electricity generating companies, KEMA and the Dutch Fly Ash Corporation has resulted in 100 % utilisation of all coal combustion by-products in the Netherlands, mainly in so-called 'bound' applications; that is, in cement and concrete products. The 'sustainable society' to which the Dutch government is dedicated implies not only reuse of residues, but also preventing soil, groundwater and surface-water pollution. To fulfil this ambition, the Building Materials Decree (BMD)[46] was implemented in 1999. This decree provides criteria for both the application and the reuse of stony and granular materials as building materials. No distinction is made between primary, secondary and waste materials. The BMD is applicable if such materials are used in construction, where they are in contact with rain, surface and groundwater – for example, in embankments, road construction, outer walls of buildings, foundations and roofs – and also if they are isolated from rain and water. Standards, methods for testing and certification schemes were introduced for implementation of the BMD in the construction industry. The BMD sets limit leaching values for a number of elements and compounds, obtained by standardised laboratory leaching tests, for different conditions of application. The limit leaching values were derived as shown below.

In this decree, leaching limits are based on the principle that, as a consequence of leaching from building materials, any increase in concentration of any compound in the underlying soil should not be higher than 1 % relative to the concentration in a standard soil over 100 years, calculated over the first 1 m depth of soil. Thus, numerical values depend on the thickness of the application used and the effective infiltration into the soil. Two categories are distinguished: Category 1 assumes 300 mm per year of effective infiltration, compatible with the average yearly rainfall in the Netherlands, and Category 2 allows only 6 mm per year of effective infiltration that can only be obtained when protective measures are taken.

Prescribed formulae convert the laboratory leaching results (NEN 7343[47] for granular materials and NEN 7345[48] for monolithic materials, Section 3.3.3) into soil load values over 100 years. In this extrapolation, correction values are applied for temperature, wetting time and changes in diffusion coefficients as a result of depletion. The calculated load values are compared with the threshold values and on such results a decision on the application is made.

In the last column of Table 3.2, threshold load values for soil have been recalculated into emission values applicable to the standard column leaching test for Category 1 Building Materials applied at a depth of 1 m. Leaching data obtained from surveys by RIVM[49] and KEMA[50] for Dutch coal fly ashes are also shown in this table. It is shown that leaching of Mo, ranging from 0.35 to 10 mg kg^{-1}, is above the limit of the BMD, at 0.24 mg kg^{-1}, in all the analysed ashes. For Se and Sb, both with the very low limit of 0.039 mg kg^{-1}, the limits are exceeded in the majority of cases. Furthermore, Cr, V and SO_4^{2-} very often exceed the leaching limits. Conversely, it must be noted that elements often considered problematical, such as As, Cd and Pb, do not normally show leaching values above the limits. From these values, it follows that the use of coal fly ash in 'unbound' applications such as granular building materials in embankments and stabilisation layers is not possible. However, the evaluation of leaching from 'bound' applications shows generally no conflict with the threshold recipient values of the BMD.

Table 3.2 The leaching behaviour of Dutch coal fly ashes, determined according to the standard leaching test NEN7343[47] (L/S = 10), expressed in mg kg[-1]. Reproduced by permission of Elsevier Limited.[82]

	RIVM/RIZA, 1993[49]				KEMA, 1995[50]				BMD height = 1 m
	Average	Min.	Max.	n	Average	Min.	Max.	n	
As	0.263	0.022	0.987	12	0.07	0.014	0.28	26	0.87
Ba	2.166	0.032	3.200	5	6.1	1.3	22.8	26	4.2
Cd	0.004	0.000	0.011	17	0.0008	0.0004	0.0040	26	0.029
Co	0.018	0.010	0.025	2	0.012	0.006	0.060	26	0.35
Cr	1.729	0.160	4.673	17	1.3	0.2	3.6	26	0.92
Cu	0.080	0.007	0.685	7	0.015	0.009	0.050	26	0.58
Mo	5.81	0.350	15.260	16	5.9	1.3	10.3	26	0.24
Ni	0.064	0.010	0.676	10	0.03	0.01	0.36	26	0.95
Pb	0.051	0.004	0.109	9	0.22	0.02	3.80	26	1.6
Sb	0.203	0.004	0.370	5	0.059	0.010	0.191	25	0.039
Se	1.335	0.010	3.700	6	0.96	0.04	4.80	26	0.039
V	3.969	0.350	17.490	9	1.7	0.1	9.9	26	1.4
Zn	0.213	0.065	1.017	5	0.068	0.033	0.300	26	3.3
SO_4^{2-}	3088	1342	3842	6	1529	85	5160	26	1122

With the results for coal fly ashes in mind it becomes clear that, according to the BMD, Municipal Solid Waste Incinerator ashes and Incinerated Sewage Sludge ashes also do not qualify for 'unbound' applications.

3.3.3 Leaching Tests

The previously mentioned CEN TC292 proposes three levels of leaching tests, to be implemented depending on the pre-existing level of knowledge on the behaviour of the material under certain conditions.[41] These levels are as follows:

- Characterisation tests: to understand the short- and long-term leaching behaviour and the parameters influencing this behaviour.
- Compliance tests: for regulatory control once the characteristics of an evaluated material have been established.
- On-site verification: quick control to verify that material meets the specifications.

Van der Sloot et al.[41] give an excellent review of chemical and physical parameters and conditions that may influence leaching results. The most important physical parameters are the particle properties (size, shape, homogeneity, porosity and permeability) and the imposed physical system (flow rate, temperature, hydrogeological conditions and the time frame of interest). Chemical factors that control release mainly deal with the composition of the material and the percolating liquid as well as possible interactions – that is, constituents, speciation, pH, Eh, equilibrium or kinetic control, sorption, and the presence of inorganic or organic complexation compounds – and with the possible influence of available biological factors.

During leaching, the fluid/particle boundary layer may form a resistance to the diffusion of dissolved matter from the particle surface to the bulk liquid. In laboratory leaching procedures, this resistance is tackled in different ways. In column leaching tests, the particulate matter is stationary and a leachant flows through and around it, carrying away the dissolved constituents. When agitation is used in batch leaching tests, the fluid is caused to flow past particles and accelerates dissolution. In batch leaching scenarios without agitation, only diffusion and Brownian motion permit the transport of the dissolving constituents.

Bearing all these variables in mind, one can imagine that worldwide a very large number of different leaching/extraction tests have been developed. This is a result of the many different types of materials to be tested under different environmental conditions. However, many such tests are variations of the same basic principle, with small modifications and adjustments to suit specific materials and conditions. Furthermore, the purpose for which a test is carried out is of great importance. Distinction must be made between regulatory requirements, specific quality criteria, impact assessment, scientific evaluation of leaching and management tools for daily practice.

An overview of available leaching test procedures for granular materials and monolithic materials is available in tabular form, with a thorough discussion of the interpretation of the results from such tests.[41] It follows that pH and the L/S ratio are the variables that most affect leaching. The tests most often used in connection with combustion residues are discussed below.

USA TCLP (Toxicity Characteristic Leaching Procedure), Method 1311
This is a standard batch leaching test developed by the US EPA for classification of waste.[51] It consists of a single batch equilibrium based test carried out at L/S = 20 for 24 hours. Very alkaline materials are leached with an acetic acid solution of pH = 2.88, and other materials with a fixed quantity of acetic acid solution buffered at pH = 4.93 with NaOH. The resulting pH is normally around 5 except for very strong alkaline materials, for which higher pH values may be found. This may lead to very variable results for elements that show high pH dependent solubility, especially for strong alkaline fly ashes with a high buffer capacity.

EU EN 12457 A specific memorandum[52] was issued by the EU on leaching from coal combustion residues, from which, with others, the proposal for a European standard leaching test was developed by the CEN TC292 committee. In 1996 the first draft European standard single-batch equilibrium-based leaching test (prEN 12457) was issued, designed for the leaching of granular waste materials and sludges.[53,54] The final version EN 12457 (in four parts) was accepted in August 2002.[42–45]

EN 12457 was specifically developed as a compliance test. The granular material must have a particle size < 10 mm and the test is carried out with demineralised water in one of four ways, depending on the properties of the waste and the rationale of the testing:

- For materials with a high solid content and particle size below 4 mm: one-stage leaching at $L/S = 2$ for 24 hours, with agitation (Part 1).
- For materials with particle size below 4 mm: one stage leaching at $L/S = 10$ for 24 hours, with agitation (Part 2).
- For materials with a high solid content and particle size below 4 mm: two-stage leaching at $L/S = 2$ and $L/S = 8$ for 6 and 18 hours, respectively, with agitation (Part 3).
- For materials with particle size below 10 mm: one-stage leaching at $L/S = 10$ for 24 hours, with agitation (Part 4).

The required particle size may or may not be obtained by size reduction and the pH during testing is dictated by the material.

EU CEN/TS 14405 A technical specification describing an up-flow percolation test was issued by the EC CEN Technical Committee 292 in June 2004.[55] This test was set up on the basis of the Dutch NEN7343 leaching test.[47]

EU DRAFT prEN 14997 A draft norm was issued in July 2004 by the same committee, dealing with a test to establish the influence of pH on leaching with continuous pH-control.[56] This test is to be applied during leaching behaviour tests for the characterisation of waste.

EU CEN/TS 14429 A test similar to the previous one, but regarding the influence of pH on leaching with initial acid/base addition, was issued in October 2005.[57]

Germany DIN 38414 S4 This is a single-batch, equilibrium-based standard leaching test widely used for regulatory purposes in Germany and Austria, but also in other countries.[58] The material should be $< 10\,mm$ and the test is run with demineralised water for 24 hours at $L/S = 10$, under shaking or slow rotation. The pH is dictated by the material, but the method allows a second and third extraction for sparingly soluble components. This method was designed for sludges and sediments and is also considered applicable for solids and pastes.

Netherlands NEN 7341 This is an availability test for the assessment of maximum leachability, designed for waste management regulations.[59] The material, if necessary, is ground to $< 125\,\mu m$ and then extraction is carried out in two steps, each at $L/S = 50$ for 3 hours, the first step at pH = 7 and the second step at pH = 4. The pH is kept constant during leaching by feedback control and the addition of HNO_3 or NaOH. The combined extract of the two steps is analysed as one sample.

Netherlands NEN 7343 This is the Dutch standard column test for the leaching of granular materials.[47] Demineralised water acidified to pH = 4 with HNO_3 percolates upwards through a 5 cm diameter column filled with the material (< 4 mm) to a height of 20 cm. The total L/S ratio is 10 and the flow is adjusted so that the total test duration is approximately 21 days. During the test, seven eluate fractions are taken at prescribed cumulative L/S ratios. The test is used for regulatory purposes to simulate leaching from waste in the medium term (up to 100 years), but does not take into account aging effects and mineral changes.

Netherlands NEN 7345 This dynamic diffusion tank leaching test was designed for monolithic materials and stabilised/solidified waste materials.[48] Thus a monolithic sample of the material (smallest dimension > 40 mm) is placed in demineralised water acidified to pH = 4 using HNO_3. The sample must be completely immersed and have free contact with the water on all sides. The total amount of water should be approximately five times the volume of the sample. The total leaching time is 64 days and the water is refreshed at prescribed times to yield eight leachates for analysis.

Netherlands NEN 7349 This serial batch leaching test is used for regulatory purposes in the Netherlands and is considered to represent leaching over a very long time scale or at very high imposed L/S ratios.[60] Five successive extractions for 23 hours each are carried out with granular material (< 4 mm). The leachant is demineralised water acidified with HNO_3 to an initial pH of 4 and with L/S = 20 for each extraction (i.e. the total accumulated L/S equals 100). This test has now been replaced by the European test EN 12457.[42–45]

Other Countries The Scandinavian countries use a column test known as the Nordtest method. This test is similar to the Dutch NEN 7343, but with a slower flow and an accumulated L/S of only 2.[61] The French (AFNOR X 31-210[62]), as well as the Japanese (JST-13), standard batch leaching tests resemble the German DIN 38414 S4 test. More details on these and other tests can be found in Van der Sloot.[41]

3.3.4 Comparison of Leaching Test Results with Field Observations

Questions arise whether the predictions from the laboratory tests are satisfactory, what the leaching is like under field conditions and how much rainwater will be evaporated. Accordingly, the conversion of results from the Dutch standard leaching tests into field conditions is important and formed a strategic research area at KEMA, where the leaching of unbound by-product and bound by-products were tested under field conditions in the open air. The first by-product (coal fly ash) was tested in lysimeters, and moulded products – that is, coal fly ash (CFA) in clay bricks, limestone bricks and concrete – were tested in purpose-built walls.

Leaching of Unbound Coal Fly Ash in Lysimeters

Results from leaching the fly ash powder in the lysimeter are compared with the laboratory tests. According to the BMD, the leaching behaviour at $L/S = 10$ has to be determined by the Dutch column leaching test (NEN 7343). If the by-products fail to meet the limits, there are two options: immobilisation or storage. The results of the column leaching test at $L/S = 1$ determines the appropriate storage conditions.

All the results discussed[63] concerned an alkaline coal fly ash (CFA), a blend of various ashes and representative of CFA produced in the Netherlands. It appears that the set of Dutch leaching tests at different L/S ratios – that is, a column test at $L/S = 10$, a cascade test at $L/S = 100$ and an availability test at $L/S = 100$ – correlate well with each other. Exceptions are Na and Cl, which show higher leaching values in the cascade test than in the availability test. Also, the European leaching test (CEN-test at $L/S = 10$)[55] gives comparable results with respect to the Dutch column test, although for some elements, K, Cr $<$ Na, B, Mg, Ca $\ll SO_4^{2-}$, the CEN test shows higher leaching. The EPA test differs from the cascade test, in the different pH regimes used. There are indications that the leaching tests themselves influence the system, causing some elements to become more mobile and others less mobile.

Water retention characteristics indicate that moisture from depths of 3.5 m can be transported to the surface, and thus can be subject to evaporation; therefore, lysimeters were built at heights of 0.95 and 3.80 m.[63] Maximum saturation, under laboratory conditions as well as under field conditions, is about 53 %. After more than eight years of operation, evaporation from the shallow lysimeter was 65 % and from the deep lysimeter 48 %. This means that the percolation assumed in the Dutch Materials Buildings Decree (BMD) of 300 mm per year gives a good prediction, but it overestimates percolation for an application that is 1 m thick.

After more than eight years, the L/S ratio obtained for the shallow lysimeter is 2.2 and that for the deep lysimeter is about 0.75. The amount of leaching of the elements was determined in two ways: via freshly obtained point samples and by integrated tank samples collected over a defined period. It appears that, in general, both approaches give the same results. By comparison of the lysimeter results with the laboratory results of the column test, it appears that:

- the elements Al, As, B, Ca, Cr, Mo and Si show similar behaviour;
- for the elements Ba, Ca, Cu, Fe, Mg and Sb, leaching in the lysimeter is lower;
- for the elements Mo, S, Se and V, leaching in the lysimeter is higher;
- for the elements Cr, K and Na, leaching in the lysimeter is far higher than in the column test.

Cr and Mo are mentioned twice because their behaviour in the shallow and the deep lysimeter is different. It seems that the leaching in the shallow lysimeter most resembles the curve of the column test, but that the cumulative leaching curve in the deep lysimeter is steeper than those of the column test and of the shallow lysimeter. This phenomenon is found for As, Cr, K, Mo, Na and Se. The continuing high pH value ensures that the concentrations of Be, Cd, Co, Cu, F, Ni, Pb, Sb, Tl and Zn still lie mostly below their lower limits of detection.

Chemical analyses of the pore water give a better insight into the leaching mechanism. It appears that after a dry–wet cycle, the peak leaching of SO_4^{2-}, Ca, Mo, K, Na and Mg starts again.

Leaching of Moulded Coal Fly Ash in Bricks and Concrete[63]

Using coal fly ash as a raw material in bricks and concrete, the leaching of As, Mo and SO_4^{2-} increases compared to that of bricks and concrete without coal fly ash. Vanadium also shows increased leaching in bricks, but not in concrete. The leaching of chromium is, however, rather complicated; in one clay, it shows an increase in leaching after adding CFA, whereas in another and also in concrete, it shows a decrease. In all cases, leaching stays well below the limits laid down in the Building

Materials Decree (BMD), so this allows coal fly ash to be used as a building material.

The results of the diffusion test extrapolated to the 5.5 years field test with the bricks in place give a reasonable prediction of the leaching of molybdenum and vanadium; however, it overestimates the leaching of arsenic and barium and underestimates the leaching of chromium and sulfate, with respect to the half-brick wall. In the field test, leaching appeared to be strongly influenced by the type of wall; that is, half-brick or hollow-brick wall with a roof. The BMD does not consider this.

3.3.5 The Removal of Metals from Ash in a Slurry

Introduction

Leaching of elements such as As, B, Cd, Cr, Mo, Se, Sb and V from coal fly ash could inhibit the potential reuse of the ash. Also, in a situation that leads to disposal, it may require the ash to be classified in a category implying increased disposal costs. Although dry treatment is preferred for practical and operational reasons, it is often not possible to remove or immobilise the leachable components without the use of water. Also, for environmental reasons, this is not an ideal option, as the problematic components will be transferred into an even more mobile liquid stream. However, when good and economically sound options for the waste-water cleaning are available, this may be the only option to improve the environmental quality of the residue. When the ash is subsequently to be used in a wet form, avoiding drying costs, it may even become a feasible option.

The environmental impact of fly ash either used in applications or disposed of in landfills depends largely on the mobility of the polluting components of the ash. The mobility of heavy metals from coal fly ash has been studied extensively, and from such studies it is clear that mobility of metals depends on their distribution among and within fly ash particles.[64–73] Dudas and Warren[74] proposed a sub-microscopic model for ash particles in which an aluminosilicate matrix is covered with a thin layer enriched in species containing heavy metals, which have condensed on the particles during cooling. Since these species are located on the surface of the particles, the mobility of the heavy metals thus depends on the amount present in this layer. The distribution of trace elements between the matrix and the surface layer of particles from Dutch fly ashes has been investigated by KEMA.[75] They found that more than 80 % of the Se in the ash is located in this layer, As, Cd, Mo and Zn between 40 and 80 %, and Cu, Pb, Sb and V between 20 and 40 %. It follows that the inner part of the particles, the Si–Al matrix, is relatively pure.

Sequential extraction tests are used to determine the behaviour of elements in soil and the affinity of elements for the different components of the soil.[76] Querol and co-workers[72,73] used a modified scheme on fly ash applying sequential extractions with distilled water; ammonium acetate at pH 7 and pH 5, an hydroxylamine/ammonium acetate buffer, and finally hydrofluoric/perchloric/nitric acid attack. Although large variations between ashes have been found, in general, S, B, Mo, Cd, Se and V are among the elements showing highest mobility, and Ba, Co, Ni, Pb and Zn among those with low mobility. This is in general agreement with the findings of KEMA. Such studies, together with analyses of different ash fractions after separation,[73] shed light on the way in which these elements are distributed among the different ash components. The most mobile elements show affinity with calcium oxides and sulfates (As, B, Cd, Mo, S, Se and Sb) and unburned carbon (Se). Less mobile are those with affinity for iron oxides (Co, Cr, Cu and Ni), and the elements with the lowest mobility show affinity to silicates and the glass fraction (Cr, Cu, Pb, V and rare earths).

Apart from the distribution of the elements in ash, for some elements the oxidation state has an influence on both mobility and toxicity. Se has been reported in fly ash mainly as Se(IV) and As as As(V).[68,70]

Possible Options for Removal Processes: Forced Leaching

The fact that a significant portion of harmful components of fly ash is present in a relatively mobile form in the outer layer of the ash particles results in the potential of considerable leaching when the material is exposed to percolating water over a prolonged time. Leaching conditions such as pH, Eh, temperature, time and water chemistry influence the intensity of leaching and standard leaching tests have been developed for certain conditions. As shown in Section 3.3.1, standards for maximum allowable leaching according to such tests have been set that are valid for certain applications. A number of elements from fly ash consistently show leaching in excess of such standards, thus limiting possible applications.

However, the same facts could be used as a challenge, in that they offer possibilities to remove the leaching elements by forced leaching, without however changing the physical properties of the ash or extracting the major elements. The simplest way of doing this is to percolate water through the ash, as in the natural environment. The only important parameter to change is the leaching time, which should be reduced from the scale of years to hours. This could be achieved by operating either at elevated temperatures, or by using higher liquid : solid ratios for short times compared to the natural environment or by adding chemicals to the water that enhance extraction. Processes following these principles are referred to as 'washing'. To be successful, rather high percentages of removal of the targeted constituents are required, so that the ash becomes significantly depleted in these constituents. However, in addition, the washed residue must also show improved leaching characteristics, as these two properties are not necessarily related. Finally, the washed ash must comply with the imposed standards, so that the ash is allowed to be used for the intended applications.

A clear disadvantage of this process is that the environmental problem of the polluting elements is transferred from a solid into an even more mobile dilute liquid form. So the environmental problem is now shifted from a 'mountain' to a 'lake'. This means that the resulting process solutions must be cleaned and if possible recycled together with the reagents used. Technically, this seems not to be a major problem, but to achieve it economically is another issue. However, the main disadvantage of washing ash is that it yields a wet product, whereas normally fly ash is sold as dry powder. Drying is an energy inefficient and expensive process and this puts a further economical constraint on the operation. However, where fly ash is applied in a wet environment, the drying step could be avoided by directly applying the residue as a slurry.

More advanced methods of extraction and possible options for the future are based on the principles of 'dry cleaning'. Extraction using supercritical CO_2 and volatile extractants do not result in large quantities of polluted solutions and a wet product. Moreover, such technologies may be more selective and offer better opportunities for recovery of the value of the extracted metals. Therefore, they are explained in more detail in the next two sections.

For a washing process to be feasible, it should be carried out on very large scale, such as normally encountered in the mineral processing industry. Such unit operations are capable of processing large quantities of material at a relatively low cost per tonne. Processes for large scale treatment of polluted soils and residues have already been proposed by various authors,[77–79] mainly by treatment with inorganic acids, but these could probably just as well be applied to the extraction of trace metals from combustion residues.

A versatile, successful and economically feasible process for removing impurities from fly ash may have more impact on the operation of a coal-fired power plant than one might think at first sight. For impurities that can be removed post-combustion, there will be no need for a pre-combustion restriction. This means that the plant has more freedom in choosing fuel, so inexpensive lower-quality coals become an option, whereas before they could not have been used due to the required quality of ash. Moreover, this also gives more freedom in co-firing with other fuels such as biomass or pet-coke.

Removal with Water

Querol *et al.*[80,81] used open and closed leaching systems at room temperature and at 90 °C to extract soluble major and trace elements from several Spanish fly ashes that showed a broad range of compositions both for major, minor and trace elements. The main purpose of the study was to reduce the concentration of heavy metals to make them more versatile for novel applications. One such application was the synthesis of zeolites, where an increased Si–Al content, as well as a decreased trace metal content, is beneficial, as the latter can contaminate the process solutions and/or the zeolite product. Therefore, the process was aimed at extracting the minor components Ca, Fe and sulfate as well as the hazardous trace metals.

Two types of extraction behaviour were found, one dominated by the dissolution of free lime, resulting in a high pH (11.6–12.1), the formation of portlandite and major extraction of Ca; and the other dominated by the dissolution of anhydrite, accompanied by a lower pH (8.2–9.4) and major extraction of sulfate. The ashes dominated by free lime dissolution are expected to maintain a high pH even when applying a high L/S ratio in a closed system, which will thus reduce the mobility of most heavy metal cations. Therefore, an open-system leaching procedure was also applied. Percolating water through a fly ash filled vessel was expected to decrease the buffer capacity of the lime at lower total water consumption, so that cationic metals could be extracted with greater ease. In addition to ambient temperature leaching, the closed-system leaching was also carried out at 90 °C.

The leaching trends found were consistent with the dissolution of small particles or the inhomogeneous coatings on the surface of the ash solid phases rather than with the dissolution of a homogeneous phase of the bulk composition. Differences between open- and closed-system leaching were not significant, except for As and V. This shows that the process is mainly solubility controlled. However, for As and V in a lime-dominated fly ash, initial leaching is low and a concentration peak in the leachates is found at a later stage. This has been shown to be consistent with precipitation of stable phases – that is, $Ca_3(AsO_4)_2.6H_2O$ and $Ca_2V_2O_7$ – due to the high pH and Ca concentrations in the initial leaching liquids. In this case, the open system at higher water inputs results in higher extraction yields compared to the closed system. Raising the temperature showed only enhanced leaching for silicate-affiliated elements, showing that this increased the degradation of the glass matrix.

For most elements to reach a low and constant leaching rate, L/S ratios of 50–100 are required and in the case of Ca, SO_4^{2-} and Mo, this value may even go up to 200. This means that in order to fully decontaminate the ash, large amounts of water must be used, and consequently must be purified. The leaching process itself may be simple and inexpensive, as stated by the authors, but the subsequent decontamination of the water and disposal of the residues thereof may incur significant costs.

The only elements for which an extraction yield above 5 % was consistently found were Ca, SO_4^{2-}, B, Mo and Se (Table 3.3). For As, Cr and V in the case of a very high L/S (≥ 250), these percentages may reach 5 % and higher. The study did not evaluate whether this resulted in improvement of the environmental quality as determined by the application of standard leaching procedures.

Table 3.3 Extraction yields (% of bulk content) obtained by leaching four different fly ashes in open and closed systems with water at ambient temperature (modified after Querol[81]).

	Closed	Open	
L/S	50	50	250
Ca	5–18	3–15	6–26
SO_4^{2-}	42–76	40–55	52–82
B	11–73	9–57	21–67
Mo	25–66	24–45	33–60
Se	8–62	4–28	4–32

Nugteren et al.[82,83] studied the removal of water-soluble contaminants in a counter-current washing system. Using an alkaline fly ash similar to the free lime dominated ash type, as studied by Querol, comparable extraction yields were found for Ca (approx. 15 %), SO_4^{2-} (approx. 40 %), Mo (approx. 30 %) and Se (20–30 %), but with a significantly reduced water consumption, as for the overall process the effective L/S is 9.

The solid–liquid separation step remains a major concern in the process, but not only because of the costs involved in the final drying operation. There is yet another important implication for the environmental quality of the residue. Metals present in the residual moisture adhering to the residue after the last filtration step will precipitate during drying. These precipitates are important, since they must be considered as highly mobile and will therefore be released during the initial stage of any subsequent leaching – as, for example, in a leaching test. Assuming 20–25 % moisture in the residue after filtration, analyses of the last filtrate indicate that for Se and Mo, approximately 0.04 and 0.05 mg kg^{-1} remain in the residue in this form. For Se, this is equal to the maximum allowable leaching by the standard leaching test according to the Dutch BMD. This shows that it will be very difficult to reach such strict limits even if large portions of the elements in question are removed.

The problem of decontamination of the water has also been studied at least partly.[84] Dissolution of free lime from the ash gives the resulting solution a pH $= 12-12.5$. This solution can be cleaned by bubbling CO_2 through it, forcing the precipitation of $CaCO_3$ accompanied by co-precipitation of Se(IV). Precipitation of ettringite seems the best option to remove sulfate prior to calcite precipitation. Mo can be removed successfully using anion exchange resins. The advantage of this route lies in the fact that the cleaned solution has a very low salt content and can be recycled in the process. The cost of this operation has not yet been evaluated, but technically it seems feasible.

Removal with Extraction Agents

Extraction of contaminating components may be enhanced by adjusting the pH and by the addition of solvents and chelating agents in the washing water. In principle, the anion-forming elements are best extracted at high pH and the cations at low pH. Large-scale extraction operations in the mining industry use solvents or lixiviants such as strong acids or bases (HCl, H_2SO_4, ammonia or NaOH), or salts (NaCN for gold extraction). Application of such lixiviants for fly ash decontamination will result in the dissolution of a very large portion of the ash, including the glass matrix, and result in substantial release of Si and Al. Therefore, extraction agents that are more specific and do not attack the glass matrix would be more appropriate. Nugteren et al.[82,83] used extraction agents that were selected from those used in soil science,[76] and from a leaching study with chelating agents to examine the mobility of coal fly ash metals in the lungs.[69] From those studied, H_4EDTA, an oxalate buffer and a citrate buffer gave the highest extraction yields. Remarkably, if oxalic acid or citric acid is used as a single lixiviant, a significant amount of the glass matrix is attacked, resulting in Si and Al extraction. The same applies when using ammonium oxalate and ammonium citrate. However, when buffers of ammonium oxalate with oxalic acid or ammonium citrate with citric acid are used in a mole ratio of 1.75 : 1, the extraction of heavy metals is high, while the extraction of Si and Al is very low. A certain minimum concentration of the extraction agent is required to achieve reasonable extraction efficiency. Increasing the concentration generally results in an increase of extraction of heavy metals, but often also in an even larger increase in extraction of the major elements Ca, Al and/or Si. An example of this is illustrated for the citrate buffer at constant L/S = 5 and a reaction time of 1 hour in Figure 3.2.

By varying the L/S ratios at constant concentrations and reaction times, it was found that for the extraction of trace elements, a ratio L/S = 5 was most appropriate. Increasing L/S further only resulted in higher extraction of major elements, without increasing the amount of extracted trace elements. Complexation reactions are generally fast, and reaction times of 5–15 minutes have proved sufficient to obtain maximum extraction for most heavy metals. The continuous extraction of Ca, and at a later stage Mg, slowly increases the pH of the solutions, which may result in re-precipitation of certain elements that were previously extracted. This is the same effect as found by Querol for As and V (see above), but appears here also for Se and Sb (Figure 3.3). Such re-precipitation may also occur in some cases for Cr, and the major elements Si and Al and must be avoided, since the precipitated phases may be soluble under the low-pH conditions required for the standard column leaching tests and thus influence environmental quality certification. If this happens, leaching during the column tests may even turn out higher than for the nontreated ash.

Modelling the extraction solutions with MINTEQA2[85] revealed that Cr_2O_3, $Cr(OH)_3$, $MgCr_2O_4$, V_2O_4 and $VO(OH)_2$ show positive saturation indices above pH 4–5 and therefore may be responsible for the decrease of Cr and V in these solutions. High concentrations of Ca would suggest that Ca-arsenates, -selenates and -vanadates might precipitate, as was found by Querol, but MINTEQA2 results show that none of these phases have positive saturation indices at pH levels below 10. This is because the high concentration of the chelating agents in the solution competes for Ca ions with the formation of stable Ca-complexes or, in the case of the oxalate buffer, forms precipitates.

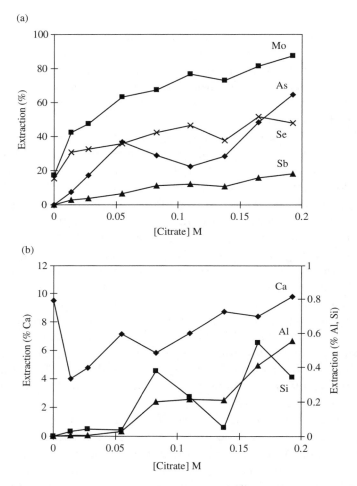

Figure 3.2 The extraction of (a) trace elements and (b) major elements with the citrate buffer at L/S = 5 and t = 1 h. Reproduced by permission of John Wiley & Sons Limited.[83]

Pre-washing with water was carried out to study the effect of removing free CaO from the ash prior to treatment with the extraction reagents. The alkalinity of the ash is thus diminished and the pH in the presence of these reagents will increase more slowly, making the reagent more efficient. This was indeed observed, since the total quantities of trace elements removed increased when using the same reagent concentration or remained about the same when using a 2–5 times lower concentration. However, the formation of Ca-complexes was not avoided as intended, since equal or even higher amounts of Ca were

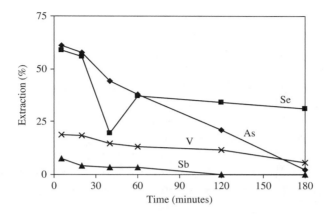

Figure 3.3 The decrease in extraction of As, Se, Sb and V with time due to re-precipitation; $0.05 \, mol \, dm^{-3}$ H_4EDTA, $L/S = 5$. Reproduced by permission of John Wiley & Sons Limited.[83]

still extracted in this second extraction step compared to the extraction from fresh fly ash. It was found that 60–90 % of the SO_4^{2-}, Mo and Se, and 15–20 % of Cr, Sb and V could be extracted under the best conditions with the reagents used. In Table 3.4, a selection of removal values is given, including the amounts removed with water by pre-washing and those removed during post-extraction washing to remove residual extractant (total $L/S = 6$ with fresh water).

Table 3.4 Removal percentages for selected elements obtained during washing experiments with different extractants.

	Molarity	Ca	SO_4^{2-}	Cr	Mo	Se	Sb	V
Water four-step, $L/S = 9$		15.0	37.9	—	27.1	18.5		
EDTA, 10 min	0.05	38.6	69.8	12.1	92.6	62.0	8.9	21.4
Oxalate buffer, 2 h	0.06/0.105	2.3	80.2	10.9	75.9	72.5	20.3	9.1
Citrate buffer, 1 h	0.06/0.105	15.0	78.6	19.0	108.9	46.8	21.4	18.5
Combined with water								
EDTA	0.04	45.5	78.2	8.9	94.5	46.7	6.4	18.0
Oxalate buffer	0.02/0.035	11.7	95.1	7.4	84.3	47.4	9.2	12.3
Oxalate buffer	0.05/0.0875	11.7	95.8	10.1	85.2	42.9	18.4	15.1
Citrate buffer	0.01/0.0175	25.1	63.6	5.8	61.0	40.5	6.7	4.9
Citrate buffer	0.025/0.044	35.3	80.0	10.4	93.6	57.3	19.9	13.0

Using extraction reagents, the concentrations of dissolved elements in the moisture left after the last filtration were even higher than those with water washing alone. Moreover, remnants of the extracting reagents may also cause further release of metals during the subsequent leaching test. So, a method of complete removal of the mobile metals and residual extractants from the washed fly ash has to be developed. Standard leaching column tests (NEN7343)[47] were carried out on the washed ashes and showed a considerable decrease in leaching for most elements; however, these did not improve when water and extraction washing were combined. Nevertheless, in most cases the values still do not comply with the requirements for granular building materials according to the Dutch BMD. A large part of the leached elements originates from the moisture left behind in the treated ash and this underlines the need to improve the solid–liquid separation.

Currently, the benefit of the technically feasible process of washing fly ash to improve the environmental quality has not yet been ascertained, as the results achieved do not show compliance with the required standards (at least in the Netherlands); however, significant improvement may be expected from further optimisation and improved solid–liquid separation. The cost of the process may be further reduced by decreasing reagent consumption, increasing recycling and improving the efficiency of decontamination of the washing waters. Experience from the mining industry has shown that processes that look rather expensive at first sight may well be carried out at much lower cost when undertaken at high throughputs. Economic feasibility mainly depends on the extent of the avoided costs, which is the difference between the market price for the treated ash and the nontreated ash.

3.3.6 The Removal of Metals by Supercritical Fluid Extraction (SFE)

Introduction

The solvating power of supercritical fluids makes them useful in processes such as the decaffeination of coffee and in the extraction of many important industrial chemicals, including medicinal compounds, natural oils and flavours, and even organic pollutants. Recently, studies on extraction of metals from various different matrices have shown the potential for SFE to be applied to the decontamination of polluted materials.[86,87]

A supercritical fluid is a substance above its critical temperature and pressure. Under these conditions it does not condense or evaporate to form a liquid or a gas, but is a fluid with properties changing continuously from gas-like to liquid-like as the pressure increases. Increasing the pressure increases the density and thus the solvating power of the fluid. Supercritical fluids can replace liquid solvents in many processes, such as metal extractions from solids, counter-current multistage separations, chromatographic separations and chemical reactions. One compound, CO_2, has so far been the most widely used for SFE because of its convenient critical temperature, cheapness, nonexplosive character and nontoxicity. Generally, it is applied to the extraction of nonpolar organic compounds, but addition of small amounts of modifiers, such as the lower alcohols, extends the use of supercritical CO_2 to many polar compounds. It has several advantages over conventional organic solvents, which include gas-like diffusivities and liquid-like densities.

By combining the solvating power of supercritical CO_2 and the metal ion complexing power and selectivity of organic ligands, a clean alternative to conventional liquid–liquid and liquid–solid extractions can result. A selective carrier molecule permeates through the sample, combines with the targeted metal and deposits it in a concentrated form in a collection vessel.

By carefully designing ligands that are capable of recognising targeted ions, SFE will gain importance as a separation and extraction technology.[88,89] In practice, the ligand is dissolved in supercritical CO_2 and pumped through an extraction cell in which the sample matrix is placed. The ligand molecules react with the target metals to form soluble metal–ligand complexes that emerge with the CO_2.

SFE Technology

At the Delft University of Technology Laboratory for Process Equipment, research has been carried out on a closed circuit in which samples were placed in an extraction vessel through which supercritical CO_2 mixed with a ligand was pumped. On emerging, the fluid is depressurised, resulting in the formation of gaseous CO_2 and a liquid or solid metal/ligand complex that is collected. The clean CO_2 is then cooled and stored, ready to be recycled. Using a 12 l vessel capable of tilting and rotation, experiments on sand spiked with Zn, Cu, Pb and Cd have shown that extraction efficiencies of up to 60 % can be obtained.[90,91] Four different ligands were tested and the commercial thiophosphinic acid Cyanex 302© showed the highest extraction for Zn, Cu and Cd, while other ligands seem to be more selective for other metals. Further improvements are envisaged where the metal stripping from the ligands is performed at supercritical conditions so that the CO_2 and ligand may be recycled without depressurising, thus saving energy.

Supercritical fluid extraction has only recently been developed, and it is foreseen that by synthesising highly selective and efficient ligands it will be technically possible to extract the leachable portion of trace elements from combustion residues and thus produce a residue that complies with environmental legislation together with a metal concentrate for potential sale. Supercritical fluids will have advantages over conventional organic solvents because they are considered as 'clean' solvents, free from the environmental concerns of disposal, handling and toxicity. Moreover, supercritical technology does not require expensive filtration and drying steps.

Results with MSWI Residues

The removal of Cu, Cd, Pb, Mn and Zn from MSWI fly ash was investigated using supercritical fluid extraction with CO_2/Cyanex 302.[92,93] The main results are as follows:

- the procedure was successfully demonstrated with 2 kg samples of ash;
- the metals investigated were Zn, Pb, Ba, Cu, Mn, Sb, Ni, Cd, Cr, V, Mo and Co;
- percentage extraction varied between 10 and 50 %, the highest being Cd = 46 % and Cu = 51 %, using Cyanex 302 as extractant;
- pre-treatment of the ash with water increased the total extraction of Cd, Zn, Pb and Mo;
- longer extraction times combined with lower ligand concentrations also increased the extraction efficiency;
- standard column leaching tests (NEN-7343)[47] showed that leachability of the residues decreased for Zn, Pb, Mo, V, Mn and Ni, but increased for Sb, Cr and Cd, while Ba and Co remained unchanged.

3.3.7 Purification of Combustion Ashes Using the SERVO Process

Introduction

The SERVO process[94] consists of heating the feed, in this case combustion ashes, in a carrier gas, normally nitrogen, containing a suitable volatile organic reagent capable of forming volatile compounds with the metals that are required to be removed from the feed. In several cases for example, Cu, Co and Ni – the metal may be recovered by direct reduction of the metal complex in the vapour phase, regenerating the extractant for recycle.[95] Where this is not possible, the metal complex can be decomposed by adsorbing in a suitable reagent for example, mineral acid – and the extractant recovered by extraction by conventional means. The most common reagents suitable for transition metals are β-diketones and Schiff base compounds, and in this case the extractants used were 2,4-pentanedione (Hacac), bis(pentan-2,4-dionato)propan-1,2-diimine[96] (H$_2$pnaa) and tetra-*iso*propyldithioimidophosphinate[97] (Hprps). The ash samples were contained in a heated extraction vessel through which the volatile extractant was passed in a stream of carrier nitrogen gas. The emerging gas and metal complex vapours were condensed in a receiver and recovered. The temperatures of the carrier gas and reaction vessel, which ranged between 200 and 280 °C, were chosen from thermal gravimetric studies of the reagents and metal complexes to maximise volatility and minimise decomposition.

Results with Combustion Residues[98]

Fly Ash from the Puertollano Coal-fired Power Station   The extraction of metals from fly ashes depends on their speciation. Thus in the case of this fly ash, where experimental studies showed that the metals are largely in the silicate phase, very little extraction was observed, with only Zn extracted to any appreciable extent: 5.2 % (Hprps) and 6.5 % (H$_2$pnaa). This poor extraction results from the use of weakly acidic extractants that cannot disrupt the silicate phase.

Rotterdam Municipal Waste Incinerator Fly Ash   This ash was from the same source as used for the SFE and was initially sieved and particles in the range 1.4–2 mm used. The diverse nature of the starting material produces a heterogeneous fly ash, with particles having different physical properties. Scanning electronic microscopy showed that some particles were burned plant material, with high carbon content (75 %) and no trace metals, while others were more like characteristic fly ash, with high concentrations of Zn and Cu. A major presence of chlorine in all the particles suggests that the wastes initially contained some PVC material. Speciation studies showed that most of the metals are easily leachable in the first three stages of the BCR procedure[100], except for Ni and Fe, which seem to be mainly contained in the silicate phase. With the highest concentration, Zn poses a significant risk, and as most of the Zn (60 %) and Cd (71.5 %) are easily leached by $0.11 \, \text{mol} \, \text{dm}^{-3}$ acetic acid, they would leach under acid rain conditions.

Results of exhaustive extraction with the SERVO process showed (Figure 3.4) that the extractant Hprps provides the greatest extraction. It is particularly interesting to note that both Cd (81 %) and Pb (75 %) are readily removed by this sulfur donor ligand. This extraction of 'soft' acids is confirmed by the extraction of both Cu (83 %) and Zn (69 %), whereas the 'hard' acids, Ni (27 %) and Fe (9 %), have much lower values. These results fit reasonably well with the speciation results, which indicate that the leachable fraction of the metals should be as

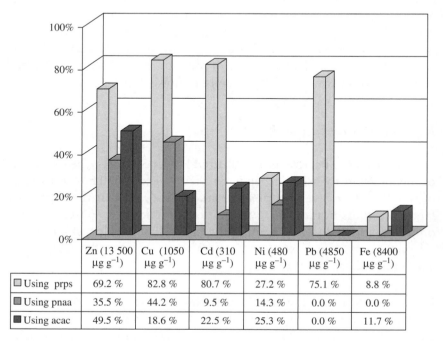

	Zn (13 500 µg g⁻¹)	Cu (1050 µg g⁻¹)	Cd (310 µg g⁻¹)	Ni (480 µg g⁻¹)	Pb (4850 µg g⁻¹)	Fe (8400 µg g⁻¹)
☐ Using prps	69.2 %	82.8 %	80.7 %	27.2 %	75.1 %	8.8 %
▨ Using pnaa	35.5 %	44.2 %	9.5 %	14.3 %	0.0 %	0.0 %
■ Using acac	49.5 %	18.6 %	22.5 %	25.3 %	0.0 %	11.7 %

Figure 3.4 SERVO extraction results for Rotterdam waste incinerator fly ash.

follows: Cd, 82 %; Pb, 79 %; Cu, 81 %; Zn, > 90 %; Ni, 60 %; Fe, 12 %. The only element where the observed extraction is much less than indicated by the speciation studies is Ni.

Orimulsion Ash The major components of Orimulsion combustion ash are $MgSO_4$, MgO, V_2O_5, $NiSO_4$, and residual C.[99] BCR speciation results showed that Ni and V are both easily leachable in steps 1, 2 and 3, leaving no residue. This ease of leaching is one of the major problems with the disposal of this material. The high concentration and easy leachability of Ni and V should favour extraction by the SERVO process, and this was confirmed by studies on pelletised Orimulsion ash with Hprps, extracting Ni (81.6 %) and V (80.0 %). Lower extraction was found with H_2pnaa (42.7 % and 22.8 %, respectively) and Hacac (63.9 % and 15.0 %, respectively).

Conclusions

The effectiveness of the SERVO process depends critically on the speciation of the elements in the feed. Thus, the process is most successful on the Rotterdam MWI and Orimulsion ashes, where the elements are readily leached, and the worst results are found with the Puertollano power plant coal fly ash, where the contaminants are present in the silicate phase.

In considering the overall feasibility of both this process and supercritical fluid extraction, the cost of extractants and the ability to recover and recycle these is vital to the overall economics. It has been shown in an earlier study[96] that the cost of recovering Ni from a laterite ore, using the SERVO process at laboratory scale with H_2pnaa as extractant, is approximately the same as the costs using other commercial processes. However, in this case, the Ni concentration in the ore is relatively high and Ni recovery by vapour phase hydrogen reduction and recycle of the extractant was proven. In addition, the favoured extractant H_2pnaa is readily synthesised from relatively low cost commercially available starting materials. In the case of the extractant Hprps, the best extractant for a number of the more toxic metals, this is not the case, as the synthesis is complicated and requires expensive raw materials.

Therefore, on the evidence of this preliminary study, the SERVO process does not currently provide a feasible process to purify combustion fly ashes except in favourable cases where the metals are readily leached and low-cost extractants can be used.

3.4   Health and Safety

3.4.1   Introduction

Concentrated fine particulate material in air is normally experienced as 'dust'. Intuitively, dust is regarded as a nuisance, as it has direct negative effects on respiration and on eyes, and it makes one dirty. At low concentrations in air, dust is not visible. Nevertheless, airborne particles, known as aerosols, are present and may have an effect on health through inhalation and contact with skin and eyes. Aerosols are always present (even in ultra-clean rooms) and typical concentrations in ambient air range from 0.04 mg cm^{-3} in rural areas to 0.15 mg cm^{-3} in heavily populated urban areas.[101] Aerosols have two main origins: natural and anthropogenic. Examples of natural aerosols are soil dust generated by wind action, salt crystals from seawater, volcanic dust, smoke from forest fires and so on. Anthropogenic aerosols originate mainly from industrial and transport activities: soot particles from car engines and particles from wear on tyres and brakes, and particles that leave the stacks of factories, among which are airborne particles from combustion. However, a large proportion is also generated in the atmosphere by gas-to-particle conversion processes.[101]

Particle sizes vary from the ultra-fine nanoparticles (< 0.1 μm) to fine particles (0.1–1 μm) and coarse particles (> 1μm) up to about 50 μm. Larger particles deposit very quickly and are only of importance very near to local point sources. If aerosol particles contain toxic components, it is understandable that concern is raised about the health risks from exposure. Ultra-fine and fine particles pose the highest health risks, as they are inhaled deeply. According to recent studies, 6 % of total mortality in Austria, France and Switzerland can be attributed to airborne particles.[102] While these very small particles dominate the atmospheric number concentrations (10 000–30 000 cm^{-3}), they amount to a negligible contribution to the atmospheric particle mass concentration.

The size of the finer particles from combustion residues is in the range of aerosols and such particles are known to contain traces of (toxic) heavy metals and are suspected of containing carcinogenic compounds. Therefore, governmental healthcare bodies have undertaken studies to evaluate the risks involved in relation to the modes and the extent of exposure. KEMA in the Netherlands has made a comprehensive study of all aspects of health issues involved with the generation of energy in coal-fired power plants.[103] These findings form an important source for the discussion below.

3.4.2   Sources of Airborne Particles

When the combustion of coal was first used to generate electricity, the fly ash initially left the stack with the flue gases, became airborne and was deposited in an area around the source, depending on the prevailing climatic conditions. Thus, people living in the direct vicinity of power stations experienced this first hand by deposition on agricultural land (not always felt as negative), on vegetables and garden furniture, and as dirt on other personal belongings. Clearly, they also inhaled particles.

In modern coal-fired power stations (Chapter 1), fly ashes are collected by particulate control equipment (normally electrostatic precipitators, ESP). Figure 3.5 shows a typical flow sheet of the particulate materials in a modern power station, indicating the ratios of flows and the concentrations of the flows important to health aspects. From this figure, it is shown that the ESPs remove most of the ash (99.75 %), as a fraction termed pulverised fuel ash (PFA). Subsequently, the flue gas desulfurisation unit (FGD), with the main purpose of removing SO_2, also extracts about 80 % of the remaining particulate material, which is simply called fly ash. This brings the overall ash removal to near

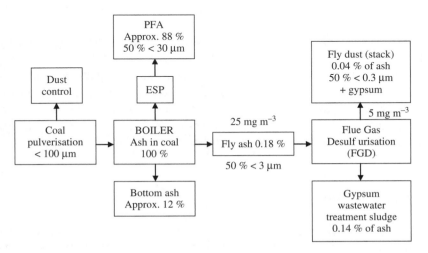

Figure 3.5   The distribution of ash flows in a modern coal-fired power station (Reproduced by permission of KEMA[103]).

99.96 %. Hence, about 0.04 % of the total ash is released through the stack in a concentration of about 5 mg cm^{-3} into the surrounding atmosphere, which in urban areas has a normal background concentration of approximately 0.1 mg m^{-3}.[101] The aerosol leaving the stack is called 'fly dust' because it contains only about 50 % true fly ash particles, the balance being gypsum particles released from the FGD unit.

From Figure 3.5, it becomes clear that the sources of aerosol emission are the systems for coal pulverisation, the handling of PFA and the stack gases. From this it is easy to understand that two groups of people may eventually be affected by exposure to aerosols generated by this industry: firstly, power station employees and workers handling and transporting PFA; and, secondly, people living in the direct vicinity of such plants.

The composition of PFA is discussed at length in Chapter 1, but it must be noted here that from the studies done by KEMA,[103] it has become clear that the composition of the 0.04 % of fly ash that leaves the stack may be different from the overall PFA composition due to differential concentration processes occurring during the passage of materials through the plant. The concentrations of major and minor elements are similar in all ashes, but the concentrations of most trace elements, especially As, Cd, Mo, Ni, Pb, Sb, Se and Zn, increase markedly from bottom ash < pulverised fuel ash < fly ash < fly dust. This is partly an effect of the decreasing particle sizes, combined with the tendencies of trace elements to be located on the surface of the particles; that is, an increasing surface to volume ratio with decreasing particle size. As indicated in Figure 3.5, the median size decreases by a factor of 10 from PFA > fly ash > fly dust. However, this effect is partly cancelled out by the fact that the ash particles in the fly dust are diluted by gypsum particles. For practical reasons, a worst-case assumption is made, namely that the metal concentrations in fly dust are taken to be equal to concentrations assuming that fly dust consists only of fly ash particles. However, exceptions have to be made for As and Se, because their concentrations have proven to be higher following the FGD system.[103]

3.4.3 Exposure

As noted above, the people susceptible to exposure to pulverised fuel ash, fly ash and/or fly dust originating from coal-fired power stations are coal-fired power station workers and those involved in transportation or processing of PFA, and people living near coal-fired power stations. Health effects may occur if such people are exposed to the above combustion residues.

Three possible mechanisms of exposure can be distinguished:

(1) *Skin and eye contact.* Concentrations of heavy metals in PFA are sufficiently low to exclude health effects on workers in power stations via this mechanism.[103] Moistened ash may be irritating in eyes and on the skin by raising the pH to very basic values. Rinsing with water can counter such effects, as is the case in the event of incidental high exposure rates during maintenance or accidents, although personal protection equipment should be used in such cases. The level of exposure through skin and eye to airborne PFA or flue dust experienced by people living nearby is sufficiently low as to be insignificant.

(2) *Oral intake.* Oral intake is possible via the hands (smoking, drinking, and eating), via consumption of crops on which ash has settled and via ash-containing phlegm. Just like grains of sand, ash particles pass through the gastro-intestinal tract in about 1–2 days and will leave the body. Due to the low pH in this tract, leaching of heavy metals is a possibility, and thus uptake of such metals through the intestinal wall into the bloodstream. Provided that adequate personal hygiene standards are maintained, intake via the hands can be considered negligible even by power plant workers. Washing of crops prior to consumption is also considered adequate for discounting this route. Small children form an exceptional group, as they often swallow earth and have less well regulated hygiene habits.

(3) *Inhalation.* Inhalation is the most important mechanism of exposure to airborne particles of PFA and flue dust. The level of penetration of particles into the respiratory tract depends on the particle size. Particles deposited in the upper tract (nose and windpipe) are removed via its surface layer (mucus), which is continuously being replaced. Small particles penetrating deeper into tracheae and alveoli are removed by pulmonary macrophages, which have a neutralising and storing capacity. Storage capacity is limited and excessive inhalation may lead to accumulation. Leaching becomes less important than in the gastro-intestinal tract because of the neutral and stable pH (7.3) in the lungs.

Chronic Non-Specific Lung Diseases (CNSLD – asthma, bronchitis, pulmonary emphysema and allergies) are due to irritation of the respiratory tracts, which may occur as a result of exposure to any kind of airborne dust, including PFA and flue dust.

3.4.4 Occupational Health and Safety

From experiments with laboratory animals, it has become clear that through inhalation of PFA limited amounts of some heavy metals may be adsorbed by the body.[104,105] However, no specific effects of such exposure to heavy metals from PFA are known. Risk analysis shows that exposure to heavy metals from PFA remains below the strict Threshold Limited Values (TLVs) for such substances implemented by the Dutch government.[106] Exposure to heavy metals has been measured in practice by KEMA, and it was shown that maximum concentrations of the individual elements are below 1 % of the TLVs for such elements, with the exception of vanadium (16 %).[107] Normal background concentrations proved to be major contributions to the total exposure.

Some compounds of As, Cd, Cr(VI), Ni and Be, which may be present in PFA, are suspected carcinogens. According to EU guidelines, a substance must be considered carcinogenic when the carcinogenic compounds therein have a concentration higher than 0.1 %. In PFA, such concentrations are in the range of 0.01–0.0001 % (Chapter 1.3.2).

Epidemiological research has shown that exposure to PFA gives no additional health risk when compared to exposure to other 'normal' dust and that no carcinogenic effects were found in animal organs after two years of exposure to PFA dust.[108] More recent work showed no negative influence on DNA[109] and confirmed that weak effects on health were comparable to the effects of other nonharmful dusts.[110]

Despite the above evidence, KEMA, in a very broadly based research project, reviewed the issue of occupational health and safety of pulverised fuel ash.[103] A Dust Assessment Methodology (KEMA-DAM®) was developed with the purpose of easily determining whether TLVs are likely to be exceeded in the event of exposure to PFA. Measurements indicate that, under normal operating conditions in a power station, the concentrations of inhalable PFA (PM_{50}) vary between 0.1 and 7 mg m^{-3} and concentrations of respirable PFA (PM_4) vary between 0.1 and 2.3 mg m^{-3}. Definitions of the different size fractions used in a health-related context are given by ISO standard 7708.[111] The TLVs for nonharmful or 'nuisance dust' are 10 mg m^{-3} for the inhalable fraction and 5 mg m^{-3} for the respirable fraction. For PFA, which contains on average 55 % inhalable particles and 5 % respirable particles, a concentration of $10\,\text{mg}\,\text{m}^{-3}$ of inhalable particles means the presence of approximately $1\,\text{mg}\,\text{m}^{-3}$ of respirable particles, so that in practice conformation with the TLV for inhalable particles is sufficient.

In devising the KEMA-DAM® model, whenever assumptions were required, the worst-case scenario was always taken (that is, maximum exposure of $10\,\text{mg}\,\text{m}^{-3}$ inhalable fraction, maximum concentrations, summation of harmful compounds, traces present in harmful speciation, traces as harmful as if present in pure form, safety factor for synergetic effects etc.). Table 3.5 summarises the results of the assessment of PFA using the KEMA-DAM® model. The composition of inhalable PFA has been derived from average coal compositions used in the Netherlands in 1999, taking into account the relative enrichment factors

Table 3.5 PFA exposure tested against Dutch TLV values using KEMA-DAM® (Reproduced by permission of KEMA[103]).

Component	Composition inhalable PFA fraction		Allowable concentration according to TLV value at $10\,\text{mg}\,\text{m}^{-3}$ PM$_{50}$ dust	Fraction of TLV value	
	Average	Max.		Average	Max.
Carcinogenic substances (mg kg^{-1} dry matter)					
As	39	100	2500	0.02	0.04
Be [a]	11	21	200	0.06	0.11
Cd	1.2	5.4	500	< 0.01	0.01
Cr(VI) [a,b]	13	18	2500	0.01	0.01
Ni [a]	135	291	10000	0.01	0.03
Major and minor elements (% dry matter)					
Al	14	16	53	0.27	0.30
Ca	2.3	3.2	36	0.06	0.09
Fe	4.6	6.2	50	0.09	0.12
Mg	0.8	1.0	60	0.01	0.02
Si	24	25	100	0.24	0.25
Ti	0.9	1.0	60	0.01	0.02
Trace elements (mg kg^{-1} dry matter)					
B	321	533	312500	< 0.01	< 0.01
Ba	1457	3443	50000	0.03	0.07
Co	49	72	2000	0.02	0.04
Cr	127	183	50000	< 0.01	< 0.01
Cs	8	10	176991	< 0.01	< 0.01
Cu	100	175	100000	< 0.01	< 0.01
F	311	1231	250000	< 0.01	< 0.01
Hg	0.3	1.9	5000	< 0.01	< 0.01
Mn	348	770	100000	< 0.01	0.01
Mo	21	58	500000	< 0.01	< 0.01
Pb	102	197	15000	0.01	0.01

Table 3.5   (Continued)

Component	Composition inhalable PFA fraction		Allowable concentration according to TLV value at $10\,\mathrm{mg\,m^{-3}}$ PM_{50} dust	Fraction of TLV value	
	Average	Max.		Average	Max.
Sb	7.3	14	50 000	< 0.01	< 0.01
Se	19	50	10 000	< 0.01	< 0.01
Sn^c	17		200 000	< 0.01	< 0.01
Te^c	< 3		10 000	< 0.01	< 0.01
Tl^c	< 7		10 000	< 0.01	< 0.01
U	10	19	20 000	< 0.01	< 0.01
V	249	397	1 000	0.26	0.41
W	13	74	100 000	< 0.01	< 0.01
Zn	258	1164	403 226	< 0.01	< 0.01
Summation of fractions of trace elements				0.42	0.75

[a] Genotoxic carcinogenic.
[b] Assumed $Cr(VI) = 0.1 \times Cr_{total}$.
[c] Assumed concentration in inhalable PFA equal to concentration in total PFA.

for the elements in inhalable PFA (determined from measurements). The maximum value represents the mean plus twice the standard deviation from these values. The allowable concentration in the fourth column represents the concentration in inhalable PFA at which the TLV (based on a time weighted average during 8 hours of exposure) will be reached. When this happens, the fraction of TLV in the last two columns will be unity. As can be seen from the table, when exposed to 10 mg m^{-3} of inhalable PFA dust, the TLVs are not exceeded for the considered elements, nor for the total sum of the trace elements. The major elements are seen to reach relatively high fractions of TLV, but this is understandable because their concentrations are high and their TLVs, or the TLVs of their oxides, are equal to the TLV of PM_{50} dust. However, as long as the PM_{50} dust concentration is no higher than $10\,\mathrm{mg\,m^{-3}}$, these values will never exceed 1. The metals Be and V are the only trace elements showing fractions above 0.1 (10 % of the TLV) due to the very low TLVs allowed for those elements. Be is probably incorporated and immobilised in the silicate crystal structures of the refractory accessory coal minerals and as such does not occur in the speciation on which the TLV for Be is based. Taking the sum of the fractions for the trace elements according to $\Sigma_{i,n} C_i / TLV_i$, where C is the concentration in the inhalable fraction of the PFA and TLV is the allowable concentration in the inhalable fraction for $10\,\mathrm{mg\,m^{-3}}$, the values still remain < 1. This is really a worst-case approach, since it assumes that the different traces have similar effects on the same organs, which in practice will probably not be the case.

The conclusion of this research shows that PFA can be qualified as a nuisance dust and that workers are not exposed to extra health risks if the concentration of the inhalable fraction (PM_{50}) of PFA in the air is kept below the TLV for nuisance dust ($10\,mg\,m^{-3}$).

In the event of a worst-case accident – for example, the release of 1000 kg of PFA within an hour – the TLV for inhalable PFA dust may be exceeded for some hours within a distance of 700 m from the centre of the source. However, in practice, workers will not be exposed for more than an hour, as they will quickly take precautionary measures.

Coal grinding operations at the power plant also contribute to the dust levels in the air. However, coal dust generally is quite coarse and deposits quickly, but it does contain a considerable amount of quartz that is known to cause silicosis. This is why a special TLV, the so-called No Effect Level (NEL) is set for coal dust at $2\,mg\,m^{-3}$ mainly for implementation in coal mines, where quartz contents are generally higher due to destruction and removal of wall rocks.[109] Measurements and calculations indicate that such levels are not attained in a power plant.[103] The detailed discussion of problems associated with coal dust is beyond the scope of this book.

3.4.5 Public Health

People living near a coal-fired power station with an open pulverised fuel ash storage facility may be exposed to airborne pulverised fuel. Current European guidelines (Directive 96/62/EC)[112] describe the framework for air quality assessment and management, and advise maximum yearly average concentrations of fine dust (PM_{10}) in outside air of $40–60\,\mu g\,m^{-3}$, with a highest daily average ($P_{99.7}$) of $100–150\,\mu g\,m^{-3}$. Follow-up Directive 1999/30/EC[113] regulates the maximum average yearly concentration for PM_{10} at $40\,\mu g\,m^{-3}$ with a P_{90} (35 days) of $50\,\mu g\,m^{-3}$. These values were implemented in 2005. Plans are also included for the implementation in 2010 of a maximum yearly average concentration of $20\,\mu g\,m^{-3}$ and restricting the maximum daily average concentration of $50\,\mu g\,m^{-3}$ to seven days (P_{98}).

In the United States, separate norms exist for PM_{10} and $PM_{2.5}$. For PM_{10}, the maximum allowable yearly concentration (averaged over three years) is $50\,\mu g\,m^{-3}$ with a P_{99} of $150\,\mu g\,m^{-3}$ and for $PM_{2.5}$ a yearly average concentration limit of $15\,\mu g\,m^{-3}$ with a P_{98} of $65\,\mu g\,m^{-3}$.[114] Keeping in mind what was stated in the introduction (Section 3.4.1), it will be noticed that background values in air in urban areas are already near the above-mentioned allowable concentrations for clean air.

KEMA used the KEMA 3D MODEL® to determine the atmospheric concentrations of airborne PFA particles for hypothetical scenarios of temporary open storage of PFA and of an accident leading to a release of 1000 kg of PFA over a short time.[103] At the perimeter, 500 m around the centre of an open PFA storage site, where the public may be exposed, the annual average concentration of particulate PFA under normal operating conditions is $2.6\,\mu g\,m^{-3}$ with a maximum hourly peak during the year of $31\,\mu g\,m^{-3}$, mainly depending on meteorological conditions. Taking into account the normal background in urban areas, this result shows that PFA particles only contribute 6.5 % to the total PM_{10} in the air. If number concentrations are considered, this figure is even less than 1 % of total PM_{10} (estimated at 0.0001 %). Hence it is concluded that an open ash storage site does not lead to exceeding the recommended limit values for PM_{10}.

Following an accident as envisaged above, the concentration of PM_{10} could increase up to $3.5\,mg\,m^{-3}$ (hourly average at peak) 500 m away from the source, which could mean that the recommended maximum daily or weekly average limits may be exceeded up to 1 km from the source. However, given the composition of PFA, damage to health is not likely to occur, although Chronic Non-Specific Lung Disease (CNSLD) sufferers may experience irritation and shortage of breath.

Deposition of particulate matter that is not a particularly striking colour is generally perceived by the public to be a nuisance if more than $1\,g\,m^{-2}$ per month is deposited. At 500 m from an open PFA storage facility, the deposition is calculated to be between 0.4 and $1.9\,g\,m^{-2}$ per month depending on weather conditions, whereas farther away from the source the deposition quickly decreases below the $1\,g\,m^{-2}$ critical level. In case of an accident as described above, the nuisance area may expand up to a distance of 1 km from the source. However, the amounts deposited are generally not sufficient to constitute a health hazard; nor is there a danger of soil pollution.[103] Cumulative deposition of $10\,g\,m^{-2}$ per year over a period of 100 years, assuming that all PFA remains in the top 25 cm, results in a soil consisting of 0.2 % PFA. Assuming that trace element concentrations in the deposited PFA equal the concentrations found in the respirable fraction (PM_4), which is again a worst-case scenario, the increase in trace element concentrations in an average sandy soil, for most elements, will be less than 1 %. The highest figure found is 6 % for Ba, but for some elements even a decrease is found.

As noted above, small children form an exceptional group, as they may swallow anything that they come across. The worst-case scenario is taken whereby children playing in a sandbox located 500 m from an open ash storage site swallow at one instance the total ash deposited during a year on 1 dm^2 of sand surface (i.e. 1/100 of 10 g, which equals 0.1 g). Assuming again that the ash has the composition of the respirable fraction and taking into account the known leaching behaviour under acid conditions, the total intake of major and trace elements can be calculated. Comparing the results with the Acceptable Daily Intake (ADI) values, it is found that except for As (41 %) and Ba (6 %), all other elements show values of less than 5 % of the ADI. This leads to the conclusion that together with the normal intake through food and drinks, the intake through deposited PFA taking the worst-case scenario would not lead to health hazards.

The contribution of fly dust from the stack of a modern coal-fired power station to the normal background fine dust in the air has been calculated as no more than 0.03 % and as such is negligible. Similarly, deposition of this fly dust in an area of 40 × 40 km around the source is 1.7 mg m^{-2} per year and is far from being a nuisance; nor does it has any effect on trace element concentrations in the soil on which it is deposited.

It was shown in Chapter 1 that PFA originating from the co-firing of the secondary fuels currently used differs very little from PFA originating from the firing by coal alone as long as the secondary fuel proportions are no higher than 10 % by dry mass. Therefore, applying the same models and analyses described above, KEMA found that, in the case of co-firing (< 10 % of the tested secondary fuels), there is no reason to believe that the TLVs will be exceeded.[103] Special attention was given to chromium and phosphorous because of possible complications with their speciation. Details concerning Cr(VI) are discussed in Chapter 3.4.6. Taking, as a worst-case scenario, Cr(VI) as 10 % of total Cr, this would not lead any TLV to be exceeded. Phosphorous in PFA, even from co-firing with a P-rich secondary fuel, has so far always been detected as phosphate and not as the more dangerous P_2O_5, and as such there is no danger of exceeding TLVs.[115]

It must be noted that around a coal-fired power station coal dust from the grinding operation does contribute to the dust levels in the air as well. Since coal dust generally is quite coarse and deposits quickly on the site (see Chapter 3.4.4), annual average concentrations as low as

$0.5 \, \mu g \, m^{-3}$ (TSPM), with a maximum daily average of $9 \, \mu g \, m^{-3}$, have been measured at 500 m from the source.[103] This has hardly any effect on the recommended limit values for PM_{10}.

3.4.6   Special Issues Regarding Health and Safety

Quartz   Quartz is of interest because it is known to cause black lungs – or, more precisely, pneumoconiosis or silicosis. Apart from this, it has recently been shown that quartz is a human carcinogenic material at concentrations above a certain threshold, and causes the serious malignant condition Progressive Massive Fibrosis (PMF).[116,117] Evidently, only the respirable fraction of quartz will have such effects, but in addition to causing this condition the surface of the quartz needs to be fresh, as it is believed that surface radicals trigger the reaction.[118] Weathering and the presence of other substances can inhibit their formation.

In the classic handbook for occupational lung diseases,[119] it is stated that PFA dust does not promote fibrogenic action in the lungs and that after prolonged exposure of laboratory animals to PFA dust, lung diseases have never been found. These findings have been confirmed by later research.[107,120,121]

Despite these results, KEMA has recently again reviewed the health implications of quartz in PFA.[103] It was found that approximately 50 % of the quartz present in the coal is vitrified during the combustion process. Most of the remaining quartz is found in the nonrespirable fraction and only about 1 % of the quartz is contained in the respirable fraction. The concentration of quartz in the respirable fraction is probably no higher than 0.1 %. Computer Controlled Scanning Electron Microscope analyses (CCSEM) of 11 000 particles in four different ash samples indicated that between 60 and 86 % of the quartz grains were embedded in vitreous ash particles and were therefore not biologically available at the surface.[122] Figure 3.6 illustrates the nature of the quartz grains in fly ash particles. This could be one of the reasons for the absence of fibrogenic properties of quartz. Another reason could be that the temperature increase to above 1200 °C during combustion has an effect on the surface properties of the quartz.[123]

In the working area of a power plant, directly underneath the ESP, the respirable atmospheric quartz concentration in the air under normal operational conditions has been measured at $0.5 \, \mu g \, m^{-3}$, which is very low compared to the Dutch TLV of $75 \, \mu g \, m^{-3}$. Combined with the fact

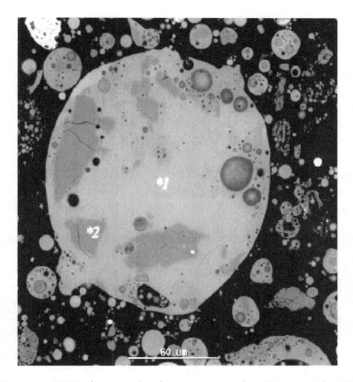

Figure 3.6 An SEM photograph of a cross-cut of a PFA particle. The basically vitreous particle (*1) contains embedded quartz grains (*2) (Reproduced by permission of KEMA[103]).

that no health effects have been reported following exposure to quartz from PFA, this leads to the conclusion that the presence of quartz in PFA does not lead to any health hazard.

Chromium (VI) The most important natural forms of chromium are Cr(III) and Cr(VI). Cr(III) is an essential element for human life (requirement 30–130 ng kg^{-1} body weight per day), but too high an intake can be harmful since Cr(III) is cytotoxic. However, Cr(VI) is not only cytotoxic but also a genotoxic carcinogen. Therefore, two TLVs exist for chromium. Since Cr(VI) is readily reduced to Cr(III) in the gastro-intestinal tract due to prevailing Eh – pH conditions, the dangerous exposure route for Cr(VI) is through inhalation. Indeed, lung cancer is the main cancer caused by Cr(VI). The main sources of Cr in the air are cigarette smoke, welding smoke (mainly indoor) and, outdoors, the chromium industry, wood preservation and oil refinery emissions. Since PFA particles do contain Cr as a trace element, the risk of exposure to power station workers and people living in the vicinity of power stations has been re-evaluated by KEMA.[103]

The average concentration of total Cr in PFA is about $130\,mg\,kg^{-1}$ (see also Chapter 1). It was found, by X-Ray Absorption Fine Structure Spectrometry (XAFS) analyses at the University of Kentucky, that between 4 and 9 % of the Cr present in (co-firing) PFA is available as Cr(VI).[124] Measurements further indicate that once in the atmosphere, the aerosol-bound Cr(VI) has the tendency to be reduced to Cr(III), although in one publication the opposite oxidation of Cr(III) to Cr(VI) was reported under extreme conditions in the atmosphere involving light and the presence of Fe(III).[125] Taking 10 % as a worst-case scenario, this would mean that the average concentration of Cr(VI) in PFA is $13\,mg\,kg^{-1}$. Taking the TLV for nuisance dust $(10\,mg\,m^{-3})$ as a maximum concentration of PFA in the air, this would mean that workers are exposed in the worst case to $0.13\,\mu g\,m^{-3}$, which is only 0.5 % of the TLV for Cr(VI). In practice, under normal operating conditions, this value would be much lower. For people living in the vicinity of the power plant, who experience exposure as a result of stack emissions and airborne dispersion, the levels of Cr(VI) are calculated to be below 1 % of the maximum acceptable level in clean air of $2.5\,ng\,m^{-3}$.

There is no danger of leaching of Cr(VI) into the environment from deposited PFA. Although Cr(VI) seems the only form that leaches readily from PFA, after leaching it reacts immediately to form insoluble Cr(III) compounds.

Radiation Radioactivity from coal combustion residues has been considered as a potential hazard to workers and public health. However, calculations made by KEMA,[103] assuming an average fly ash composition, have shown that if a worker spends his whole working time 25 m from an ash storage site, he would be exposed to 0.016 mSv of radiation per year. Taking 1 mSv as the occupational radiation limit, this represents only 1.6 %. The same worker experiencing the TLV for inhalable dust $(10\,mg\,m^{-3})$ would be subjected to 5.6 % of the limiting radiation value, although this would be mainly internal radiation through inhalation. For people living near a coal-fired power plant, the calculated values range from 0.004 % to 0.1 % of the recommended maximum exposure from nonnatural radiation sources, taken as 0.1 mSv per year. This means that radiation doses associated with exposure to PFA are well below the relevant limits.

Organic Compounds In principle, any combustion process that takes place in the presence of chlorine can lead to the formation of dioxins.

However, the dioxin levels in stack gases and PFA particles are extremely low because of nearly complete combustion, high combustion temperatures, low chlorine levels and the presence of SO_3, a suspected inhibitor. KEMA measured $1.5\,pg\,I\text{-}TEQ\,m^{-3}$ in flue gas and $< 1\,pg\,I\text{-}TEQ\,g^{-1}$ in PFA, < 0.1 % of the generally accepted limit values.[103] The most dangerous dioxin species, 2,3,7,8-TCDD, has not been reported in fuel ashes.

Polycyclic aromatic hydrocarbons (PAHs) are products of incomplete combustion. However, flue gases from coal-fired power plants only contain a few micrograms of PAHs per cubic meter, of which half is removed by the gas desulfurisation system. The average concentration of carcinogenic PAHs (those with four or more benzene rings) in the emitted flue gases is estimated at approximately $26\,ng\,m^{-3}$.[103] Once they reach ground level, the concentration has decreased by dilution down to $0.1\,pg\,m^{-3}$, which is far below the accepted exposure levels of $2300\,pg\,m^{-3}$ in rural areas and $8600\,pg\,m^{-3}$ in urban areas. PAHs were not detected in PFA. However, even if the concentration were assumed to equal the lower limit of detection, the contribution from dispersed PFA in air would be no more than $0.4\,pg\,m^{-3}$.

3.4.7 Products from Combustion Residues

It should be evident from the above discussions that during manufacture of (novel) products from combustion residues, the concentrations of these residues are in general lower than at the primary site where they were produced. This means that if the same precautions and measures are taken as at the power plants, workers and people living near such sites are in no significant danger.

Chapter 5 considers more in detail the occupational and public health problems related to products from combustion residues and the requirements for measures taken for such operations.

References

1   BSI, *BS3892. Pulverised fuel ash for use in concrete.* British Standards Institute: London (1965).
2   BSI, *BS3892. Part 1: Pulverised fuel ash for use as a cementitious component in structured concrete.* British Standards Institute: London (1982).
3   BSI, *BS3892. Part 2: Specification for pulverised fuel ash for use as a type 1 addition.* British Standards Institute: London (1996).
4   BSI, *BS3892. Part 3: Specification for pulverised fuel ash for use in cementitious grouts.* British Standards Institute: London (1997).

5 BSI, *BS EN 450. Fly ash for concrete – definitions, requirements, and quality control*. British Standards Institute: London (1995, 2005).

6 BSI, *BS EN 206. Concrete – Part 1: Specification, performance, production and conformity*. British Standards Institute: London (2001).

7 Sear, L.K.A., *Properties and use of coal fly ash: A valuable industrial by-product*. Thomas Telford: London (2001).

8 CUR, *Vliegas in cement, toeslag en beton. CUR Report 96-6*. Stichting CUR, Gouda, the Netherlands (1996).

9 CUR, *Poederkoolvliegas verkregen door bijstoken voor mortel en beton. CUR Report C99-1*. Stichting CUR, Gouda, the Netherlands (1999).

10 CUR, *Toepassing van poederkoolvliegas in mortel en beton. CUR Aanbeveling 94*. Stichting CUR, Gouda, the Netherlands (2003).

11 ASTM, *C618-03. Standard specifications for coal fly ash and raw or calcined natural pozzolan for use as a mineral admixture in concrete*, American Society for the Testing of Materials (2003).

12 Chatterjee, K. and Stock, L.M., *Towards organic desulfurization. The treatment of bituminous coals with strong bases*, Energy Fuels, 5: 704 (1991).

13 Wills, B.A., *Mineral Processing Technology*, 5th ed. Pergamon Press (1992).

14 Keppeler, J.G., *Carbon burn-out, an update on commercial applications*. In Proc. Int. Ash Utilization Symposium (on CD), 22–24 October, Kentucky (2001).

15 Tranquilla, J.M. and Maclean, J.H., *Microwave carbon burnout (MCB) gas by-products and deportment of specific metallic elements*. In Proc. Int. Ash Utilization Symposium (on CD), 22–24 October, Kentucky (2001).

16 Tondu, E. (STI), *Commercial separation of unburned carbon from fly ash*. In SME Annual Meeting, 11–14 March, Phoenix, AZ (1996).

17 Bittner, J.D. and Gasiorowski, S.A., *STI's six years of commercial experience in electrostatic beneficiation of fly ash*. In Proc. Int. Ash Utilization Symposium (on CD), 22–24 October, Kentucky (2001).

18 Jiang, K.X. and Stencel, J.M., *The influence of ash particles interactions during pneumatic transport, triboelectric beneficiation*. In Proc. Int. Ash. Utilization Symposium (on CD), 22–24 October, Kentucky (2001).

19 Lockert, C.A., Lister, R. and Stencel, J.M., *Commercialization status of a pneumatic transport, triboelectrostatic system for carbon/ash separation*. In Proc. Int. Ash Utilization Symposium (on CD), 22–24 October, Kentucky (2001).

20 Kawatra, S.K. and Eisele, T.C., *Removal of unburned carbon from fly ash by froth flotation*. In SME Annual Meeting, 11–14 March, Phoenix, AZ (1996).

21 Groppo, J.G. and Brooks, S.M., *Methods of removing carbon from fly ash*. US Patent 5,456,363 (1995).

22 Groppo, J.G., Robl, T.L. and McCormick, C.J., *Method for improving the pozzolanic character of fly ash*. US Patent 5,817,230 (1998).

23 Groppo, J. and Robl, Th., *Ashes to Energy – The Coleman power plant project*. In Proc. Int. Ash Utilization Symposium (on CD), 22–24 October, Kentucky (2001).

24 Hwang, J.Y., Liu, X. and Zimmer, F.V., *Beneficiation process for fly ash and the utilization of cleaned fly ash for concrete application*. In Proc. 11th Int. Symp. on Use and Management of Coal Combustion By-products. January, Orlando, Fla (1995).

25 Hwang, J.Y., *Wet process for fly ash beneficiation*. US Patent 5,047,145 (1990).

26 Hamley, P., Lester, E., Thompson, A., Cloke, M. and Poliakoff, M., *The removal of carbon from fly ash using supercritical water oxidation*. In Proc. Int. Ash Utilization Symposium (on CD), 22–24 October, Kentucky (2001).

27 Höller, R., Wirsching, F. and Hamm, H., *Removal of ammonia from fly ash*. European Patent #0135148 (1984)

28 Bittner, J., Gasiorowski, S. and Hrach, F., *Removing ammonia from fly ash*. In Proc. Int. Ash Utilization Symposium (on CD), 22–24 October, Kentucky (2001).

29 Gasiorowski, S.A. and Hrach, F.J., *Method for removing ammonia from ammonia contaminated fly ash*. US Patent 6,077,494 (2000).

30 Majors, R.K., Hill, R., McMurry, R. and Thomas, S., *A study of the impact of ammonia injection on marketable fly ash including quality control procedures*. In Proc. 1999 Conf. on Selective Catalytic and Non-catalytic Reduction for NO_x Control. US DOE-FETC. pp. 11–13 (1999).

31 Minkara, R.Y., *Chemical treatment of ammoniated ash*. In Proc. Int. Ash Utilization Symposium (on CD), 22–24 October, Kentucky (2001).

32 Blondin, J. *et al.*, *A new approach to hydration of FBC residues*. In Proc. 12th Int. Conf. on Fluidised Bed Combustion. San Diego (1993).

33 Blondin, J. and Anthony, E.J., *Use of FBC ashes in no-cement mortars*. In Proc. 14th Int. Conf. on Fluidised Bed Combustion, Vancouver, Canada (1997).

34 Sikkinga, J., Segal, H.R., Driel, C.P. van and Kerkdijk, C.B.W., *Fly ash and AFBC ash upgrading by HGMS. Report phase I*. FDO-Holec (1983).

35 Kruger, R.A., *Recovery and characterization of cenospheres from South African power plants*. In Proc. Ninth Int. Ash Use Symposium, 3: 1–20 (1991).

36 Ghafoori, N., Honaker, R. and Sevim, H., *Processing, transporting, and utilizing coal combustion by-products*. In Proc. Int. Ash Utilization Symposium, 18–20 October, Kentucky, pp. 101–114 (1999).

37 Gurupira, T., Jones, C.L., Howard, A., Lockert, C., Wandell, T. and Stencel, J.M., *New products from coal combustion ash: selective extraction of particles with density < 2*. In Proc. Int. Ash Utilization Symposium (on CD), 22–24 October, Kentucky (2001).

38 Eymael, M.M.Th. and Cornelissen, H.A.W., *Processed pulverized fuel ash for high strength concrete*. Waste Management, 16: 237–242 (1996).

39 Galloway, R.L., *A History of Coal Mining in Great Britain*. Macmillan: London (1882).

40 Wallis, S.M., Scott, P.E. and Waring, S., *Review of leaching test protocols with a view to developing an accelerated anaerobic leaching test. AEA-EE-0392*. AEA Environment & Energy, Harwell, Didcot, Oxon, UK (1992).

41 Sloot, H.A. van der, Heasman, L. and Quevauviller, Ph., *Harmonisation of Leaching/Extraction Tests*. Studies in Environmental Science 70. Elsevier: Amsterdam, the Netherlands (1997).

42 CEN, *EN 12457 Characterisation of waste; Leaching; Compliance test for leaching of granular waste materials and sludges; Part 1: One stage batch test at a liquid to solid ratio of 2 l/kg for materials with high solid content and with particle size below 4 mm (without or with size reduction)*. Comité Européen de Normalisation (2002).

43 CEN, *EN 12457 Characterisation of waste; Leaching; Compliance test for leaching of granular waste materials and sludges; Part 2: One stage batch test at a liquid to solid ratio of 10 l/kg for materials with particle size below 4 mm (without or with size reduction)*. Comité Européen de Normalisation (2002).

194   LIMITATIONS OF COMBUSTION ASHES

44   CEN, *EN 12457 Characterisation of waste; Leaching; Compliance test for leaching of granular waste materials and sludges; Part 3: Two stage batch test at a liquid to solid ratio of 2 l/kg and 8 l/kg for materials with high solid content and with particle size below 4 mm (without or with size reduction).* Comité Européen de Normalisation (2002).

45   CEN, *EN 12457 Characterisation of waste; Leaching; Compliance test for leaching of granular waste materials and sludges; Part 4: One stage batch test at a liquid to solid ratio of 10 l/kg for materials with particle size below 10 mm (without or with size reduction).* Comité Européen de Normalisation (2002).

46   BMD, *Dutch Building Materials Decree (Bouwstoffenbesluit).* Staatscourant (1999).

47   NEN7343, *Leaching characteristics of solid earthy and stony building and waste materials. Leaching tests. Determination of the leaching of inorganic constituents from granular materials with the column test.* Netherlands Normalization Institute, Delft (1995).

48   NEN7345, *Leaching characteristics of building and solid waste materials – Leaching tests: Determination of the leaching behaviour of inorganic components from shaped building materials, monolithic and stabilized waste materials.* Netherlands Normalization Institute, Delft (1995).

49   Aalbers, Th.G. *et al.*, *Milieuhygiënische kwaliteit van primaire en secundaire bouwmaterialen in relatie tot hergebruik en bodem – en oppervlaktewaterbescherming.* RIVM report no. 771402006. RIVM/RIZA (1993).

50   KEMA, *Databank spoorelementen. Deel 8: Uitloging kolenreststoffen.* Report no. 64234-KES/WBR 95-3123 (1995).

51   Anon, *US EPA, Part 261, Appendix II – Method 1311 Toxicity Characteristic Leaching Procedure (TCLP),* Federal Register, 55: 11 863–11 877 (1990).

52   CEN, *Leaching and soil/groundwater transport of contaminants from coal combustion residues.* Report EVR 14054EN. Comité Européen de Normalisation (1992).

53   CEN, *Characterisation of waste. Leaching. Compliance test for leaching of granular waste materials. Determination of the leaching of constituents from granular waste materials and sludges.* Draft European Standard prEN 12457, CEN/TC 292. Comité Européen de Normalisation (1996).

54   CEN, *Characterisation of waste – Methodology guideline for the determination of the leaching behaviour of waste under specified conditions.* Draft PrENV, CEN/TC 292. Comité Européen de Normalisation (1996).

55   CEN, *Characterisation of waste – Leaching behaviour tests – Up-flow percolation test (under specified conditions).* CEN Technical Specification 14405, CEN/TC 292. Comité Européen de Normalisation (2004).

56   CEN, *Characterisation of waste – Leaching behaviour tests – Influence of pH on leaching with continuous pH-control.* Draft European Standard prEN 14997, CEN/TC 292. Comité Européen de Normalisation (2004).

57   CEN, *Characterisation of waste – Leaching behaviour tests – Influence of pH on leaching with initial acid/base addition.* CEN Technical Specification 14429, CEN/TC 292. Comité Européen de Normalisation (2005).

58   DIN 38414 Teil 4. *Deutsche Einheitsverfahren zur Wasser, Abwasser- und Schlammuntersuchung; Schlamm und Sedimente (Gruppe S).* Beuth-Vertrieb, Deutsche Industriele Norm, Berlin (1984).

59 NEN7341, *Leaching characteristics of solid earthy and stony building and waste materials. Leaching tests. Determination of the availability of inorganic components for leaching.* Netherlands Normalization Institute (NNI), Delft (1995).

60 NEN7349, *Leaching characteristics of building and solid waste material, leaching tests, determination of the leaching of inorganic components from granular materials with the cascade test.* Netherlands Normalization Institute, Delft (1995).

61 Nordtest, *Solid waste, granular inorganic material: Column test. Nordtest method NT ENVIR 002.* Espoo, Finland (1995).

62 AFNOR X-31-210. *Déchets: Essai de lixiviation X31-210. AFNOR T95J.* Association Française de Normalisation, Paris, France (1988).

63 Meij, R., *Private communication* (2004)

64 Sakai, S., Mizutani, H., Takatsuki, H. and Kishida, T., *Leaching test of metallic compounds in fly ash of solid waste incinerator.* J. Japan Society of Waste Management Experts, **6**: 225–234 (1995).

65 Fällman, A.-M. and Hartlén, J., *Leaching of slags and ashes – controlling factors in field experiments versus in laboratory tests.* In Environmental Aspects of Construction with Waste Materials, J.J.J.M., Goumans, H.A. van der Sloot and T.G. Aalbers (eds), Elsevier: Amsterdam, the Netherlands, pp. 39–54 (1994).

66 Garavaglia, R. and Caramuscio, P., *Coal fly ash leaching behaviour and solubility controlling solid.* In Environmental Aspects of Construction with Waste Materials, J.J.J.M., Goumans, H.A. van der Sloot and T.G. Aalbers (eds.), Elsevier: Amsterdam, the Netherlands, pp. 87–102 (1994).

67 Janssen-Jurkovičová, M., Hollman, G.G., Nass, M.M. and Schuiling, R.D., *Quality assessment of granular combustion residues by a standard column test: prediction versus reality.* In Environmental Aspects of Construction with Waste Materials, J.J.J.M., Goumans, H.A. van der Sloot and T.G. Aalbers (eds), Elsevier: Amsterdam, the Netherlands, pp. 161–178 (1994).

68 Wadge, A. and Hutton, M., *The leachability and chemical speciation of selected trace elements in fly ash from coal combustion and refuse incineration.* Environ. Pollut., **48**: 85–99 (1987).

69 Harris, W.R. and Silberman, D., *Time-dependent leaching of coal fly ash by chelating agents.* Environ. Sci. Technol., **17**: 139–145 (1983).

70 Hoek, E. van den and Comans, R.N.J., *Speciation of As and Se during leaching of fly ash.* In Environmental Aspects of Construction with Waste Materials, J.J.J.M., Goumans, H.A. van der Sloot and T.G. Aalbers (eds), Elsevier: Amsterdam, the Netherlands, pp. 467–476 (1994).

71 Fernandez-Turiel, J.L., de Carvalho, W., Cabanas, M., Querol, X. and Lopez-Soler, A., *Mobility of heavy metals from coal fly ash.* Environ. Geol., **23**: 264–270 (1994).

72 Querol, X., Juan, R., Lopez-Soler, A., Fernandez-Turiel, J.L. and Ruiz, C.R., *Mobility of trace elements from coal and combustion wastes.* Fuel, **75**: 821–838 (1996).

73 Querol, X., Fernandez-Turiel, J.L. and Lopez-Soler, A., *Trace elements in coal and their behaviour during combustion in a large power station.* Fuel, **74**: 332–343 (1995).

74 Dudas, M.J. and Warren, C.J., *Submicroscopic model of fly ash particles.* Geoderma, **40**: 101–114 (1987).

75 Meij, R. and Krijt, G.D., *Databank spoorelementen deel 3 poederkoolvliegas.* KEMA report no. 63597-KES/WBR 93-3113 (1993).

76 Batley, G.E., *Trace Element Speciation: Analytical Methods and Problems*. CRC Press: Boca Raton, Fla (1989).

77 Rulkens, W.H. and Assink, J.W., *Extraction as a method for cleaning contaminated soil: possibilities, problems and research*. In Natl. Conf. Manage. Uncontrolled Hazard Waste Sites, 31 October – 2 November, Washington, DC, pp. 576–583 (1983).

78 Muller, G. and Rietmayer, S., *Chemische Entgiftung: das alternative Konzept zur problemlosen und endgultigen Entsorgung Schwermetall-belasteter Baggerschlamme*. Chem-Ztg., **106**: 289–292 (1982).

79 Katsuura, H., Inoue, T., Hiraoka, M. and Sakai, S., *Full-scale plant study on fly ash treatment by the acid extraction process*. Waste Manage., **16**: 491–499 (1996).

80 Querol, X., Umaña, J.C., Alastuey, A., Bertrana, C., Lopez-Soler, A. and Plana, F., *Extraction of water soluble impurities from fly ash*. Energy Sources, **22**: 733 (2000).

81 Querol, X., Umaña, J.C., Alastuey, A., Ayora, C., Lopez-Soler, A. and Plana, F., *Extraction of soluble major and trace elements from fly ash in open and closed leaching systems*. Fuel **80**: 801–813 (2001).

82 Nugteren, H.W., Janssen-Jurkovičová, M. and Scarlett, B., *Improvement of environmental quality of coal fly ash by applying forced leaching*. Fuel **80**: 873–877 (2001).

83 Nugteren, H.W., Janssen-Jurkovičová, M. and Scarlett, B., *Removal of heavy metals from fly ash and the impact on its quality*. J. Chem. Tech. Biotechnol., **77**: 389–395 (2002).

84 Nugteren, H.W., Suñé Llop, E., Janssen-Jurkovičová, M. and Scarlett, B., *The removal of oxyanions of Se, Mo and B from fly ash leachates*. In Proc. REWAS'99 Global Symposium on Recycling, Waste Treatment and Clean Technology, San Sebastian, Spain, 5–10 September, **3**: 2367–2375 (1999).

85 *MINTEQA2. A geochemical assessment model for environmental systems*. EPA/600/3-917021. US Environmental Protection Agency, Athens, GA (1991).

86 Glennon, J.D., O'Connell, M., Leahy, S., Mehay, H. and Smith, C.M.M., *Efficient removal of chromium from leather waste and flyash using supercritical fluid extraction technology*. In Proc. REWAS'99: Global Symposium on Recycling, Waste Treatment and Clean Technology, San Sebastian, Spain, 5–10 September, **3**: 2329–2336 (1999).

87 Kersch, Chr., Roosmalen, M.J.E. van, Woerlee, G.F. and Witkamp, G.J., *Extraction of heavy metals from fly ash and sand with ligands and supercritical carbon dioxide*. In Proc. of the 5th Int. Symp. on Supercritical Fluids, Atlanta, GA (2000).

88 Glennon, J.D., Hutchinson, S., McSweeney, C.C., Harris, S.J. and McKervey, M.A., *Molecular baskets in supercritical CO_2*. Anal. Chem., **69**: 2207–2212 (1997).

89 Glennon, J.D., Hutchinson, S., Walker, A., Harris, S.J. and McSweeney, C.C., *New fluorinated hydroxamic acid reagents for the extraction of metal ions using supercritical CO_2*. J. Chromatogr. A, **770**: 85–91 (1997).

90 Kersch, C., Woerlee, G.F. and Witkamp, G.J., *An experimental study on the supercritical fluid extraction of heavy metals from sand*. In Proc. 6th Meeting Supercritical Fluids Chemistry and Materials, Nottingham, UK, pp. 591–596 (1999).

91 Kersch, C., Woerlee, G.F. and Witkamp, G.J., *Extraction of heavy metals from solid matrices with ligands and supercritical carbon dioxide*. In Proc. Metal Separation Technologies beyond 2000, Hawaii, pp. 161–165 (1999).

92   Kersch, Chr., Roosmalen, M.J.E. van, Woerlee, G.F. and Witkamp, G.J., *Extraction of heavy metals from fly ash and sand with ligands and supercritical carbon dioxide.* Ind. Eng. Chem. Res., 39: 4670 (2000).

93   Kersch, Chr., Van der Kraan, M., Woerlee, G.F. and Witkamp, G.J., *Municipal waste incinerator fly ash: supercritical fluid extraction of metals.* J. Chem. Tech. Biotechnol., 77: 256–259 (2002).

94   Cox, M., Duke, P.W. and Gray, M.J., *Winning metal from an ore.* Patent GB 213 5984 A (1984).

95   Cox, M., Duke, P.W. and Gray, M.J., *Extraction of metals by the direct thermal attack of organic reagents.* In Extraction Metallurgy'85, 9–12 September, Inst. Mining and Metallurgy, London, pp. 33–41 (1985).

96   Pichugin, A.A. and Cox, M., *Direct recovery of nickel from Greek laterite ore using SERVO process.* In Proc. Nickel/Cobalt'97 Vol. II, Pyrometallurgical Fundamentals and Process Development, L.A. Levrac and R.A. Bergmann (eds), Metallurgical Society of CIM, pp. 125–136 (1997).

97   Cupertino, D.C., Keyte, R.W., Slawin, A.M.Z., Williams, D.J. and Woolings, J.D., *Preparation and single crystal characterization of $^iPr_2(S)NP(S)^iPr_2$ and homoleptic $[^iPr_2(S)NP(S)^iPr_2]^-$ complexes of zinc, cadmium and nickel.* Inorg. Chem., 35: 2695 (1996).

98   Allimann-Lecourt, C., *Extraction of metals from contaminated materials, sediment samples and industrial solid waste using a novel technology (SERVO process).* Ph.D. thesis, University of Hertfordshire (2004).

99   Vitolo, S., Seggiani, M., Filippi, S. and Brocchini, C., *Recovery of vanadium from heavy oil and Orimulsion fly ashes.* Hydrometallurgy, 57: 141–149 (2000).

100 Sahuquillo, A., Lopez-Sanchez, J.F., Rubio, R., Rauret, G., Thomas, R.P., Davidson, C.M. and Ure, A.M., *Use of a certified reference material for extractable trace metals to assess sources of uncertainty in the BCR three-stage sequential extraction procedure.* Anal. Chim. Acta, 382: 317–327 (1999).

101 Hinds, W., *Properties, Behavior and Measurement of Airborne Particles*, 2nd edn. John Wiley and Sons, Inc: (1999).

102 Kunzli, N., Kaiser, R., Medina, S., Studnicka, M., Chanel, O., Filliger, P., Herry, M., Horak, F. Jr, Puybonnieux-Texier, V., Quenel, P., Schneider, J., Seethaler, R., Vergnaud, J.-C: and Sommer, H., *Public-health impact of outdoor traffic-related air pollution: a European assessment.* The Lancet, 256: 795–801 (2000).

103 Meij, R., *Status report on the health issues associated with pulverised fuel ash and fly dust.* V2.1. KEMA report 50131022-KPS/MEC 01-6032 (2003) (summary report in English. Detailed reports in 10 volumes in Dutch).

104 Rothenberg, S.J., Seiler, F.A., Hobbs, C.H., Casuccio, G.S. and Spangler, C.E., *Isolation and characterisation of fly ash from rat lung tissue.* J. Tox. and Env. Health., 27: 487–508 (1989).

105 Srivastava, V.K., Chauhan, S.S., Srivastava, P.K., Shukla, R.R., Kumar, V. and Misra, U.K., *Placental transfer of metals of coal fly ash into various fetal organs of rat.* Arch. Toxic, 64: 153–156 (1990).

106 Hoeksema, H.W., *Working conditions for fly ash workers and radiobiological consequences of living in a fly ash house.* In Proc. Int. Conf. on Ash Technology and Marketing (Ash Tech '84) (1984).

107 Meij, R., Te Winkel, B.H. and Overbeek, J.H.M., *Gehaltes aan macro- en spoorelementen in inhaleerbaar en respirabel poederkoolvliegas.* KEMA report no. 65031.200-KST/MAT 97-65034 (in Dutch) (1997).

108 Vink, G.J. and Arts, J.H.E., *Potential health effects of coal fly ash and risks of occupational exposure.* MT TNO report 87/117b (1987).

109 Borm, P.J.A., *Toxicity and occupational health hazards of coal fly ash (CFA). A review of data and comparison to coal mine dust.* Annals of Occupational Hygiene, **41**: 659–676 (1997).

110 Borm, P.J.A., Knapen, A.M., Schins, R.P.F., Herwijnen, M. Van, Schilderman, P.A.E.L., Maanen, J. Van, Smith, K. and Aust, A., *In vitro effects of coal fly-ashes: Hydroxyl radical generation, iron release, and DNA damage and toxicity in rat lung epithelial cells.* Inhalation Toxicology, **11**: 1123–1141 (1999).

111 ISO, *Air quality – Particle size fraction definitions for health related sampling.* ISO standard 7708 (1995).

112 Council Directive 96/62/EC of 27 September 1996 on ambient air quality assessment and management. EU (1996).

113 Council Directive 1999/30/EC of 22 April 1999 relating to limit values for sulphur dioxide, nitrogen dioxide and oxides of nitrogen, particulate matter and lead in ambient air. EU (1999).

114 Smith, I.M. and Sloss, L.L., $PM_{10}/PM_{2.5}$ – *emissions and effects.* IAE Coal Research Report no. CC/08. ISBN 92-9029-312-8 (1998).

115 Meij, R. and Weijers, H.M., *Het vóórkomen van fosfor in kolenreststoffen.* Rapportnummer 99533011-KST/MAT 99-6549 (in Dutch) (1999).

116 IARC, *Monograph on the Evaluation of the Carcinogenic Risks of Chemicals to Humans, Volume 68: Silica, Some Silicates, Coal Dust and Para-aramid Fibrils.* IARC Working group on the evaluation of carcinogenic risks to humans, Lyon, France (1996), WHO Publication, IARC Press (1997).

117 Castranova, V., *From coal mine dust to quartz: mechanisms of pulmonary pathogenicity.* Inhalation Toxicology, **12** (Supplement 3): 7–14 (2000).

118 Donaldson, K. and Borm, P.J.A., *The quartz hazard: a variable entity.* Annals of Occupational Hygiene, **42**: 287–294 (1998).

119 Parkes, W.R., *Occupational Lung Disorders*, 2nd edn (reprinted 1993). Butterworths London (1993).

120 Pigott, C.H., *Results from macrophage test and haemolysis tests.* ICI Central Toxicology Lab., Macclesfield, UK (1983).

121 Maanen, J. van, Borm, P.J.A., Knapen, A.M., Herwijnen, M. van, Schilderman, P.A.E.L., Smith, K., Aust, A., Tomatis, M. and Fubini, B., *In vitro effects of coal fly-ashes: Hydroxyl radical generation, iron release, and DNA damage and toxicity in rat lung epithelial cells.* Inhalation Toxicology, **11**: 1123–1141 (1999).

122 Meij, R., Nagengast, S.J. and Te Winkel, B.H., *The occurrence of quartz in coal fly ash particles.* Inhalation Toxicology, **12** (Supplement 3): 109–116 (2000).

123 Fubini, B., Zanetti, G., Altilia, S., Tiozzo, R., Lison. D. and Saffiotti, U., *Relationship between surface properties and cellular responses to crystalline silica: studies with heat-treated cristobalite.* Chem. Res. Toxicology, **12**: 737–745 (1999).

124 Meij, R., *Het vóórkomen van chroom in kolenreststoffen.* Report no. 99533011-KST/MAT 99-6544 (in Dutch) (1999).

125 Zhang, H., *Light and iron(III)-induced oxidation of chromium(III) in the presence of organic acids and manganese(II) in simulated atmospheric water.* Atmospheric Environment, **34**: 1633–1640 (2000).

4

Novel Products and Applications with Combustion Residues

4.1   Introduction

Henk Nugteren
Delft University of Technology, The Netherlands

4.1.1   The Use of Secondary Resources

Finding outlets for the ever-growing amounts of solid wastes and industrial residues, together with saving of primary raw materials, are some of today's key issues in the protection of the environment and moving towards sustainable development. Proposed solutions generate a wide range of concerns and initiatives world-wide.

One of the possibilities of saving natural resources is to replace primary raw materials by secondary raw materials. The accrued benefits will be increased if those secondary raw materials are made of or can be recovered from waste or residue streams, thus also reducing the amount of solid residues produced.

The primary raw materials that are most suitable to be replaced by secondary raw materials are those for which generally only a few simple specifications between rather broad margins must be fulfilled. Such raw materials are normally relatively low-cost materials. The majority

Combustion Residues: Current, Novel and Renewable Applications   Edited by Michael Cox,
Henk Nugteren and Mária Janssen-Jurkovičová   © 2008 John Wiley & Sons, Ltd

of these are classified as *industrial minerals*, and hence if replaced by secondary raw materials, these would be termed *secondary industrial minerals*.

Considering the uses of industrial minerals that may be replaced by suitable secondary raw materials, it is noted that most fall into the category of construction materials, ceramic and glass raw materials and fillers, filters, adsorbents and immobilisers. The first categories are favourable because of the large bulks involved; the last categories are particularly interesting because of the highly specialised nature of the products, together with their additional environmental benefits.

The definition of an industrial mineral[1] is:

> any rock, mineral, or other naturally occurring substance of economic value,
> exclusive of metallic ores, mineral fuels, and gemstones; one of the non-metallics

Hence, industrial minerals are natural minerals that have a *use* in industry, and thus for secondary industrial minerals the same definition applies, with the difference that their origin is from a secondary source. This secondary source may be an industrial process, from which the secondary industrial mineral can just be a by-product or a residue, used directly – for example, fly ash from power generation in a coal-fired facility, to be used in the cement/concrete industry – or to be used after industrial upgrading through an appropriate technology – for example, zeolites made from coal fly ash, to be used as absorbents.

Industrial minerals are often listed and described either by uses or by commodities. Since in this book the commodity can be briefly summarised as being 'combustion residues', the approach of classifying uses seems the most appropriate. In Table 4.1, a general list is given of uses of industrial minerals as applied in textbooks for classification and description purposes. Application of combustion residues in the cement and concrete industry is well-established, and lightweight aggregates have also been manufactured on an industrial scale. A few areas, such as 'ceramic raw materials', 'insulating materials', 'roofing' and 'fillers, filters and absorbants' are under development and examples of research in these fields are given in the following sections. The main task for the future will be to further develop new potential areas for possible applications, as given in italics in the table, and even to extend the number of potential areas by exploring creative ideas. As the list does not pretend to be complete, novel applications may also be found beyond this list.

A striking point is that 'slags' figure as such on the list. This means that although blast furnace slags do not fit the definition of (primary)

Table 4.1 Industrial minerals classified by use.[1] Opportunities for combustion residues are given in bold for established outlets, underlined for those under development and in italics for potential uses.

Abrasives
Ceramic raw materials
Chemical industry raw materials
Construction materials
   Aggregates
      Crushed stone
      Lightweight aggregates
      Sand and gravel
      Slags
   Cement raw materials
   Dimension and cut stone
   Gypsum and anhydrite
   Insulating materials
   Roofing
Electronic and optical uses
Fertiliser minerals
Fillers, filters and absorbents
Fluxes
Foundry sand
(Gem materials)
Glass raw materials
Mineral pigments
Refractories
Well drilling fluids

---

industrial minerals, they have already been used so extensively and over such a long period that their use is commonplace and thus slags are included in the list of industrial minerals. Strictly speaking then, 'slags' should be listed as a commodity rather than a use. This illustrates how important and specific the use of slags has become through the years.

Coal fly ash may already deserve a similar position as far as usage in the cement and concrete industry is concerned. But there are obviously more possible uses for combustion residues in general. There are of course physical, chemical, environmental, economical and societal constraints, but the future for new uses and markets is still bright.

The key factor in acceptance of an industrial mineral is its technical suitability and the ability to supply a constant-quality product, as guaranteed by specifications and certification. This may apply even more strongly to secondary industrial minerals, because of customer perception problems. Waste-derived materials are often considered inferior by definition and therefore all mentioned constraints require equal attention.

Further, as for (primary) industrial minerals, the market value of secondary industrial minerals, once technically accepted, is highly dependent on the source location and the industry location. Slag is an important secondary industrial mineral in areas where steel plants are located, but some distance away from the source of supply primary industrial minerals may be used instead.

4.1.2 Standards and Specifications

The long history of use of ash in cement and concrete has resulted in clear specifications for the ash, mainly because all the relevant aspects have been studied and the relationship between the ash properties and product quality are known.

Nevertheless, even in this established field new developments are taking place, as interdependent relationships between the properties of ingredients, process variables and product properties are still being explored, as shown below in Section 4.2.

However, the maturity of development is very different in other areas of ash utilisation, especially in the field of novel applications, where relationships have not yet been defined and hence standards and specifications do not exist.

As an example, the case can be made for fly ash as a filler in polymers (Section 4.8). Fillers in the polymer industry are sold in various grades, each according to prescribed quality criteria. Even for simple fillers such as talc, well-documented grades are on the market. Such quality criteria will necessarily have to apply to fly ash fillers for polymers. It is illusory to think that a fly ash without any specifications could be accepted as a polymer filler. Batches of certified fly ash meant for the cement or concrete industry are not appropriate for filler applications, as they are unlikely to meet the properties relevant to fillers. So, new specifications will have to be agreed, probably even for each particular polymer product. Such specifications, agreed with the users, will be necessary for all the novel products presented in this chapter.

Returning to the example of the fillers in polymers, it can be seen that in the majority of the studies referred to in Section 4.8, the focus has been on the properties of fly ash filled polymers as compared to the unfilled polymer or to the polymer utilising commercial available fillers. This is of course important, but another aspect, the properties of the combustion residue itself, has received very little attention. Most authors provide limited general information about the residues used, such as 'C class fly ash, ca 0.1 μm with 17 % hollow particles',[2] but others do not give any data at all.[3] The work at MTU[4] and by Kruger[5–7] are exceptions, not only because they better document the properties of the fly ash, but also because they study the influence of ash properties on the properties of the final product by using different fractions of beneficiated ash.

The main goal of the production industry is to make a high-quality product and, to add further complication, many different products, according to certain product specifications PS_1 to PS_j, as shown in the matrix on the right in Figure 4.1. Such a product can be obtained by formulating a number of ingredients $I_1 \ldots I_n$, as shown in the first matrix. The ingredients must be processed under certain process conditions $C_1 \ldots C_k$ to yield the required product. So, the correct formulation (mix of ingredients) processed under the correct conditions will result in the required quality for the product. However, this will only work out well if the ingredients are all of good quality and therefore each ingredient I_i has its own set of specifications $I_iS_1 \ldots I_iS_m$. For the polymer industry, normally more than one of the ingredients is a filler. For such a

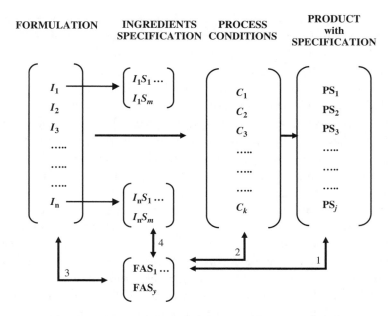

Figure 4.1 The dependency of the specifications of ingredients (filler) on product specifications (quality), process conditions, formulation and specifications of other ingredients. Reproduced by permission of Wiley VCH. Nugteren, H.W., Coal fly ash: from waste to industrial product. *Part. Part. Syst. Charact.*, 24: 49–55 (2007).

filler, the effects of all significant properties on the final product quality have been quantified, so the allowable ranges for the filler properties have been determined and this defines the filler specifications.

Assume that a filler currently in use – for example, kaolinite – has to be replaced by a combustion residue filler; for example, fly ash. It should be clear that the specifications of the kaolinite cannot be simply translated to the fly ash. Moreover, the fact that properties can behave interdependently and that some fillers possess properties that others lack completely – for example, highly spherical fly ash particles versus the plate-like kaolinite particles – makes it difficult to transfer the specifications of one filler to another. So for the fly ash a new set of specifications $FAS_1 \ldots FAS_y$ will have to be developed, focusing on the fly ash properties that may have an influence on the quality of the product, as shown by arrow 1 in Figure 4.1. A full set of ashes that is statistically sound to cover all combinations of possible influence factors should be tested for the effects on the final product. The results will have to be compared with the product quality requirements, and hence allowable ranges for measurable fly ash specifications should emerge.

The problem with the studies referred to so far is that if the impact of variations in filler properties on product properties or quality is not studied systematically, this will not lead to specifications for fly ash as a filler in polymer. Kruger has initiated this process by looking at the effect of using two different fractions of the same ash as a filler in rubber and has compared the characteristics of such rubbers with those made with conventional fillers. But what he also noticed was that the addition of fly ash had an influence on the curing time; that is, on one of the process conditions. So, influences on the set of process conditions must also be studied in a systematic way (arrow 2) and this may further limit or expand the specifications for the filler.

New developments in cement show that negative influences on product quality may be compensated by changes in formulation (Figure 4.1, arrow 3).[8] Furthermore, the replacement of one filler by another one may also have an influence on the specifications of other ingredients (arrow 4). In the case of the polymer example, the sphericity of fly ash particles will decrease the viscosity of the paste and therefore may allow other ingredients to have less influence on this property. Conversely, the strength may decrease because of the absence of the plate-like kaolinite particles, so that another ingredient must compensate for it by changing its specifications.

The influences indicated by arrow 1 can be studied by testing a full set of different fly ashes for a given application. This may require many experiments to be carried out, but in principle it is rather straightforward. The economic benefit lies simply in the difference in cost of the exchanged ingredients. Full exploration of the influences depicted by arrows 2, 3 and 4 is much more complicated, but the economic benefits may be important as well, as processing costs or saving of other ingredients may be involved. Moreover, on the basis of these relationships, it may be possible to replace one ingredient by another – for example, fly ash – whereas this would not have been possible if only the influences governed by arrow 1 had been taken into consideration.

In addition to this, fly ashes may be beneficially pre-treated, as examples in this chapter will show (size fractions, removal of minerals and surface treatment). This will of course increase the cost of the fly ash, but, again using the example of fillers, looking at the current world market prices for polymer-grade mineral fillers, there seems room for such operations.

The above procedure of establishing standards and specifications is rather tedious, costly and product specific. Therefore, it should not be expected that an industry will undertake such studies on its own while its current ingredient materials are well established, used for a long time, and widely available at acceptable quality and price. So, in order to promote fly ash for novel applications, such as fillers in polymers, fly ash researchers will have to take the initiative. In the case of initially positive results, the target industry must be convinced of the possibility for the replacement of an ingredient by a combustion residue. Of course, technical assistance from such industry experts is required during further research activities and development studies.

4.1.3   Outline of This Chapter

In this chapter, a number of selected novel products with their applications are treated in detail. It is not intended to cover the whole area of all the research undertaken towards new outlets for combustion residues all over the world. However, the most well-known and striking developments in which partners of the former EU thematic network PROGRES were involved, together with some other outstanding examples, form the base for this chapter.

The novel products described show that not only are conventional residues used for new applications, but that the new generation of ashes also provides new opportunities for making different novel products. Examples are the use of incinerated sewage sludge ash (ISSA, Section 4.3) for the production of heavy clayware ceramics, and the conversion of coal gasifier slags into glass polyalkenoate cements (Section 4.6).

For some products, the chemical and mineralogical properties and characteristics of the combustion residues are the determining factors for successful use (zeolites, Section 4.4; and geopolymers, Section 4.9), whereas for other applications the physical properties seem more important (fire retardants, Section 4.7; fillers, Section 4.8). However, when allowable limits for usage are set, both chemical and physical properties may play a role, as is the case for cement and concrete (Section 4.2). Due to the different stages of development of the various products, a great variation in degree of determination of the required properties exists between the described products. For cement and concrete applications, the requirements are set out in very specific national and European guidelines (see Chapter 2 and Section 4.2); for zeolites, recent research has shown chemical and mineralogical constraints for either direct conversion or SiO_2-extracted pure zeolites (Section 4.4). For glass polyalkenoate cements, the Si : Al ratio is of prime interest, but for fire retardants and geopolymers specific requirements have not yet evolved. The section on fillers shows that in many preliminary investigations only very little attention is drawn to such requirements, as often only 'inert fine spherical particles' are required.

As shown above, successful applications may only be expected when specific requirements for the chemical and physical limits of the combustion residue to be used are established. When fully developed, this will lead to standards and procedures for certification.

References

1 Lefond S.J. (ed.), *Industrial Minerals and Rocks*, 5th edn, Society of Mining Engineers of the American Institute of Mining, Metallurgical, and Petroleum Engineers, Inc.: New York (1983).

2 Şen, S. and Nugay, N., *Tuning of final performances of unsaturated polyester composites with inorganic microsphere/platelet hybrid reinforces*. European Polymer Journal, 37: 2047–2053 (2001).

3 Chamberlain, A. and Hamm, B., *The viability of using fly ash as a polymer filler*. In 56th Annual Techn. Conf. of Soc. of Plastics Engineers, 3: 3415–3417 (1998).

4 Huang, X., Hwang, J.Y. and Gillis, J.M., *Processed low NO_x fly ash as a filler in plastics*. In Proc. 12th Int. Symp. on Use and Management of Coal Combustion By-products (CCBs), American Coal Ash Association, Orlando, Fla (1997).

5 Kruger, R.A., Hovy, M. and Wardle, D., *The use of fly ash fillers in rubber*. In Proc. 1999 International Ash Utilization Symposium, Lexington, Kentucky, pp. 509–517 (1999).
6 Kruger, R.A., Hovy, M. and Wardle, D., *The use of fly ash fillers in rubber*. In Proc. PROGRES Workshop on Novel Products from Combustion Residues, H.W. Nugteren (ed.), Morella, Spain, pp. 295–303 (2001).
7 Kruger, R.A., Jubileuszowa Miedzynarodowa Konferecja Popioly z Energetyki, Warszawa, 14–17 pazdziernika 2003.
8 Dhir, R.K., McCarthy, M.J. and Magee, B.J., *Impact of BS EN 450 on concrete construction in the UK*. Construction and Building Materials, **12**: 59–74 (1998).

4.2   New Developments in Cement and Concrete Applications

Rod Jones and Michael McCarthy
University of Dundee, Scotland

Types of combustion residues involved
The current use of combustion residues in cement and concrete is mainly restricted to coal fly ash (PFA) conforming to standards such as BS3892 and EN450. The aim of this section is to explore fly ash use at or beyond the property limit requirements of these standards and other recent developments with the material in concrete and to examine other combustion residues in this application. The materials covered are:

- coarse fly ash;
- conditioned and lagoon fly ash;
- co-combustion fly ash;
- silica fume;
- cement kiln dust; and
- incinerator fly ash and sewage sludge ash.

State of development
The effect of including such materials on the properties of cement and concrete has received varied coverage in the literature. In general, it has been found that they can be incorporated in concrete with some modifications to the mix composition. In some cases, the materials considered are being used in construction; in others, they are at the investigation stage.

4.2.1   Introduction

As indicated in Chapter 1, power plant operation, quality of coal, combustion characteristics, boiler design and the ash removal system all influence the characteristics and quality of ash produced. Fly ash has been used in construction for around 70 years and this has grown over the past 20 years. While there is a wide range of use between countries, on average around 50 % of fly ash produced is employed in this application.

In Chapter 2.4, many aspects of the traditional use of fly ash in concrete are described, different types of fly ash reviewed, and the development of standards for use in cement and concrete discussed. There, the influence of key fly ash characteristics on the properties of concrete was extensively treated. In this section, newer developments, looking to extend fly ash use, are reported, and coverage of other combustion residues and new-generation ashes such as those arising from co-combustion is included (Section 4.2.4). This section does not consider the environmental implications associated with the application of the materials in concrete, which in some cases could influence or preclude their use, and the reader is referred to relevant published technical information on this.

4.2.2 Fly Ash Conforming to the EN 450 Standard

Traditionally, fly ash used in the UK as a cement component in concrete has tended to be that at the finer end of the particle size range. The introduction of the European standard, EN 450, covers the use of coarser material in this role (up to 40.0 % retained on a 45 μm sieve, compared to 12.0 % in BS 3892, Part 1). It is likely that this will affect the degree to which fly ash acts as a cement component and, thus, have implications for concrete construction. The effects of fly ash fineness (from different sources) on concrete cube strength are shown in Figure 4.2.[1-3] This demonstrates that with increased coarseness of fly ash (increasing 45 μm sieve retention), there was a gradual reduction in strength, which increased with cement content in concrete. The workability of concrete, measured in terms of slump, also gave some reductions with coarser fly ash.

To accommodate these effects on concrete strength, minor adjustments to the mix proportions through the w/c ratio were examined. It was found that by reducing the water content and using a superplasticising admixture to maintain workability, it was possible, with coarser material, to match strength development of finer fly ash, over test ages to 180 days.

Following the minor changes required to the concrete mix constituent proportions, to allow for the effects of coarser fly ash to match 28 day strength, other aspects of concrete performance including, engineering and durability properties were measured. The results demonstrate that, at equal strength, coarser fly ash had little or no effect on these properties compared to finer fly ash concrete (see Table 4.2).[1-3] Overall, this

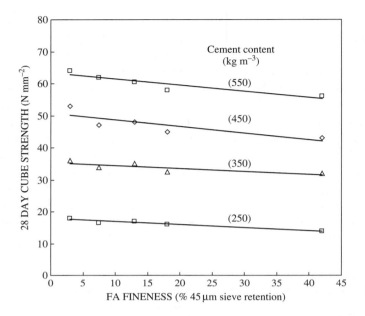

Figure 4.2 The influence of fly ash fineness on the 28 day compressive strength of concrete containing 30 % fly ash.[1-3]

Table 4.2 A comparison of selected engineering and durability properties of BS 3892-1 and BS EN 450 FA (30 % fly ash) concrete.[1-3]

Property	Design strength ($N\,mm^{-2}$)	Concrete with fly ash to	
		BS 3892, Part 1	BS EN 450
Modulus of elasticity[a] ($kN\,mm^{-2}$)	40	23	23
Creep coefficient[b]	40	1.11	1.13
Ultimate drying shrinkage[c] ($\times 10^{-6}$)	40	492	408
Carbonation depth[d] (mm)	35	17.0	17.5
Chloride diffusion[e] ($cm^2\,s^{-1} \times 10^{-9}$)	50	4.9	4.3

[a] At 28 days to BS 1881, Part 121.
[b] At 28 days loading (0.4 fcu).
[c] At 20 °C, 55 % RH.
[d] Enriched CO_2, 20 °C, 55 % RH (30 weeks).
[e] Two–compartment cell.
[a-c] Fly ash fineness BS 3892, Part 1, 3.3 %; BS EN 450, 29.7 %.
[d,e] Fly ash fineness BS 3892, Part 1, 3.5 %; BS EN 450, 35.0 %.

suggests that fly ash with a wider range of particle sizes can be used with minor modifications to concrete mixes to match the performance of that containing finer fly ash.

4.2.3 Conditioned and Lagoon Fly Ash

The majority of national standards covering fly ash as a cement component in concrete require that it is kept dry, or near dry, before use, with typical moisture levels of < 3.0 % given (stored and delivered in a dry condition in EN 450–1). As a result, in most cases, only material stored in dry conditions – that is, in bags or silos – can be used in this application, while material that has water added at low levels (known as conditioned fly ash, with around 10–20 % moisture) or high levels (as a slurry), and is stored in stockpiles or lagoons, cannot.

Studies of low-lime fly ashes[4–8] have found that with water added, fly ash agglomerates, as indicated by changes in fineness (see Figure 4.3[4,8]) and particle size distribution data. Tests for mineralogy suggest that this was due to fly ash combination with water, which was also reflected in LOI data (giving minor increases: see Figure 4.3). These effects were time dependent and were influenced by the free lime present, but were not affected by other fly ash properties.

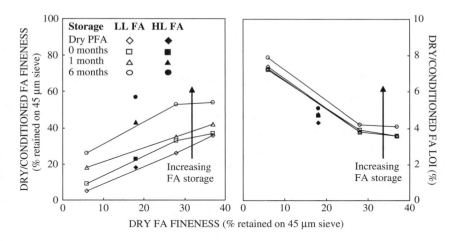

Figure 4.3 The influence of moisture (10 %) and storage period on the fineness and LOI of low-lime (LL) and high-lime (HL) fly ash.[4,8]

In using conditioned fly ash (i.e. immediately after moistening, or that stored for various periods thereafter) as a cement component in concrete, allowances for the moisture present in the ash can be made, by adjusting the fly ash and water contents of the concrete mix. Following this approach, concrete with conditioned fly ash gave both lower and higher workabilities and compressive strengths compared to those with dry fly ash, with performance generally poorer with longer-term fly ash storage.[5] Procedures for assessing conditioned fly ash for use in concrete and a mix design technique[6,8] have been developed. By these means, it was possible to match strength and other aspects of performance to that of dry fly ash concrete,[6] although the economics of this approach, in practice, would have to be assessed carefully.

Investigations of fly ash in an excess of water, or lagoon fly ash,[7] indicate that this type of storage had a slightly greater impact on concrete performance than that of conditioning. This may reflect fine particle loss under these conditions; although conditioned and stockpiled fly ash typically shows more agglomeration.[7] It may be necessary for processing with this material to separate it into specific fractions, prior to use in concrete.

Full-scale trials with conditioned fly ash indicate satisfactory concrete performance, if the material is handled as for fine aggregate.[8] It is possible that difficulties may arise in handling moistened fly ash, in particular when passing it through hoppers, the design of which may be an issue in full-scale concrete production.

4.2.4   Co-combustion Fly Ash

Recent developments in electricity generation have seen the introduction of co-combustion, which replaces part of the coal fired in power stations with organic waste materials or harvested biomass (co-fuels), reducing CO_2 production. A description of the most frequently encountered co-combustion ashes is given in Chapter 1.5. Issues relating to the environmental impact of ash produced from this process have recently received attention.[9]

The particle size and chemistry of co-fired ashes have been described in several studies,[10-12] with various co-fuels (harvested biomass and wood-based). These indicate some agglomeration or coarsening and variations in chemistry; for example, alkali and calcium content between fly ashes. A review of research carried out in Europe[13] suggests small changes in chemical composition for a range of co-combustion fly ashes (with co-fuel levels up to 20 % by mass of the fuel) compared to their references. Other work, reported in Chapter 1.5.1, also found minimal effects on chemical composition using the KEMA TRACE MODEL® for a large number of co-fuels, when used up to 10 % by mass. Limited studies examining the impact of co-combustion fly ash on fresh and strength properties of concrete suggest minor influences on these.[14]

The use of co-combustion fly ash in concrete is covered in the revision of EN 450, published in 2005, see Chapter 2.4.3, page 98.

4.2.5 Silica Fume

Silica fume is recovered from the gaseous waste streams of the ferro-silicon/silicon alloy industries. Early uses were made in Scandinavia in the 1950s.[15] However, silica fume started to find wider application as a cement component in concrete construction in the 1970s and 1980s. It is frequently used in high-strength/high-performance concrete. A significant amount of research has been carried out on silica fume and its use in concrete, and many papers have been published. The reader is, therefore, directed to several reviews for detailed information on silica fume[15–20] and this section provides a summary of the main issues raised in these and other publications.

Silica fume normally comprises extremely fine spherical particles. The very high surface area of the material ($\geq 15\,000\,\mathrm{m^2\,kg^{-1}}$ to BS EN 197-1) means that it is amongst the finest of the solid components used in concrete, and for handling purposes it is supplied in slurry or amorphous densified form. Chemically, it has a very high silica content ($\geq 85\,\%$ by mass to BS EN 197-1).

The material is typically included as a cement component at levels of up to 10 %, although this can be higher, and given its particle size, tends to influence the fresh properties of concrete. Hence, superplasticising admixtures are ubiquitous with the material in concrete to enable control of workability. Increased cohesiveness and reduced bleeding of concrete are frequently obtained with the material.[21] Silica fume can aslo have implications in relation to plastic shrinkage of concrete[22] and need for effective curing practices.

The high fineness and pozzolanicity of silica fume means that it can be used to enhance compressive strength compared to Portland cement (PC) concrete of equal w/c ratio, with differences tending to increase with silica fume level – see, for example, Figure 4.4.[16] Strength influences both at early ages and in the longer term have been noted, but as with other pozzolanic materials, depend on the concrete mix proportions and curing environment conditions.[23–25] Most other engineering properties with silica fume in concrete at equal w/c ratio (including flexural strength, modulus of elasticity and creep strain) are generally similar or improved compared to PC concrete. In the case of drying shrinkage,

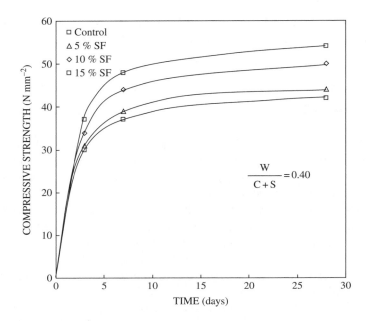

Figure 4.4 The influence of the silica fume level on the strength development of concrete to 28 days.[16]

both lower and higher shrinkage strains for silica fume concrete have been noted compared to reference concretes.[20,25]

The permeation characteristics of concrete can be significantly improved through the use of silica fume,[25,26] reflecting its physical packing and chemical reaction product influences, reducing the pore size volume and interconnectivity of the concrete microstructure. It has also been found to enhance several aspects of durability, including abrasion[25,27] and chloride resistance.[28,29] With other durability properties, including carbonation,[29] and freeze/thaw resistance,[30] depending on silica fume level and mix proportions, issues including initial curing and the inclusion of air-entraining admixtures may be important in matching PC concrete performance.

Other developments have seen the use of silica fume in ternary blend cements in combination with PC and other by-product materials such as fly ash and ground granulated blast furnace slag (GGBS) [31–33] to give enhanced performance.

The use of micronised fly ash as an alternative to silica fume can provide a similar effect in terms of strength development.[34] This micronised fly ash did not, however, reach the fineness of silica fume and, therefore, the strength improvement was not as great.

4.2.6   Cement Kiln Dust

Cement Kiln Dust (CKD) is derived from the manufacture of Portland cement. It is collected from the kiln exhaust gases and its properties are influenced by the characteristics of the production process. CKD particles are generally angular in shape, and while they may have a range of sizes, the majority are typically $< 50 \, \mu m$.[35,36] Chemically, CKD differs from PC, due to incomplete calcination and volatile components present.[36] While its composition may vary between sources, its main components typically include silica, alumina calcium oxide, and it may contain sulfate, alkalis and chloride.[35-38]

Studies examining concrete with CKD as a cement component at low levels (6 % by mass cement)[38] have found that its influences on 90 day compressive strength depended on the source of CKD, and w/c ratio, with similar or lower strength noted compared to PC concretes. Other work testing CKD,[39] covering a range of levels, found that both compressive and tensile strength of concrete decreased with increasing CKD level (see Figure 4.5),[39] although this behaviour was modified slightly with different cement types (blast furnace slag cement and sulfate-resisting cement).

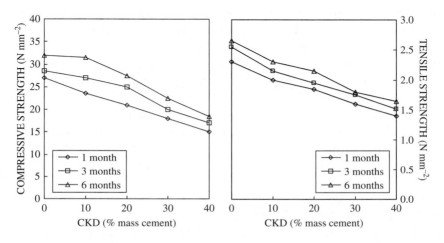

Figure 4.5   The influence of CKD level in cement on the compressive strength of concrete (w/c = 0.5, water cured at 20 °C).[39]

The freeze/thaw resistance of CKD concrete can be influenced by material source, with both similar and poorer performance found, compared to reference concrete.[38] These type of performance variations have also been noted with regard to chloride ingress/chloride-induced corrosion for different CKD.[37,38]

CKD can be used in ternary cements with PC and either GGBS or fly ash and, given its chemical composition, may contribute as a pozzolanic activator.[40] A potential limitation of CKD is its chloride content, which may represent a risk with respect to corrosion of steel in reinforced concrete. Similarly, sulfate levels could also be an issue with regard to its use in concrete.

4.2.7   Municipal Solid Waste Ash and Sewage Sludge Ash

The incineration of municipal waste leads to the production of municipal solid waste fly ash (MSWA), while the combustion of waste from water treatment gives incinerated sewage sludge ash (ISSA). The general characteristics of these residues are given in Chapter 1.5.3. A summary of the properties of interest for possible applications in cement and concrete is given below.

MSWAs normally comprise a range of particle sizes (from $< 10\mu m$ to around 200 μm) and types (including[41] spheres, flakes, prisms, needles and dust agglomerates). Their chemical composition may also be variable, but the main components are silica, alumina and calcium oxide, and they may contain sulfate alkalis and chloride.[41-44] Particles of sewage sludge ash tend to be in the range of 2.5 to 250 μm and to compromise smaller glassy material together in a porous structure.[45] The main components of these ashes are silica, alumina, calcium oxide, iron oxide, sulfate and phosphate, although again these vary between sources.[45-48]

There is disagreement in the literature on the impact that MSWA and ISSA have on the workability of concrete, with studies suggesting different performances compared to PC concrete controls.[46-50] This appears to depend on the characteristics of the ash and the levels used in cement.

It has been suggested that the strength contribution of MSWA is likely to be due to hydration of lime present[50] and it has been found that by limiting MSWA levels in cement, comparable or higher compressive strengths to PC concretes can be obtained.[50,51] However, there may be reductions in compressive strength with increasing MSWA content

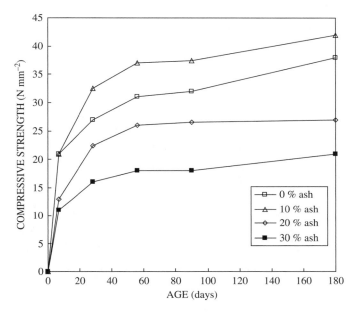

Figure 4.6 The influence of the MSWA level on the strength development of concrete.[50]

(see Figure 4.6),[50] which seem to reflect ash reactivity and influences on microstructure.[52] It should be noted that the strengths achieved with MSWAs depend on the nature of the ashes and their method of production.[51]

It appears that at relatively low ISSA levels in cement, compressive strengths may be similar or increase compared to PC concrete, but reduce progressively as ash levels are increased to 20 % (Figure 4.7),[48] similar to those found for MSWA in Figure 4.6. These types of effect have been found in other work[47] (at w/c ratio = 0.6) but with greater strength reductions noted with increasing ISSA levels compared to the reference concrete.

Studies examining the effectiveness of MSWA in cement on freeze/thaw resistance of concrete have found that this depends on the characteristics of the ash used, with some ashes giving improved performance and others poorer compared to PC concrete.[51] As with other properties, the level of MSWA in cement influences this aspect of performance. In studies[48] examining the permeation properties of ISSA

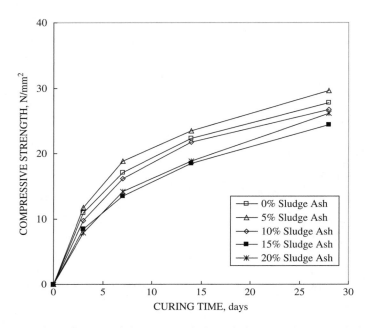

Figure 4.7 The influence of the sewage sludge ash level on the strength development of concrete (w/c = 0.50, water cured at 20 °C).[48]

concrete, water absorption and permeability decreased up to a certain ISSA level in cement and then tended to increase – that is, similar to strength.

Given the alkali contents associated with MSWA, ash combination with PC and fly ash or GGBS to create cementitious blends, represent an option for the material. However, the chloride and/or sulfate content in these ashes could be an issue with regard to durability and would need to be determined, among other properties e.g. environmental, in evaluating suitability of the material for use in concrete construction applications. The chemistry of ISSA suggests that it could be used cementitiously, but again would depend on individual ash compositions.

4.2.8   Acknowledgements

Acknowledgement is given to Professor R.K. Dhir, Dr N.A. Henderson, Dr M.C. Limbachiya and Dr J.E. Halliday of the Concrete Technology Unit, University of Dundee for providing information used in this Section.

References

1 Dhir, R.K., McCarthy, M.J. and Magee, B.J., *Impact of BS EN 450 PFA on concrete construction in the UK.* Construction and Building Materials, 12: 59–74 (1998).

2 Dhir, R.K. and McCarthy, M.J., *Use of BS EN 450 fly ash in concrete.* Concrete Journal, 32: 26–29 (1998).

3 McCarthy, M.J. and Dhir, R.K., *Towards maximising the use of fly ash as a binder.* Fuel Journal, 78: 121–132 (1999).

4 McCarthy, M.J., Tittle, P.A.J. and Dhir, R.K., *Characterisation of conditioned PFA for use as a cement component in concrete.* Magazine of Concrete Research, 51: 191–206 (1999).

5 McCarthy, M.J., Tittle, P.A.J. and Dhir, R.K., *Influences of conditioned PFA as a cement component in concrete.* Magazine of Concrete Research, 52: 329–343 (2000).

6 McCarthy, M.J., Tittle, P.A.J., Dhir, R.K. and Kii, H.K., *Mix proportioning and engineering properties of conditioned PFA concrete.* Cement and Concrete Research, 31: 321–326 (2001).

7 McCarthy, M.J., Tittle, P.A.J. and Dhir, R.K., *Lagoon PFA: feasibility for use as a binder in concrete.* Materials and Structures, 31: 699–706 (1998).

8 McCarthy, M.J. and Tittle, P.A.J., *Stockpile and pond stored fly ash for use in structural concrete.* In Proc. Int. Fly Ash Symposium, Kentucky, pp. 571–578 (2001).

9 van der Sloot, H.A. and Cnubben, P.A.J.P., *Verkennende evaluatie kwaliteitsbeïnvloeding poederkoolvliegas: bijstoken van biomassa in een poederkoolcentrale of bijmenging van biomassa-assen met poederkoolvliegas.* ECN Report ECN-C—00-058, ECN Publicatie SF, Petten, the Netherlands (2000).

10 Robinson, A.L., Junker, H., Buckley, S.G., Hornung, A. and Hornung, U., *Interactions between coal and biomass when co-firing.* In Proc. 27th Int. Symposium on Combustion, Boulder, Colorado, 2–7 August, Combustion Institute, Pittsburgh, PA (1998).

11 Drift, A. van der and Olsen, A., *Conversion of biomass, prediction and solution methods for ash agglomeration and related problems.* ECN Report ECN-C—97-042. ECN Publicatie SF, Petten, the Netherlands (1997).

12 Zygarlicke, C.K., McCollor, D.P. and Toman, D.L., *Ash interactions during the cofiring of biomass with fossil fuels.* In Proc. 26th Int. Tech. Conf. Coal Utilities and Fuel Systems, Gaithersburg, MD (2001).

13 Van Den Berg, J.W., Vissers, J.L.J., Hohberg, I. and Wiens, U., *Fly ash obtained from co-combustion-state of the art on the situation in Europe.* In Fédération Internationale du Béton, FIB Symposium, Concrete and Environment, Berlin (2001).

14 McCarthy, M.J., Brindle, J.H. and Dhir, R.K., *A review of co-combustion fly ash for use in concrete construction.* In Proceedings of International Symposium – Recycling and Reuse of Waste Materials R.K. Dhir, M.D. Newlands and J.E. Halliday (eds). Thomas Telford: London, pp. 447–458 (2003).

15 Fédération Internationale de la Précontraint, *Condensed Silica Fume in Concrete. State of the Art Report.* Thomas Telford: London (1988).

16 Mehta, P.K., *Condensed silica fume.* In Cement Replacement Materials, R.N. Swamy (ed.). Surrey University Press, UK, pp. 134–170 (1986).

17 The Concrete Society, *Microsilica in concrete.* Technical Report No 41, The Concrete Society, Slough, UK (1993).

18   Malhotra, V.M. and Mehta, P.K., *Pozzolanic and Cementitious Materials*, Vol. 1. Advances in Concrete Technology. Gordon and Breach: New York (1996).

19   American Concrete Institute, *Guide for the use of silica fume in concrete*. Report ACI 234–96 (1996).

20   Fidestol, P. and Lewis, R., *Microsilica as an addition*. In *Lea's Chemistry of Cement and Concrete*, 4th edn, P.C. Hewlett (ed.). Arnold: London (1998).

21   Khayat, K.H., Vachon, M. and Lancot, M.C., *Use of blended silica fume cement in commercial concrete mixtures*. ACI Materials Journal, **94**: 183–192 (1997).

22   Branch, J., Hannant, D.J. and Mulheron, M., *Factors affecting the plastic shrinkage of high strength concrete*. Magazine of Concrete Research, **54**: 347–344 (2002).

23   Wild, S., Sabir, B.B. and Khatib, J.M., *Factors influencing strength development of concrete containing silica fume*. Cement and Concrete Research, **25**: 1567–1580 (1995).

24   Carette, G.G. and Malhotra, V.M., *Long-term strength development of silica fume concrete*. ACI Special Publication, SP 132–55, American Concrete Institute (1992).

25   Dhir, R.K., Limbachiya, M.C., Henderson, N.A., Chaipanich, A. and Williamson, G., *Use of unfamiliar cement to ENV 197-1 in Concrete*. Concrete Technology Unit Report CTU/ 1098, University of Dundee (1999).

26   Bayashi, Z. and Zhou, J., *Properties of silica fume concrete and mortar*. ACI Materials Journal, **90**: 349–356 (1993).

27   Laplante, P., Aitcin, P.C. and Vezina, D., *Abrasion resistance of concrete*. ASCE Journal of Materials in Civil Engineering, **3**: 19–28 (1991).

28   Jones, M.R., *Performance in chloride bearing exposures*. In *Euro-cements, Impact of ENV 197 on concrete construction*, R.K. Dhir and M.R. Jones (eds). E & F N Spon: London, pp. 149–167 (1994).

29   Gjorv, O.E., *Effect of condensed silica fume on steel corrosion in concrete*. ACI Materials Journal, **92**: 591–598 (1995).

30   Cohen, M.D., Zhou, Y. and Dolch, W.L., *Non-air-entrained high strength concrete – is it frost resistant?* ACI Structural Journal, **89**: 406–415 (1992).

31   Ozyildirim, C., *Laboratory investigation of low-permeability concretes containing slag and silica fume*. ACI Materials Journal, **91**: 587–594 (1994).

32   Thomas, M.D.A. and Shehata, M.H., *Use of multi-component cementitious systems in high performance concrete*. ACI Special Publication 189 (SP189-17), American Concrete Institute pp. 295–319 (2000).

33   Jones, M.R. and Magee, B.J., *A mix constituent proportioning method for concrete containing ternary combinations of cement*. Magazine of Concrete Research, **54**: 125–139 (2002).

34   Eymael, M.M.Th. and Cornelissen, H.A.W., *Processed pulverized fuel ash for high-performance concrete*. Waste Management, **16**: 237–242 (1996).

35   Konsta-Gdoutos, M.S. and Shah, S.P., *Hydration and properties of novel blended cements based on cement kiln dust and blast furnace slag*. Cement and Concrete Research, **33**: 1269–1276 (2003).

36   Corish, A. and Coleman, T., *Cement kiln dust*. Concrete, **29**: 40–42 (1995).

37   Helmy, I.M., Amer, A.A. and El Didamony, H., *Chemical attack on hardened pastes of blended cements: attack of chloride solutions*. Zement–Kalk-Gips, **44**: 46–50 (1991).

38   Batis, G., Katsiamboulas, A., Meletiou, C.A. and Chaniotakis, E., *Durability of reinforced concrete made with composite cement containing kiln dust*. In *Concrete*

for *Environment Enhancement and Protection,* R.K. Dhir and T.D. Dyer (eds). E & F N Spon: London, pp. 67–72 (1996).

39 Shoaib, M.M., Balaha, M.M. and Abdel-Rahman, A.G., *Influence of cement kiln dust substitution on the mechanical properties of concrete.* Cement and Concrete Research, **30**: 371–377 (2000).

40 Dhir, R.K., Dyer, T.D. and Halliday, J.E., *Activation and acceleration of Portland cement/GGBS blends using cement kiln dust (CKD).* In Proceedings of the International Congress 'Creating with Concrete', 'Modern Concrete Binders: Binders, Additions and Admixtures', R.K. Dhir and T.D. Dyer (eds). Thomas Telford: London, pp. 361–370 (1999).

41 Forestier, L. and Libourel, G., *Characterisation of flue gas residues from municipal solid waste combustors.* Environ. Sci. Technol., **32**: 2250–2256 (1998).

42 Mohamad, A.B., Gress, D.L. and Keyes, H., *A microscopic comparison of the physical and cementitious properties of coal fly ash, coal – RDF fly ash, incinerator fly ash and incinerator bottom ash.* In Proc. 11th Int. Conf. on Cement Microscopy. International Cement Microscopy Association New Orleans, LA, pp. 81–100 (1989).

43 Auer, S., Kuzel, H.-J., Pollman, H. and Sorrentino, F., *Investigation on MSW fly ash treatment by reactive calcium aluminates and phases formed.* Cement and Concrete Research, **25**: 1347–1359 (1995).

44 Andac, M. and Glasser, F.P., *The effect of test conditions on the leaching of stabilised MSWI fly ash in Portland cement.* Waste Management, **18**: 309–319 (1999).

45 Gunn, A.P., Dewhurst, R.E., Giorgetti, A., Gillott, N.L., Wishart, S.J.W. and Pedley, S., *Use of sewage sludge in construction.* CIRIA Report C608, CIRIA, London (2004).

46 Monzo, J., Paya, J., Borrachero, M.V. and Corcoles, A., *Use of sewage sludge admixtures in mortar.* Cement and Concrete Research, **26**: 1389–1398 (1996).

47 Tay, J.H., Sludge *ash as filler for Portland cement concrete.* J. Environ. Eng., **113**: 345–351 (1987).

48 Pinarli, V., Sustainable *waste management – studies on the use of sewage sludge ash in the construction industry as a concrete material.* In Sustainable Construction, Use of Incinerator Ash, University of Dundee, UK, Thomas Telford: London, pp. 415–425 (2000).

49 Tay, J.H. and Cheong, H.K., *Use of ash derived from refuse incineration as a partial replacement of cement.* Cement and Concrete Composites, 171–175 (1991).

50 Cheong, H.K., Tay, J.H. and Snow, K.Y., *Utilisation of municipal solid waste fly ash as innovative civil materials.* Proc. Symp. on Use of Incinerator Ash, R.K. Dhir, T.D. Dyer and K.A. Paine (eds), University of Dundee, UK, Thomas Telford: London, pp. 369–380 (2000).

51 Triano, J.R. and Frantz, G.C., *Durability of MSW fly ash concrete,* ASCE Journal of Materials in Civil Engineering, **4**: 369–384 (1992).

52 Halliday, J.E., *The potential for reusing wastes from thermal processes as a constituent of blended cements.* Internal Report. University of Dundee (2002).

4.3   Combustion Residues in Heavy Clayware Building Products

Michael Anderson
University of Hertfordshire, Hatfield, UK

Types of combustion residues involved
The first comprehensive study of the use of PFA in heavy clayware building products (bricks, tiles and pipes) took place over 60 years ago. Results published at that time showed that it could be successfully employed to partially (or even totally) replace the normal clay used for the manufacture of building bricks. However, the inherent problem of PFA variability has until recently prevented its large-scale adoption within this market sector. Today, the growing availability of newer combustion residues, particularly those arising from the incineration of sewage sludge (ISSA) and municipal solid waste (MSWF), offer possible application in this same role. In addition, smaller volumes of many other combustion residues now arising from the incineration of miscellaneous wastes such as paper-pulp, wood and bark chippings, food/chemical industry by-products and pyrolysis residues can also be considered for their potential suitability.

State of development
Air-classification and stockpile/blending technologies are now able to overcome the earlier obstacle of PFA variability, thereby allowing its increasing use in heavy clayware products. However, the type of clay and the manufacturing method employed considerably influences the amount that can be incorporated. Growing pressure to conserve the remaining reserves of many rapidly diminishing clay deposits is now stimulating increasing interest in the development of new products with partial replacement of the traditional clay. The newer combustion residues are actively under investigation to fulfil this role, although with varying levels of success. Thus, ISSA has been shown to offer lower firing costs when incorporated in clay bricks. However, MSWF has proved less attractive owing to its high soluble salts content that detrimentally affects the fired appearance of the host products. The other (minor) combustion residues continue to be appraised on a case-by-case basis to identify ceramic benefits that they may individually offer.

4.3.1 Introduction

Ceramic building materials fall under the general heading of 'heavy clayware', which embraces a diverse variety of fired merchandise, all with specific end uses and manufactured from a wide array of different clays, shales, marls, mudstones and siltstones – which for the current discussion will be collectively referred to as 'clays'.

Throughout the EU in recent years, there has been a steady decline in the volume demand for these products, due both to changes in the way new buildings are constructed, as well as the growth of alternative (mainly cheaper) building materials. However, heavy clayware still represents a large volume user of virgin raw materials, as illustrated in Table 4.3, which compares recent consumption figures for four separate EU member states. Thus, this manufacturing sector still potentially offers considerable scope as a future 'host' for significant quantities of a variety of combustion residues.

Table 4.3 The tonnage of clay used for heavy clayware products in selected EU member states (2001).

EU member state	Millions of tonnes
UK	8.5
The Netherlands	2.7
Germany	22.0
Italy	23.0

Heavy clayware products can be considered under three basic categories: bricks, tiles and pipes. Within each of these groupings there is a wide diversity of individual products on the market, with specific purpose applications – bricks and blocks (subsequently collectively referred to as bricks) for wall construction, beam-work and flooring; tiles for roofing, cladding and paving; and pipes for sewer construction, field drainage, chimney pots and flue linings. However, dominant among these (with an estimated 80 % of market share tonnage) are bricks and it is this product type that offers the best future opportunity for the largest volume inclusion of combustion residues. Therefore, in the current discussion, bricks will be used as the casebook example – although the results described can also be broadly related to the other categories of heavy clayware products.

The following review examines the broad ceramic possibilities for each of the common combustion residues (some new, some old) that are now available, and considers their individual suitability for use in clay brick manufacture. This review begins with an assessment of the potential role of PFA which, as previously reviewed in Chapter 2, has historically received the most attention, but which mainly due to its inherent variability, has failed to make significant impact within the commercial brick manufacturing sector. This problem is examined further in the following section.

4.3.2 PFA Variability and its Influence on Clay Brick Properties

In Chapter 2.9 (p. 126), PFA variability has already been identified as the largest obstacle in preventing its wider utilisation in the heavy clayware industry in the past. The PFA being produced at power stations is seldom homogeneous; rather, it is the product of a process which, even more so today, is constantly subjected to operating variables that significantly affects its quality over quite short periods of time. Such changes come about due to a variety of reasons broadly related to the necessary adjustments to operating schedules that directly control boiler running behaviour. Thus, during the time taken for a boiler to be brought up from cold to its optimum operating temperature, the PFA produced will frequently contain a higher than normal proportion of unburned coal and coarser, poorly formed ash particles. Furthermore, once steady-state combustion has been achieved, subsequent adjustments to the loading level of the boiler are regularly required to meet the changing demand for electricity. These corrections are made in various ways. For example, the number of grinding mills supplying pulverised coal to an individual boiler may be reduced and this will result in a change in the particle-sizing of the PFA. Under extremely low-level operating conditions, oil may be introduced into the boiler to stabilise the flame, in which case there may be some oil contamination of the ash being produced at that time. Today, the coexistence of several alternative electricity generating options (i.e. coal, gas and nuclear) in many EU member states contributes towards further variability, owing to the policy of rapidly switching between supply options to achieve maximum economy. The outcome response since the earlier practice of steady-state generation at most large coal-fired power stations has been the emergence of PFA of increasing variability. Maximum efficiency (base-load) firing conditions at which power station boilers are ideally designed to run, is no longer the way that most are now operated. Further complicating factors have also arisen over recent years, such as irregular changes in the source of coal, the fitting of 'low-NO_x' burners and the increasing practice of some power stations to burn small amounts of biomass or oil-derived by-products in combination with the coal. Additionally, regular maintenance schedules such as boiler cleaning (soot-blowing) contribute further towards increasing the level of variation in the ash being regularly produced. Consequently, today, the PFA passing out of the boilers of a typical large power station is purged from the exhaust gases by the filter system and accumulates in the collection hoppers as irregular bands and lenses of variable composition. When this situation is viewed in the context of the essential narrow tolerance requirements of PFA for clay brickmaking, it is clear that there is a high probability of it leading to problems in this manufacturing process. This is highlighted in Figure 4.8, which shows the results of particle-size variation and carbon content (determined as loss on ignition) recorded for individual PFA samples collected over a 40 day period at a large 2000 MW power station in the UK. Here, the fluctuations in carbon values are seen to be modest and consistent with what is expected for the base-load

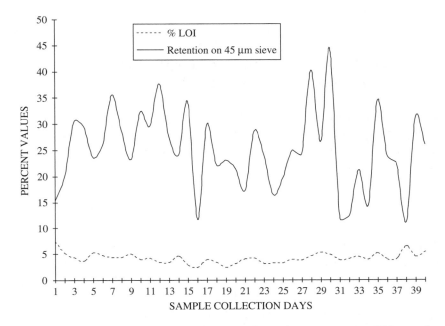

Figure 4.8 Particle sizing and LOI variability shown by power station PFA samples taken over a 40 day sampling programme (under base-load conditions).

running conditions that were in operation over the duration of sampling. However, in contrast, the particle-size composition is seen to be highly erratic from day to day. A repeat of this monitoring exercise today would doubtlessly reveal even greater variability, particularly in the case of increased carbon content fluctuation.

In the manufacture of some commercial building products such as Autoclaved Aerated Concrete (AAC) blocks and sintered lightweight aggregate, PFA possessing the level of fluctuation displayed in Figure 4.8 can be successfully used due to the ability of both processes to accommodate such levels of variability. However, within the clay brick manufacturing process and depending on the amount used, such changes in ash quality can be expected to have a dramatic effect on the fired ceramic properties of any brick product in which it is used. Earlier results[1] showed that as the replacement level of PFA is increased, the resulting bricks can be expected to yield unpredictable dimensions, varying water absorption values and fluctuating strengths. Such products would not be able to meet the relevant standards within the UK or in other EU member states on a day-to-day basis. Consequently, some remedial action is needed as a prerequisite for promoting a wider use of this by-product in commercial brick manufacture of the future. Possible approaches to tackle this problem are examined below.

4.3.3 Procedures for Achieving Acceptable PFA Consistency for Clay Brick Manufacture

As a way forward to allow PFA to be successfully used more widely in the future in heavy clayware products, the following practical approaches to overcoming the problem of inherent variability are considered appropriate.

Up-grading by Air-classification

Today, the poorer-quality PFA now being widely produced is detrimentally affecting its long-established market use in the cement and concrete industry. As a result, it has been necessary to implement upgrading procedures to achieve the level of consistency and quality specified in the relevant standards. This need has led in relatively recent times to the increasing use of mechanical processing equipment to tackle the problem. There are many different designs of apparatus appropriate to fulfil this upgrading function, but their operating principles are all broadly similar and are briefly reviewed in Chapter 3.2 (p. 146). Thus the PFA is directed into the classifier, where individual particles are subjected simultaneously to centrifugal and centripetal friction drag-forces, both of which can be adjusted to achieve a specific 'cut-point', yielding an appropriate product of consistent particle size. This 'product specification' can be maintained irrespective of the fluctuating quality of the raw PFA being drawn off upstream from the collection hoppers. Today, such classifiers are installed at many power stations, but currently these are devoted to supplying 'quality assured' PFA for use into the cement and concrete products market. However, processed ash of this same quality is equally appropriate for use (at relatively high replacement levels) in clay brick products. Unfortunately, its current selling price is linked to the price of cement, which is far higher than the combined cost involved in quarrying and preparing clays for brick manufacture. Consequently, although proven to be technically appropriate for clay bricks, due to its excessive price, its use in this role currently remains commercially impracticable.

Blending/stockpiling

The essential raw materials quality demanded by the brick industry, which has been referred to earlier in Chapter 2.9, has resulted in the requirement for brick manufacturers to pay far greater attention to all stages of their manufacturing process, and this begins with the initial winning and stockpiling of the clays from their quarries. Thus, it is now common practice at many brickworks to construct 'stockpiles' of carefully placed layers appropriate to their particular brick recipes during the drier summer months in order to last through until the following spring. Due to variations in the quality of clay existing in many brickworks quarries, such stockpiles are most frequently built up of successive horizontal layers of even thickness, having been excavated from different parts and horizons in the quarry according to a carefully pre-planned extraction schedule. The subsequent reclaiming of these stockpiles for factory use is normally undertaken by a front-end excavator, which removes this material by cutting vertically downwards through the horizontal layering, thus ensuring a homogenous blend.

Adopting this same relatively simple quality control procedure, it is proposed that PFA could be pre-prepared in a similar fashion and thereby introduced into the brickmaking process with the essential consistency required. To achieve this, it would be necessary for close collaboration between the power station and brick manufacturer involved. The envisaged procedure is shown schematically in Figure 4.9. Here, PFA from the power station either in a damp conditioned form after removal from primary silo storage, or alternatively excavated from a 'conditioned' stockpile or dewatered lagoon, could be built up into horizontally layered piles at the power station. This would later be mechanically removed (as shown) by vertical cutting, followed by transportation to the brickworks for inter-layering with the quarried clay during the routine stockpiling–building operation. When subsequently required for use (following normal brickworks reclamation procedure), the layered stockpile would be removed by vertical cutting, to then be conveyed to the beginning of the factory making operation. As an outcome of this two-phase procedure of layering at the power station and subsequent stockpile inter-layering at the brickworks, it is believed that the variability inherent in the raw PFA would be successfully controlled.

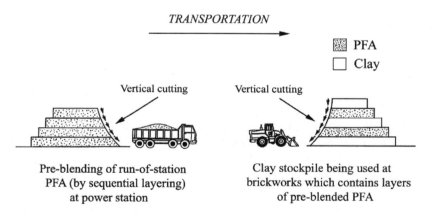

Figure 4.9 A proposed procedure for pre-blending PFA for subsequent incorporation into brickworks clay stockpiles to eliminate variability.

Having overcome this problem, consideration needs to be given to the amount of such 'homogenised' PFA that might be potentially used at a brickworks to achieve market-acceptable products. This prospect is examined in the next section.

4.3.4 How Much PFA Can be Used in Clay Bricks?

There is no definitive answer to this question, as it is dependent upon a number of interrelated variables, most important of which concerns the composition of the particular PFA proposed for use and the ceramic properties of the intended 'host' brick clay. In addition, the type of processing equipment used for forming the bricks is also very influential on the final outcome.

There are three common forming methods used for making commercial bricks, each offering particular advantages within differing manufacturing and marketing situations. The essential features of each of these and their respective merits in terms of the successful incorporation of PFA are as follows.

Vacuum Extrusion

Using this making method, the clay is mixed with water into a stiff plastic consistency and extruded by an auger under vacuum by forcing the plastic clay-mix through a tapered rectangular die, to emerge as a continuous smooth 'column'. As this column exits the auger die-mouth, it is cut into individual bricks. To reduce weight and improve thermal efficiency, such bricks are usually 'perforated' by a series of holes created during the extrusion operation. The introduction of PFA (which possesses no plasticity) into this forming process creates additional friction during the extrusion operation, as well as reducing the cohesive strength of the formed product after it has been dried prior to firing. With very plastic (clay-rich) brick bodies, levels of up to 60–70 % (dry) w/w PFA replacement can be achieved and, after subsequent drying, will be sufficiently robust to be handled and placed in the kiln without damage. However, with poorly plastic or 'short' clays (containing only a small amount of clay minerals) a reduction to around 30–35 % (dry) w/w substitution may be anticipated. Chemical products (plasticisers) are available on the market that can be added to the brick-mix to make up the lack of clay. These artificially provide additional 'smoothness' to enhance extrudability and also to improve the subsequent dry-strength of the bricks, enabling their being handling into the kiln without risk of damage. However, the price of introducing such supplements will add substantially to overall manufacturing costs.

Soft-mud Moulding

Using this method, the clay is prepared in a very 'soft' condition, with typically 25–30 % water being added to it. Subsequently lumps or 'clots' of this wet mixture are thrown (commonly by machine) into pre-sanded brick shaped moulds. These moulds are then inverted and the wet products turned out onto pallets to dry. Using this making procedure, the clay component tends to be more evenly dispersed through the individual bricks than with the extrusion process. However, at the same time, because a high level of remnant void spaces (formerly occupied by water) are created as it is dried out, this type of product is inherently more 'open' in microstructure and inevitably weaker than equivalent extruded bricks in their comparative unfired state, and consequently more prone to handling damage. Thus, the introduction of nonplastic PFA will cause even further weakening in strength. Consequently, with a plastic clay, a substitution level of 40–50 % PFA (dry) w/w is possibly achievable. However, if using less plastic clays, only 15–25 % PFA (dry) w/w is likely, if the product is still to possess enough strength to be handled into the kiln without damage after drying.

Semi-dry Pressing

Less common now than either vacuum extrusion or soft-mud moulding, this brickmaking process uses damp granulated clay with a typical moisture content of only 10–12 % (dry weight basis). A measured volume of the clay is discharged into a mould-box of the machine and pressed to form the brick, which is then ejected. There is an attractive energy saving by using this making method because of the relatively small amount of water used in the forming operation. Consequently, such bricks require only minimal drying and are frequently placed directly into the kiln. However, because the cohesive bond holding them together after pressing relies on particle surface to surface adhesion of the clay granules, the presence of nonplastic PFA has been found to interfere with this bonding process, thereby yielding weaker bricks with poor dry strength and thus prone to more damage than the standard products when being handled during transfer to the kiln. Thus, bricks manufactured by this method are likely to be able to accommodate only around 10 % PFA (dry) w/w as a maximum practical level. In this context, it should be noted that the previously reported 100 % PFA bricks developed by the British Electricity Authority (BEA) laboratories and the 80 % PFA/20 % FBA bricks developed by the West Virginia University both opted for this making method, as described in Chapter 2 (pp. 124–125). In these experimental products, clay was completely eliminated and replaced by much more effective chemical binders, allowing this entirely nonplastic material to be pressed into a cohesive shape, which after drying, hardened into robust bricks suitable for transporting safely into the kiln without damage.

4.3.5 The Future of PFA as a Commercially Acceptable Clay Brickmaking Raw Material

In reviewing the historical development of PFA in brickmaking in Chapter 2.9.2, clear lessons emerged as to the viability of PFA in this potential role in the future. Firstly, in the case of 100 % PFA bricks, the making technology was shown to be successful, particularly in the case of the West Virginia University projects where 20 % of the PFA brick-mix was replaced with FBA. Both this and the BEA product were confirmed to possess the necessary physical properties to enter the market. However, while the commercial future of the BEA brick was discounted at the time mainly for shortcomings relating to PFA quality, the West Virginia University experimental product failed to be promoted commercially due to its 'bland' visual appearance, which was considered to be unattractive (from a marketing point of view) when compared with conventional clay bricks.

In resolving this earlier problem, it is believed that there is still a commercial future for bricks made entirely from PFA/FBA if appropriate surface treatment/texturing is undertaken to enhance their rather plain appearance. Alternatively, if expressly manufactured for an end-use where the product would be 'rendered' (covered with plaster or cement) or painted over on its exposed surfaces, this difficulty would not arise. Nevertheless, with either approach, it must be stressed that an essential prerequisite for successful manufacture on a commercial scale will depend on the use of supplies of PFA that are of consistent composition. Owing to the generally poor quality of PFA now being widely produced, this emphasises the need to use a 'beneficiated' supply source equal in quality to that currently demanded for inclusion in high-quality concrete products. In the alternative situation, where PFA may be considered for introduction only as a partial clay substitute in clay bricks, a wide range of advantages exist, as indicated below:

(1) better brick-forming properties;
(2) improvements in drying rate;

(3) lower maturing temperature and reduced overall firing times;
(4) more stable unfired and fired shrinkages;
(5) greater control over fired porosity;
(6) improved frost resistance.

Yet it must be stressed that all such benefits are not available for every 'brick clay' considered in a potential 'host' role. Rather, research has shown that an essential requirement for success involves correctly matching each individual source of clay to a specific quality of PFA possessing compatible physical/ceramic properties – thereby achieving an improved brick compared with the traditional 'all clay' product.

Notwithstanding the above limitations, there are important technical and environmental arguments for expanding the use of PFA in brick-making. This embraces recognition both that the clay reserves in many traditional brick manufacturing areas are continuing to decline rapidly, while at the same time process energy costs within this manufacturing sector are rapidly increasing.

4.3.6   Applications of More Recent Combustion Residues in Heavy Clayware Products

Although PFA remains indisputably the largest 'bulk' combustion residue available for use in brickmaking, there are other more recent industrial/municipal incineration by-products for possible considera-tion in this same context. In particular, sewage sludge and municipal solid waste are being incinerated in increasing volumes on the grounds that this is the most efficient disposal route available. But the corol-lary is that the resulting ashes are becoming a major disposal concern. Their possible inclusion in bricks as a more acceptable disposal solu-tion is a relatively new avenue of investigation, but one that is now receiving increasing attention. The following brief resumé overviews the findings that have been established so far in respect of this future possibility.

Incinerated Sewage Sludge Ash (ISSA)

Today, in many EU member states, substantial quantities of sewage sludge are being incinerated using modern fluidised bed technology.[2] The resulting incinerated sewage sludge ash (ISSA) is odourless, pathogen-free and fine grained (typically approximately 20–250 μm) with good free-flowing characteristics. It possesses a low dry bulk density and individual particles generally possess an open-textured spongy structure with the capacity to absorb a considerable amount of water (for more background information, see Chapter 1 (pp. 45 and 64ff)).

Table 4.4 shows a characteristic ISSA analysis compared with a typical brick clay. The ISSA is shown to vary noticeably from the clay in its higher iron, calcium and phosphate contents, all of which in certain circumstances can assist glass formation during the firing of the bricks. This characteristic has shown itself to be of particular interest to brick manufacturers, owing to its potential to reduce firing costs.

Table 4.4 An analysis of a typical ISSA, compared with a typical brick clay (%).

	ISSA	Brick clay		ISSA	Brick clay
SiO_2	35.75	57.15	Cr_2O_3	0.16	< 0.01
Al_2O_3	11.15	9.54	HfO_2	n/d	< 0.01
TiO_2	1.00	0.63	Pb_3O_4	0.09	< 0.01
Fe_2O_3	16.90	11.41	P_2O_5	11.92	0.23
Mn_3O_4	0.26	6.06	V_2O_5	0.01	n/d
CaO	12.85	3.21	ZnO	0.32	< 0.01
MgO	1.94	1.01	ZrO_2	0.05	n/d
K_2O	1.49	0.98	SO_3	3.12	< 0.05
Na_2O	0.24	0.61	LOI[a]	4.37	9.08
BaO	0.16	0.19			

[a] LOI for ISSA is largely unburned carbon, whereas for brick clay it is largely water.

The potential application of ISSA as a partial replacement material in heavy clayware has only begun to appear in technical publications since the early 1990s.[3-14] Yet preliminary research is recorded as already having been undertaken in the development of a lightweight aggregate and a tile product from a combination of PFA and ISSA,[15] as a result of finding that both these by-products have complementary firing characteristics.

One issue of concern relating to the widespread commercial use of ISSA in brickmaking is the presence of heavy metals within this material. Raw sewage sludge frequently contains significant levels of such species, particularly where it originates from industrial areas. Moreover, once incinerated, their level of concentration in the resulting ISSA is substantially enriched. Consequently, if ISSA is to be successfully used as a future brickmaking raw material, there will be a need to address the present concerns about their presence and the subsequent likelihood of their leaching out from the fired products. This possibility has been examined independently by a number of different research centres[3,7,10,12] and the results have broadly shown that metal release from ceramic bodies submitted for leachate testing were very low, confirming that such species were either being immobilised within the glassy melt-phase or converted to low solubility metal compounds during the firing process. Further work will be required to ensure that this pattern of behaviour is consistent with all clays for which such inclusions may be considered.

ISSA has one further advantage for use in brick manufacture; that is, unlike PFA it is recognised to possess a high level of consistency on a day-to-day production basis due to the efficiency of modern fluidised bed combustion technology[16] and, consequently, at a modest replacement level of 5–15 % (dry) w/w ISSA can be considered for use by brick manufacturers on a regular basis, with the knowledge that it will not detrimentally affect the normal ceramic properties of their products.

Municipal Solid Waste Fly Ash (MSWF)

Municipal solid waste incineration (MSWI) is increasingly seen as a most convenient disposal method in many EU member states. However, although this process can reduce the bulk volume of such waste by as much as 90 %, it still leaves behind two by-products. One is hearth ash, which makes up typically 90 % mass-weight of the total residue and is currently finding new application outlets in the civil engineering field. However, the remaining 10 % mass-weight of fly ash is a significantly greater problem, because it contains concentrations of potentially leachable heavy metals (far higher than ISSA) and can consequently only be disposed of in hazardous waste landfill sites; (for more background information see Chapter 1.6.3, p. 59ff).

Table 4.5 shows a typical analysis of the MSWF sampled from a modern 'waste to energy' municipal solid waste incinerator. Apart from the presence of the heavy metals, this ash is seen to be dominantly composed of a mixture of reacted and unreacted hydroxides of calcium. This is due to the lime-scrubber installed to intercept and clean the exhaust gases arising from the combustion bed. These gases are strongly acidic due to the hydrochloric acid evolved during the incineration of large quantities of plastics among the waste. In the scrubber, a sprayed

Table 4.5   An analysis of a typical MSWF.

	MSWF		MSWF
SiO_2	2.6	Cl	25.3
Al_2O_3	2.6	SO_3	6.6
TiO_2	0.8	P_2O_5	0.6
Fe_2O_3	1.1	Br	0.2
MnO_2	0.1	ZrO_2	Trace
CaO	47.0	Cr_2O_3	Trace
MgO	0.8	SrO	Trace
K_2O	3.7	Sb_2O_3	Trace
Na_2O	3.0	CdO	Trace
PbO	0.6	NiO	Trace
ZnO	2.1	LOI	1.9
SnO_2	0.2		

mixture of lime and water chemically reacts with the exhaust gases as they come into contact. Their reaction products greatly increases the total burden of fly ash compared with that originally arising from the combustion hearth. This ash is highly deliquescent, immediately absorbing moisture once exposed to the air. Consequently, compared with PFA and ISSA it presents increased difficulties when handling or when consideration is given to introducing it into any commercial end-uses.

From a survey of published literature, the possibility of using MSWF in ceramic products appears to have received little attention. Among the few recorded studies, one approach describes the conversion of MSWF into a glass-ceramic through controlled vitrification.[17-18] Another investigation[19] reports on its use when intermixed with rock-cutting waste and fired to produce a ceramic tile product. However, one preliminary laboratory investigation is recorded to have examined the possibility of using this material in clay bricks[20] by replacing 5 % (dry) w/w of the clay within a commercial brick body. The experimental bricks produced were formed using the vacuum extrusion method and fired at the normal factory maturing temperature of the clay. From the results obtained, it was concluded that although the ash did not significantly change the physical properties of the bricks when compared with an 'all clay' control product run in parallel, they proved unattractive from a visual perspective, due to the migration of large quantities of soluble salts to their surfaces, which produced an unsightly scum coating that would severely affect commercial interest in its potential use in this role.

Miscellaneous (Small-volume) Combustion Residues

Other combustion residues arising from various miscellaneous inciner-ation processes are also now beginning to receive attention within the heavy clayware industry. Until recently, the relatively small volumes involved have led them mostly to be consigned to landfill. However, rising disposal costs and increasingly more stringent environmental legislation is leading many producers to seek alternative solutions. Collectively, these newer combustion residues are very diverse in nature; that is, derived from the paper industry, wood and bark ash, pyrolysis products (notably from sewage sludge and rubber tyres), ashes from waste-derived fuels, peat, fruit pulp, clinical waste and so on. Based on the encouraging results described earlier involving the utilisation of the more common combustion residues in clay bricks, it is likely that a number of these newer combustion residues may also prove appro-priate for inclusion in heavy clayware products. Some are likely to play a passive role, while others may improve the existing brick properties and perhaps also contribute to energy savings. In every case, the amount that can be successfully introduced will be dependent on the outcome of systematic investigations starting in the laboratory and subsequently proven at the factory. Broad guidelines giving a preliminary insight into issues likely to influence the future growth of this technology are briefly considered in the following section.

4.3.7 The Future Prospects of Combustion Residues in Clay Bricks

The optimum substitution level of individual combustion residues in clay products will always be limited by two interrelated factors: (a) where the level at which the addition begins to cause an unacceptable deterioration in the 'host' products properties inasmuch that it can no longer maintain existing specifications; and/or (b) where the level of substitution reaches a point at which although still within specifications, the appearance of the product makes it unacceptable (visually) to the commercial market. Yet it should be added that this latter situation will not always be an issue, owing to the fact that the in-service destination of clay building bricks fall within two distinct categories – those that are 'rendered' – that is, covered with plaster or cement – or alternatively coated with paint and thus hidden from view, and those that are used as 'facing' products; that is, where they are visibly exposed at the outside or inside surfaces of wall constructions. These alternative possibilities are shown diagrammatically in Figure 4.10. The former practice is very widely used in continental Europe, while the latter situation is found extensively in the UK. As a result, far greater levels of combustion residues can be potentially introduced into bricks that are to be rendered – as any surface pock-marking, blemishes or discoloration caused by the incorporation of this particular residue is hidden from view.

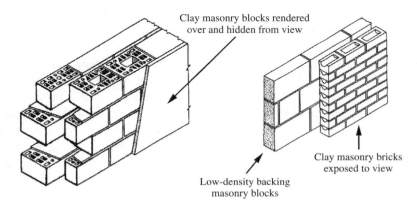

Figure 4.10 The two alternative construction procedures: left, the continental European preference (large perforated clay masonry blocks with the outer surface rendered); right, the UK preference (a backing course of low-density masonry blocks with an outer leaf of exposed clay masonry bricks).

Most pronounced among the deleterious 'visual' side-effects referred to above and shared by all three of the high-volume combustion residues (PFA, ISSA and MSWF) is the common presence of soluble salts, which can promote a manufacturing fault known as 'drier scum'. This occurs when such salts dissolve in the water added to the mix prior to forming into bricks, and then subsequently migrate to their outside surfaces during the drying stage, where they form a whitish film that still remains present after firing. This is visually unattractive and severely down-grades the potential selling price of affected products. Remedial action is possible, involving adding chemicals such as barium carbonate or barium chloride to the brick-mix containing the combustion residue. These compounds immobilise the salts and prevent their migration to the surface.[21] For those clay brick products destined for 'exposed' building situations, the use of such additives is essential to prevent this problem. However, for bricks that are rendered or painted over at their exposed surfaces, the presence of scumming is of little concern.

As a final footnote on this topic, it is necessary to briefly address the important issue of market responsibility for future clayware products containing combustion residues of this sort. Increasing opportunities are now being presented to brick manufacturers within EU member states to incorporate a widening variety of industrial and urban-derived combustion residues within their bricks. The future widespread adoption of this practice will not only contribute towards helping to extend the life of many clay deposits now nearing exhaustion, but will also provide a new income-stream for participating brick manufacturers, offering them the opportunity to reduce operating costs by entering into collaborative ventures with waste producers and/or disposal operators. However, in anticipation of such 'new' brick products increasingly entering the marketplace in the future, it is of paramount importance that their production is carried out in a responsible way and within internationally acceptable guidelines. Therefore, it will be essential to establish an audit-traceable recording procedure linked to these products, both for life-cycle referencing and in order to satisfy the growing amount of relevant environmental legislation. These requirements, which affect heavy clayware products and all other future building products incorporating combustion residues manufactured within the EU, are addressed in Chapter 5.6.

References

1 Anderson, M. and Jackson, G., *The beneficiation of power station coal ash and its use in heavy clayware ceramics*. Trans. J. Brit. Ceram. Soc., **82**(2): 50–55 (1983).
2 Hemphill, B., *Fluidised bed technology for sludge destruction*. Water Eng. & Management, Dec., 37–40 (1988).
3 Anderson, M., Skerratt, R.G. and Birchall, C., *The use of fluidised bed sewage sludge ash in brickmaking*. Proc. Conf. Aqua Enviro (University of Leeds) European Conference on Sludge and Organic Waste, Wakefield, 12–15 April, Session 4, paper 8 (1994).
4 Anderson, M., Skerratt, R.G., Thomas, J.P. and Clay, S.D., *Case study involving using fluidised bed incinerator sludge ash as a partial clay substitute in brick manufacture*. Wat. Sci. Tech., **34**(3–4): 507–515 (1996).
5 Anderson, M., *Encouraging prospects for recycling incinerated sewage sludge ash (ISSA) into clay-based building products*. J. Chem. Technol. Biotechnol., **77**: 352–360 (2002).
6 Anderson, M., Ellott, M. and Hickson, C., *Factory-scale proving trials using combined mixtures of three by-product wastes (including incinerated sewage sludge ash) in clay building bricks*. J. Chem. Technol. Biotechnol., **77**: 345–351 (2002).
7 Alleman, J.E., Bryan, E.H. and Stumm, T.A., *Sludge-amended brick production; applicability for metal-laden residues*. Wat. Sci. Tech., **22**(12): 309–317 (1990).

8   Tay, J.-H. and Show, K.-Y., *Utilisation of municipal wastewater sludge as building and construction materials*. Resources, Conservation and Recycling, **6**: 191–204 (1992).

9   Wiebusch, B. and Seyfried, C.F., *Utilisation of sewage sludge ashes in the brick and tile industry*. Wat. Sci. Tech., **36**(11): 251–258 (1997).

10 Wiebusch, B., Ozaki, M., Watanabe, H. and Seyfried, C.F., *Assessment of leaching tests on construction materials made of incinerator ash (sewage sludge):investigations in Japan and Germany*. Wat. Sci. Tech., **38**(7): 195–205 (1998).

11 Wiebusch, B., Rosenwinkel, K.-H. and Seyfried, C.F., *Utilization of sewage sludge and ashes from sewage sludge incineration as construction material*. Tile & Brick Int., **14**: (4) 267–277 (1998).

12 Schirmer, T., Mengel, K. and Wiebusch, B., *Chemical and mineralogical investigations on clay bricks containing sewage sludge ash*. Tile & Brick Int., **15**(3): 166–174 (1999).

13 Lin, D.E. and Weng, C.H., *Use of sewage sludge ash as brick material*. J. Environ. Eng., **127**(10): 922–927 (2001).

14 Pan, S.C. and Tseng, D.H., *Sewage sludge ash characteristics and its potential applications*. Wat. Sci. Tech., **44**(10): 261–267 (2001).

15 Anderson, M., Skerratt, R.G. and Birchall, C., *Ceramic materials and methods of manufacturing such materials*. United Kingdom Patent GB 2297 971 B (1997).

16 Anderson, M. and Skerratt, R.G., *A variability study of incinerated sewage sludge ash produced by a large commercial incinerator in relation to its future use in ceramic brick manufacture*. Brit. Ceram. Trans., **102**(3): 109–113 (2003).

17 Boccaccini, A.R., Kopf, M. and Stumpfe, W., *Glass-ceramics from filter dusts from waste incinerators*. Ceramics Internat., **21**: 231–235 (1995).

18 Romero, M., Rawlings, R.D. and Rincon, J.Ma., *Development of a new glass-ceramic by means of controlled vitrification and crystallisation of inorganic wastes from urban incineration*. J. European Ceram. Soc., **19**: 2049–2058 (1999).

19 Hernandez-Crespo, M.S. and Rincon, J.Ma., *New porcelainized stoneware material obtained by recycling of MSW incinerator fly ashes and granite sawing residues*. Ceramics Internat., **27**: 713–720 (2001).

20 Staffordshire University, New disposal routes for Campbell Road municipal waste incinerator flyash, Entrust funded technical investigation report on behalf of the Staffordshire Environmental Fund Ltd. unpublished (2002).

21 Williams, A.N. and Ford, R.W., *Scum prevention by using additives: laboratory tests*. Trans. J. Brit. Ceram. Soc., **81**: 88–90 (1982).

4.4   Zeolites

Xavier Querol and Natalia Moreno
CSIC, Barcelona, Spain

Types of combustion residues involved
Coal fly ash (PFA) is the combustion by-product most suitable for conversion into zeolites. However, restrictions on the chemical and mineralogical composition of the ashes set limits to the suitability for the conversion process. For the two different processes proposed, direct conversion and two-step zeolite synthesis, the requirements are different. One subsection is completely dedicated to the description and explanation of these requirements.

State of development
For direct conversion, a pilot production of 2.2 tonnes has been achieved. For the two-step SILEX process, only laboratory work has been performed. Zeolite products from both processes have been fully characterised and applications tested. The zeolitic material produced from the pilot production was tested for environmental applications at a larger scale.

4.4.1   Introduction

Natural zeolite deposits are generally associated with hydrothermal alteration and/or weathering of glassy volcanic rocks, whereas man-made or synthetic zeolites may be obtained from a wide variety of Si- and Al-rich starting materials.

The compositional similarity of fly ash to certain volcanic materials, precursors of natural zeolites, was the main reason why Höller and Wirsching studied the synthesis of zeolites from this coal by-product.[1] Although this potential application may consume only a small proportion of the produced fly ash, the final products may represent a higher added value compared with the current applications in the construction industry.

Zeolite is a crystalline aluminium–silicate mineral, with group I or II elements as counter-ions. The structure consists of a framework of $[SiO_4]^{4-}$ and $[AlO_4]^{5-}$ tetrahedra, linked to each other at the corners by sharing their oxygen atoms. These tetrahedra make up a three-dimensional network, including many voids and open spaces. The voids are responsible for the specific properties of zeolites, such as the adsorption of molecules in internal channels. The substitution of Si(IV) by Al(III) in the tetrahedra accounts for a negative charge on the structure which gives rise to the exchange of cations. As a consequence of the above structural properties, zeolitic materials have a wide range of industrial applications. These applications are mainly based on the above properties; that is, cation exchange capacity (CEC) and the ability to adsorb specific gas molecules – for example, use as molecular sieves for H_2O, SO_2, CO_2 or NH_3. Moreover, zeolites are capable of reversibly adsorbing water without any change in the chemistry or physical properties of the zeolite itself.

Since the term zeolite is used for a large group of crystalline Al–Si phases, with very different properties and applications, it should be clarified that the zeolitic material obtained from coal fly ash is used as molecular sieves and ion exchangers, and that the complex and highly pure zeolites used for catalysis and other technological uses cannot be prepared due to technological limitations and the mixture of components in the fly ash.

Table 4.6 summarises the zeolite species synthesised from coal fly ash in different studies. It is important to note that the possible industrial applications of these materials obtained from fly ashes vary considerably. Thus, there are zeolites (such as X zeolite) with a large pore size (7.3 Å) and a high CEC (close to $5\,meq\,g^{-1}$), which make this zeolite an interesting molecular sieve and cation exchange material. However, other zeolites, such as analcime, sodalite, cancrinite and kalsilite, have a small pore size (around 2–3 Å) and a low CEC, and hence have a low potential application as either a molecular sieve or ion exchanger.

Table 4.6 Zeolites and other neomorphic phases synthesised from fly ash.

High CEC, some with large channels (up to 7 Å)	
NaP1 zeolite	$Na_6[(AlO_2)_6(SiO_2)_{10}].15H_2O$
Phillipsite	$(Na, K)_{10} [(AlO_2)_{10} (SiO_2)_{22}].\ 20H_2O$
K-chabazite	$K_2 [(AlO_2)_2 (SiO_2)].\ H_2O$
Zeolite F linde	$K_{11} [(AlO_2)_{11} (SiO_2)_{11}].\ 16H_2O$
Faujasite	$Na_{12}Ca_{12}Mg_{11} [(AlO_2)_{59} (SiO_2)_{133}].\ 235H_2O$
Herschelite	$Na_4 [(AlO_2)_4 (SiO_2)_8].\ 12H_2O$
Zeolite A	$Na_{12} [(AlO_2)_{12} (SiO_2)_{12}].\ 27H_2O$
Zeolite X	$Na_{86} [(AlO_2)_{86} (SiO_2)_{106}].\ 264H_2O$
Zeolite Y	$Na_{56} [(AlO_2)_{56} (SiO_2)_{136}].\ 250H_2O$
Low CEC and small channels	
Perlialite	$K_9NaCaAl_{12}Si_{24}O_{72}.15H_2O$
Analcime	$Na_{16} [(AlO_2)_{16} (SiO_2)_{32}].\ 16H_2O$
Hydroxyl-sodalite	$Na_6 [(AlO_2)_6 (SiO_2)_6].\ 7.5H_2O$
Kalsilite	$KAl(SiO_4)$
Tobermorite	$Ca_5(OH)_2Si_6O_{16}.4H_2O$

Other species, characterised by a relatively high Al(III) : Si(IV) ratio, such as NaP1, A, X, KM and chabazite, have high CEC. Therefore, these synthesised materials may have significant applications in waste-water decontamination technology.

4.4.2   Synthesis of Zeolites

Since the studies of Höller and Wirsching,[1] many patents and technical articles have proposed different methods based on the dissolution of Al–Si-bearing fly ash phases in alkalis and the subsequent precipitation of zeolitic species.[2-16] Based on the process used for the zeolite synthesis, the methods may be classified as conventional alkaline activation (or direct conversion) and preparation of pure zeolitic products by a two-step synthesis (the SILEX process).

Direct Conversion

This methodology is based on *in situ* conversion of Al–Si-bearing species in fly ash into zeolitic species. To this end, different conversion conditions are obtained by varying the activation solution to fly ash ratio, temperature, pressure and reaction time to obtain different types of zeolite. This process has been modified by several authors as follows:

(a) the introduction of an alkaline fusion stage prior to the conversion, resulting in the synthesis of very interesting zeolites such as zeolite A;[4-6,17]
(b) microwave assistance to accelerate the process;[18]
(c) a process using salt mixtures instead of aqueous solutions, which was developed by Park *et al.*[19,20] to avoid the generation of waste water (dry or molten-salt conversion).

The direct conversion process has the following major limitations:

(a) the final product is a mixture of zeolites and residual nonreacted fly ash components;
(b) the nonconverted fly ash fractions may contain leachable amounts of environmentally unfriendly elements, such as As, Cu, Mo, Se and V, due to only a fraction of these elements being dissolved during conversion.

Using this methodology, up to 13 different zeolites have been obtained from the same fly ash (European Coal an Steel Community Contract 722/ED/079).[12] The zeolite content of the resulting materials varies widely (40–75 %), depending mainly on the activation solution: fly ash ratio and the reaction time. The design of the direct conversion process has advanced so that it has been applied by Querol *et al.* at a pilot plant scale for the production of 2.2 tonnes of zeolitic material in 8 h in a single-batch experiment.[13] Such pilot plant products have been obtained from both Teruel and Narcea fly ashes,[12,13] that have low and high glass contents, respectively (Table 4.7). The processes using microwave assistance, and the dry conversion or alkaline fusion procedures, have so far only been applied at a laboratory scale.

The main technical limitation of direct conversion is that relatively high temperatures (125–200 °C) are required to dissolve Si and Al from the fly ash at a reasonable rate. Under these conditions, many of the high-CEC and large-pore zeolites (e.g. zeolites A and X) cannot be synthesised. If the temperature is reduced, then the synthesis yield is reduced considerably and a very long activation time is required. However, KM (equivalent to phillipsite), NaP1, Na-chabazite (herschelite), K-chabazite, Linde F and other high-CEC zeolites may still be obtained with high synthesis yields in the temperature range of 125–200 °C. Thus, a product with a CEC of 2.7 meq g^{-1} has been obtained from the pilot plant experiment mentioned above.

In most studies, fly ash activation is carried out in digestion bombs or autoclaves, varying the activation agent (mainly KOH and/or NaOH), the temperature (50–200 °C), the conversion time (3–48 h), the solution concentration (0.5–5.0 mol dm^{-3}), the pressure (the vapour pressure of solution at the selected temperature) and solution to sample ratio (2–18 cm^3 g^{-1}). Following the reaction, the zeolitic material is filtered, washed with water, dried at room temperature and analysed by XRD. Determination of CEC and comparison with a pure standard zeolite CEC provides a semi-quantitative estimation of the zeolite content of the synthesised materials.

Types of Zeolite Fly ash consists of different Al- and Si-bearing phases with different solubilities in an alkaline environment (glass > quartz > mullite). During the first stages of the zeolitisation process, the composition of the process solution is mainly determined by the composition of the glass phase. Consequently, the type of zeolite species prepared is influenced by the glass composition, as the

Table 4.7 Characterisation of the fly ashes reported in this section.

	Teruel, Spain	Narcea, Spain	Meirama, Spain	Puertollano, Spain	Alkaline, Netherlands	Neutral, Netherlands	Monfalcone, Italy
Major (%)							
SiO_2	48.3	55.2	49.2	58.6	46.8	53.3	50.8
Al_2O_3	23.9	23.3	17.6	27.4	24.8	26.1	33.4
Fe_2O_3	16.0	6.9	10.4	7.3	9	7.4	6.4
CaO	5.4	4	11.8	0.8	6.8	3.1	2.4
MgO	1.0	2.5	2	1	3.7	0.6	0.8
Na_2O	0.2	0.7	0.4	0.3	1.2	0.1	0.4
K_2O	1.4	3.8	0.4	2.4	2	0.6	0.7
P_2O_5	0.2	0.3	0.2	0.1	0.7	1.5	0.3
TiO_2	0.8	0.9	0.5	0.7	0.9	1.8	2.6
MnO	0.04	0.1	0.1	0.1	0.1	0.1	0.03
SO_3	0.8	0.4	2.2	0.2	1	0.5	0.3
LOI	2.0	1.9	5.2	1.1	3	4.8	1.9
SiO_2/Al_2O_3 (wt/wt)	2.0	2.4	2.8	2.1	1.9	2	1.5
Traces ($mg\,kg^{-1}$)							
As	79	98	94	140	48	55	39
Cr	107	177	47	108	140	196	136
Mo	15	6	5	11	13	7	12
Se	3	6	7	7	16	n.d.	<5
V	206	173	154	202	325	226	455

Minerals (%)						
Quarrz	8.6	6.6	6.9	10.4	11.2	7.1	3.2
Cristobalite	< 0.3	< 0.3	4.5	< 0.3	< 0.3	< 0.3	0.5
Mullite	19.4	3.8	19.6	20.7	20.1	10.9	25.9
Others	8.7	2.6	5.8	3.5	3.7	1.1	0.5
Glass	62.7	85.6	62.5	64.7	63.1	80.1	73.1
Glass composition (%)							
SiO_2	54.6	55.4	50.9	65.5	47.4	53.7	54.4
Al_2O_3	15.9	23.9	5.4	19.4	16.4	22.8	20.3
SiO_2/Al_2O_3	3.4	2.3	9.4	3.4	2.9	2.4	2.7

$SiO_2 : Al_2O_3$ ratio in the solution is an important factor. Thus differences in mineralogical composition of two fly ashes, having similar bulk $SiO_2 : Al_2O_3$ ratios, will result in different glass $SiO_2 : Al_2O_3$ ratios and hence may give rise to different zeolitic species.

Both the temperature and the concentration of the activation agent have a very important influence on the zeolites obtained. Thus, by increasing both parameters (i.e. 200 °C and 5 mol dm^{-3}), low-CEC zeolites such as hydroxyl-cancrinite and hydroxyl-sodalite are obtained. Conversely, low temperature and concentration (i.e. < 150 °C and 0.5–3.0 mol dm^{-3}) allows the synthesis of high-CEC zeolites such as NaP1, A zeolite or chabazite. Table 4.8 summarises the zeolite species that may be synthesised from most of Spanish fly ashes as a function of temperature, concentration and solution to fly ash ratio.

Table 4.8   Zeolites and other neomorphic phases synthesised from mostly Spanish fly ash as a function of activation agent (NaOH or KOH), molarity and temperature for activation solution to fly ash ratios approximately 2 dm^3 kg^{-1}.

T (°C)	Alkali (mol dm^{-3})	Synthesised product
NaOH		
< 100	< 1.0	No activation
	2.0–3.0	Low-activation, A zeolite
150	< 1.0	No activation
	1.0	Low activation, NaP1 (herschelite)
	2.0–3.0	NaP1 (herschelite traces)
150–200	0.5	Low activation
	5.0	Low CEC phases with herschelite
200	1.0–3.0	NaP1 and herschelite
> 200	> 3.0	Analcime, hydroxyl-sodalite and -cancrinite
KOH		
< 150	1.0–2.0	Low activation
150	2.0	KM
	5.0	KM, chabazite and Linde F traces
200	2.0	KM
	5.0	Low-CEC phases (kalsilite) with minor KM

NaP1-rich zeolitic material was obtained at pilot plant scale from the Narcea fly ash (Table 4.7) by using 3 mol dm^{-3} NaOH solutions at a ratio of 2 cm^3 g^{-1}, at 125 °C in 8 h.[13,16,21] The zeolitic material obtained possessed a CEC of 2.7 meq g^{-1}, which is approximately equivalent to a zeolite content of 60 % (Figure 4.11 and Table 4.9).

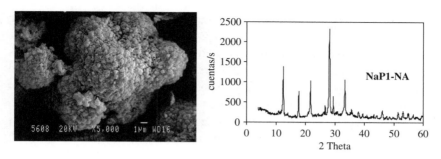

Figure 4.11 An SEM micro-photograph (left) and an XRD pattern (right) of NaP1–Na zeolitic material synthesised from Narcea fly ash at pilot plant scale.

Table 4.9 Cation exchange capacity (meq g^{-1}) of zeolitic material synthesised from fly ash (using the ISRIC method[22]). The mean CEC for Spanish fly ashes and natural commercial zeolitic products are reported for comparison.

Zeolitic product	CEC (meq g^{-1})
SILEX process	
A	5.3
A–X	4.7
X	4.3
Direct conversion	
Pilot plant NaP1	2.7
Herschelite	2.1
KM	1.9
Linde F	1.9
Analcime	0.6
Sodalite	0.3
Synthetic commercial products	
A	5.4
X	4.3
Natural commercial products	
Clinoptilolite	1.5–2.0
Fly ash	
Mean values	< 0.05

Synthesis Yield Most of the cited studies showed higher conversion efficiencies with NaOH solutions than KOH solutions at the same temperature. Although most of the zeolites reported in Table 4.6 may be obtained from a specific fly ash,[12] the reaction time needed to reach acceptable synthesis yields is inversely proportional to the aluminium–silicate glass content of the fly ash. Thus, for higher aluminium–silicate glass contents, shorter activation periods and lower solution : fly ash ratios are needed to reach high yields of zeolite. Using a low solution : fly ash ratio $(2 \, dm^3 \, kg^{-1})$ in consideration of possible industrial application, the time needed to obtain a high yield with a high glass fly ash is 6–8 h, but longer reaction times (24 h) are required to obtain similar yields from fly ashes containing high quantities of mullite and/or quartz and a low glass content.[13] The conversion efficiency also depends on the particle size distribution and on the content of the nonreactive phases, mainly hematite, magnetite and lime.

Synthesis of Pure Zeolites

Hollman et al.[23] developed a two-stage synthesis procedure consisting of an initial alkaline leaching/extraction step that was separated from the precipitation of the zeolite. The aim of this process was to produce zeolites of high purity with a high CEC, without residual fly ash components and free of leachable elements, such as As, Mo, Se and V. The EU-sponsored BRITE EURAM project SILEX (SILica EXtraction, contract BRPR-CT98-0801) significantly improved the SiO_2 extraction and, as a result, yielded a much higher material recovery in the form of a pure zeolite, with > 99 % purity.[16] In addition, the extraction residue turned out to be zeolitic material comparable in zeolite content to the direct conversion products described above. Thus this SILEX process not only has the advantage of producing pure zeolites without any residual fly ash, but also is able to produce zeolites with a large pore volume. By addition of Al, the $SiO_2 : Al_2O_3$ ratio in the synthesis solution can be varied so that species such as X and A zeolites become possible products. However, this process has only been tested at laboratory scale, and the predicted production costs seem much higher than for products obtained from the direct conversion process.[24]

Silica Extraction Extraction of silica from a fly ash is favoured by an alkaline environment[25] and the SiO_2 comes from the dominant aluminium–silicate glassy matrix, opaline SiO_2 and various crystalline phases such as mullite, quartz and feldspar. These SiO_2-bearing phases in the fly ash may be classified according to their potential solubility in alkalis as follows: glass = opaline silica > tridymite – cristobalite > quartz > feldspar > mullite.[26]

   In order to synthesise the desired zeolites, there are a number of limitations on the extracted solution, and thus for the extraction process itself:[24]

(a) The $NaO_2 : SiO_2$ ratio in the solution after combining with an Al-rich product has to be kept within the stability range of zeolites X and A. Therefore, if the $Na_2O : SiO_2$ molar ratio in the leachate is > 1.3, the extract cannot be used as a starting solution for the synthesis of high-CEC zeolites, but only for zeolites such as sodalite with a low CEC. For this reason, optimisation of the extraction process has to aim for high silica extraction yields in relatively low $NaO_2 : SiO_2$ leachates.

(b) During extraction, the immediate consumption of almost all of the liberated Al by precipitation of zeolites cannot be avoided, so that the solid residue from the extraction is converted into a zeolitised product. For such a product to be useful, it should have a relatively high CEC; thus the growth of NaP1 or KM zeolites has to be promoted in the residue. Consequently, high NaOH or KOH concentrations ($> 3\ mol\ dm^{-3}$) and high temperatures ($> 150\,°C$) have to be avoided.

Table 4.10 shows the optimal extraction conditions for selected Spanish, Dutch and Italian fly ashes, taking into account the above limitations.[24,27]

Table 4.10 Optimal extraction conditions determined for selected fly ashes (SILEX project).[24]

Fly ash	Extraction conditions				Extraction yield		SiO_2 fixed as zeolite $(g\ kg^{-1})$
	NaOH $(mol\ dm^{-3})$	$T\ (°C)$	L/S $(dm^{-3}\ kg^{-1})$	t (h)	SiO_2 $(g\ kg^{-1})$	Na_2O/SiO_2	
Alkaline	2	150	2	5	70	1.0	168
Meirama	2	150	2	4	127	0.7	199
Monfalcone	3	120	3	5	186	1.2	135
Neutral	3	120	3	10	166	1.2	245
Puertollano	3	120	3	9	405	0.6	111

Synthesis Yield As already mentioned, the synthetic process is based on the combination of the silica-rich extracts from fly ashes with alumina-rich solutions. The selected aluminium source was a waste solution from the Al-anodising industry,[24] but other Al-rich solutions might also be used. The best products were synthesised by combining SiO_2-rich extracts from the Meirama or Puertollano fly ashes (with 40 and $135\ g\ dm^{-3}$ of SiO_2, respectively) with an Al_2O_3-rich solution from a Dutch Al-anodising industry, containing $117 \, g \, dm^{-3}$ of Al_2O_3 and $86 \, g \, dm^{-3}$ of Na_2O.

Different zeolites may be obtained with this method by varying:

(a) the $Na : Si : Al : H_2O$ molar ratios of the starting solution;
(b) the gelling conditions – mainly temperature, time and stirring;
(c) the aging conditions;
(d) the crystallisation conditions – mainly temperature and time.

The synthesis conditions for zeolites X and A were optimised and the following products obtained using the two-step process:[16]

(1) Zeolite X with a purity of > 99.9 % and a CEC of $4.3 \, meq \, g^{-1}$ was synthesised (Figure 4.12) by mixing the two solutions to produce the following molar ratios: $Na_2O : SiO_2 = 1.2$, $SiO_2 : Al_2O_3 = 1.9$, $H_2O : Na_2O = 41.0$. Gelling occurred with stirring for 15 min at 60 °C, and crystallisation carried out at 90 °C for 7 h.

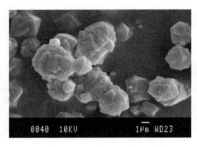 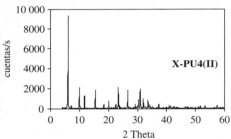

Figure 4.12 An SEM micro-photograph (left) and an XRD pattern (right) of X-PU4(II) zeolitic material synthesised from silica extracts from Puertollano fly ash and an added high-Al waste water.

(2) Zeolite A with a purity of 98 % and a CEC of $5.3 \, meq \, g^{-1}$ was synthesised (Figure 4.13) as above with the following molar ratios: $Na_2O : SiO_2 = 1.3$, $SiO_2 : Al_2O_3 = 1.7$, $H_2O : Na_2O = 41.0$. Gelling again occurred with stirring for 15 min at 60 °C, and the product crystallised at 90 °C for 8 h.

The CECs of these pure products are similar to those of commercial products, and much higher than for those obtained by direct conversion (Table 4.9). DIN 38414-S4 leaching tests were applied to the final pure zeolites and confirmed that no detectable As, Mo, Se and V was leached,

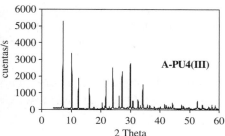

Figure 4.13 An SEM micro-photograph (left) and an XRD pattern (right) of A-PU4(III) zeolitic material synthesised from silica extracts from Puertollano fly ash and an added high-Al waste water.

unlike the products from direct conversion. Consequently, this suggests that these elements, although initially extracted together with the silica, remain in solution during the zeolite synthesis, as may be expected from the dominant anionic species present at the working pH.

The Suitability of Fly Ash for Zeolite Synthesis

During investigations carried out in the SILEX project, the suitability of ashes for either of the two proposed processes was studied based on characterisations and experiments with 23 different European fly ashes.[24,27] The bulk chemistry, the mineralogy and the chemistry of the glass fractions seem to be key factors for their behaviour during the zeolite synthesis. The mineralogical composition of the ashes was quantitatively determined by XRD using internal references[28] and the composition of the glass phase calculated by subtracting the composition of these crystalline phases.

From such studies, the following factors were shown to be important when considering the suitability of a given fly ash for direct conversion to zeolites:

(a) A high content of Al_2O_3 and SiO_2 favours conversion; that is, $Al_2O_3 + SiO_2 > 65 \%$, as determined from chemical analysis.
(b) The $SiO_2 : Al_2O_3$ ratio in the glass matrix should be in the range of 1.8–2.4.
(c) A glass content $> 80 \%$, and low concentrations of mullite ($< 10 \%$) and quartz ($< 10 \%$), are required for fast conversion; otherwise, long activation periods are needed for acceptable yields.

(d) Low concentrations of leachable metals such as As, Cr, Mo, Se and V are needed to ensure the quality of the conversion product, since these elements may be partially retained in the products.

On the other hand, the requirements for a fly ash from which at least $100\,g\,kg^{-1}$ of SiO_2 could be extracted were found to be as follows:

(a) Bulk SiO_2 content > 50 %.
(b) $SiO_2 : Al_2O_3$ ratio in the glass matrix > 2.0.
(c) Concentration of SiO_2 in the glass > 50 %. However, if the concentration is < 50 %, but opaline SiO_2 and/or cristobalite is present, these may compensate for the low SiO_2 in the glass. The occurrence of opaline silica and/or cristobalite indicates devitrification of the glass, and hence excess SiO_2 in the original glass, favourable for silica extraction.

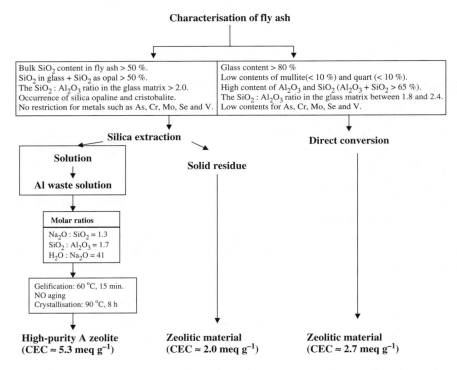

Figure 4.14 The suitability of a fly ash for silica extraction and/or zeolitisation by direct conversion. If the condition of one of the two options is not fulfilled, the ash is considered not suitable for zeolite synthesis.

(d) No technical restrictions for leachable elements such as As, Mo, Se and V, since these do not appear in the final products but remain in the process solutions. However, these elements may build up in solution and therefore restrict the possibility of recycling process water, so becoming an economic factor.

Figure 4.14 summarises the procedures to determine the suitability of a fly ash for either silica extraction or direct conversion.

4.4.3 Potential Applications of Zeolites Synthesised from Fly Ash

Simultaneously with the development of synthetic methods, intensive research was carried out on the potential application of the zeolites synthesised from fly ash. The high $Al(III)/Si(IV)$ ratio of these zeolites accounts for the high CEC of some compounds such as NaP1, A, X, KM, F, K-chabazite, Na-chabazite (herschelite) and faujasite. The high CEC, up to 5 meq g^{-1} for some pure zeolites, suggests a high potential for applications in waste-water treatment. In particular, the removal of heavy metals, radioisotopes and ammonium from solutions has been tested extensively.[7,10,12–17,21,29–38] The possibility of using these zeolitic materials as molecular sieves for flue gas purification technology has also been investigated.[32,39,40] Breck's review on *ion exchange reactions in zeolites* provides detailed information on general applications of zeolites.[41]

Some proposed environmental applications of zeolitic material from coal fly ash are discussed below.

Ion Exchange

Most of the studies on the use of zeolites derived from fly ash for waste-water treatment have been carried out using synthetic solutions under laboratory conditions. Such studies have shown that the synthesised zeolitic material possess CECs ranging from 0.3 to 5.3 meq g^{-1} (Table 4.9), and that in common with other ion exchangers there is competition between cations in solution to occupy exchangeable sites. Thus, high-Fe^{3+} or -Ca^{2+} solutions in urban and industrial waste waters may considerably reduce the uptake of ammonium.[42] High ammonium and heavy metal (Mn, Cd, Pb, Cu, Cr, Zn) removal efficiencies were also found by treating municipal landfill leachates with a 1:1 mixture of activated carbon and synthetic zeolitic material.[43]

Treatment of acid mine waters was investigated[21,37,44] using doses of synthetic zeolites from 5 to 40 g dm^{-3}, depending on the zeolitic species and the water matrix, mainly consisting of high concentrations of Ca^{2+}, Mg^{2+} or Fe^{3+}. The following tentative order for the affinity of different ions for NaP1 zeolite exchange sites was obtained:[44] $Fe^{3+} > Al^{3+} \geq Cu^{2+} \geq Pb^{2+} \geq Cd^{2+} = Tl^+ > Zn^{2+} > Mn^{2+} > Ca^{2+} = Sr^{2+} > Mg^{2+}$. These results demonstrate that NaP1 and A zeolites have a higher affinity for heavy metal ions than for Ca^{2+} and Mg^{2+}. Thus, solutions containing up to 600 mg dm^{-3} of heavy metals and up to 800 mg dm^{-3} of Ca could be treated to reduce the heavy metal content to < 0.5 mg dm^{-3}, with relatively high levels of Ca still remaining in solution. Since a pH rise (from 2.5 to 5) was induced by the zeolite addition, precipitation of metal-bearing solid phases enhanced the efficiency of decontamination. These studies concluded that acidic mine waters may be efficiently treated by a direct cation exchange treatment using low doses of zeolites (Figure 4.15). Such treatment may also be advantageous in cases such as extraction wells, where it can be used without producing solid waste precipitates in the well, as found following treatment with alkalis.

An important limitation to the application of direct conversion products for waste-water treatment is the possibility that hazardous leachable elements, such as Mo, As, Cr and V, in the residual fly ash particles may form part of the product. However, in most of the reported experiments, the content of these leachable elements in the treated solutions were low enough not to compete with the benefits of the cation uptake. If pure zeolitic material synthesised from high-Si fly ash extracts is used, this limitation does not exist.

Another potential use of synthetic zeolitic material is the immobilisation of heavy metals in contaminated soils. When phyto-remediation strategies are adopted for the final recovery of polluted soils, immobilisation of metals is necessary to avoid leaching of the metals and groundwater pollution. Synthetic zeolitic material, prepared as described above by direct conversion at pilot plant scale, was applied at different doses from 10 to 25 tonnes per hectare to experimental fields in Doñana area of southern Spain[37], polluted in March 1998 by a spill of pyrite mud containing high concentrations of heavy metals. Although land reclamation took place immediately after spillage, several restricted areas remained affected by residual pyrite mud. The applied zeolitic material was manually mixed with the soil and the soil was sampled at

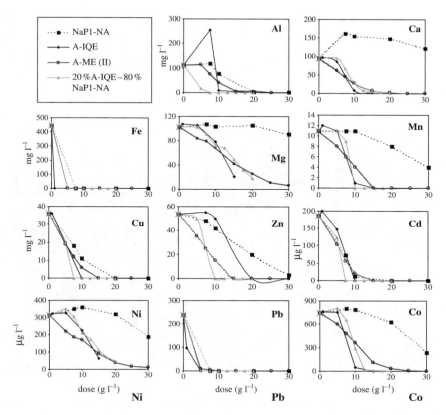

Figure 4.15 Plots of Al, Ca, Fe, Mg, Mn, Cu, Zn, Cd, Ni, Pb and Co concentrations versus zeolite dose applied in the decontamination tests of highly polluted Rio Tinto water using the following zeolitic products: (a) synthesised by direct conversion in a pilot plant (NaP1-NA); (b) a reference commercial zeolite (A-IQE); (c) synthesised by the SILEX process (A-ME(II)); and (d) a blend of (a) and (b) (20 % A-IQE+80 % NaP1-NA).

one and two years after the zeolite addition. The results showed that zeolite application considerably decreased the leaching of elements such as Cd, Co, Cu, Ni and Zn compared with a control site. The reduction in the leachable heavy metal concentration could be due either to the exchange capacity of the added NaP1 zeolite; to precipitation induced by the pH rise from 3 to 7; or to metal sorption on the surface of illite in the soil. Detailed investigations verified that the adsorption on illite did not occur in the reference soil, due to the acidic pH. However, it was shown that the pH rise caused by zeolite addition could promote adsorption on illite.[45] This study concluded that both ion exchange on the zeolitic material and adsorption on illite simultaneously occurred for zeolite doses > 15 tonnes per hectare. As a consequence of this treatment, the experimental fields with the added zeolitic product showed significant plant growth when compared with the reference field (Figure 4.16).

Figure 4.16 Experimental fields with and without the addition of the zeolitic product from the pilot experiment (direct conversion). Note that the treated fields show significant plant growth.

Similar studies have been performed to immobilise Cd in contaminated soils where drastic reduction of the leachable Cd content (from 88 to 1 %) was obtained by adding synthesised zeolitic products to polluted soils at a dose of 16 wt%.[36]

Molecular Sieves

Preliminary results on the use of fly ash zeolites as molecular sieves for flue gas treatment and separation and recovery of gases such as CO_2, SO_2 and NH_3 has been reported.[32,39,40] Adsorption capacities were determined using thermo-gravimetrical analysis or gas sorption analysis, with direct detection of desorbed gases.

Zeolite A, and in particular zeolite X, have larger channel diameters than the above gas molecules, at 4.1 and 7.3 Å, respectively, but zeolites

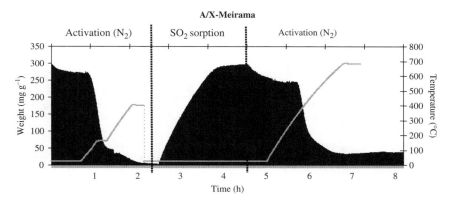

Figure 4.17 The thermo-gravimetric evolution of SO_2 sorption–desorption on a zeolitic product synthesised from the silica extracts from Meirama fly ash (mainly zeolite A). The black area indicates the variation of weight: (a) the loss of water and activation of zeolite (in the first N_2 region), following from (b) SO_2 sorption and (c) the SO_2 desorption and regeneration of zeolite (in the second N_2 activation stage). The grey line shows the variation of temperature.

such as sodalite and analcime have smaller channel diameters, around 2.3 Å, that do not allow the trapping of these gases. Other zeolites such as KM or NaP1 have a complex structure with two sizes of channel, one with a very small diameter and the other with a size close to that of zeolite A, which also makes trapping of these molecules difficult. Thus, the adsorption capacities of SO_2 and NH_3 measured for sodalite and analcime reach a maximum of only 6 mg g^{-1}.[32,40] It has also been shown that these sorption values increase a little, up to 20 mg g^{-1}, for KM and NaP1, but the highest sorption capacities for SO_2 were measured for Na-chabazite (herschelite) and zeolites A and X at 100, 300 and 380 mg g^{-1}, respectively (Figure 4.17), in agreement with the larger channels found in these zeolites. Therefore, zeolites A and X, together with Na-chabazite, are the most interesting zeolites for flue gas treatment. The first two can be obtained from the two-step silica extraction process from fly ash, but not from direct conversion, since their synthesis is strongly limited by temperature ($< 100\,^\circ$C).

It is important to note that in actual industrial applications the presence of water vapour in the flue gas may considerably reduce the gas uptake capacity of these zeolites. Consequently, the major potential applications of zeolitic material for gas treatment may be both the uptake of water vapour or SO_2 or NH_3 sorption from dry gaseous effluents.

References

1 Höller, H. and Wirsching, U., *Zeolites formation from fly ash*. Fortschr. Miner., **63**: 21–43 (1985).

2 Hemni, T., *Synthesis of hydroxy-sodalite ('zeolite') from waste coal ash*. Soil Sci. Plant Nutr., **33**: 517–521 (1987).

3 Mondragón, F., Rincon, F., Sierra, L., Escobar, J., Ramirez, X. and Fernandez, J., *New perspectives for coal ash utilization: synthesis of zeolitic materials*. Fuel, **69**: 263–266 (1990).

4 Shigemoto, N., Shirakami, S., Hirano, S. and Hayashi, H., *Preparation and characterisation of zeolites from coal fly ash*. Nippon Kagaku Kaishi, **5**: 484–492 (1992).

5 Shigemoto, N., Hayashi, H. and Miyaura, K., *Selective formation of Na-X, zeolite from coal fly ash by fusion with sodium hydroxide prior to hydrothermal reaction*. J. Materials Sci., **28**: 4781–4786 (1993).

6 Shigemoto, N., Sugiyama, S. and Hayashi, H., *Characterization of Na-X, Na-A, and coal fly ash zeolites and their amorphous precursors by IR, MAS NMR and XPS*. J. Materials Sci., **30**: 5777–5783 (1995).

7 Kolousek, D., Seidl, V., Prochazkova, E., Obsasnikova, J., Kubelkova, L. and Svetlik, I., *Ecological utilization of power-plant fly ashes by their alteration to phillipsite: Hydrothermal alteration, application*. Acta Univ. Carol. Geol., **37**: 167–78 (1993).

8 Shin, B.-S., Lee, S.-O. and Kook, N.-P., *Preparation of zeolitic adsorbent from waste coal fly ash*. Korean J. Chem. Eng., **12**: 352–356 (1995).

9 Chang, H.L. and Shih, W.H., *Conversion of fly ash to zeolites for waste treatment*. Ceram. Trans., **61**: 81–88 (1995).

10 Park, M. and Choi, J., *Synthesis of phillipsite from fly ash*. Clay Sci., **9**: 219–229 (1995).

11 Querol, X., Plana, F., Alastuey, A. and López-Soler, A., *Synthesis of Na-zeolites from fly ash*. Fuel, **76**: 793–799 (1997).

12 Querol, X., Umaña, J.C., Plana, F., Alastuey, A., López-Soler, A., Medinaceli, A., Valero, A., Domingo, M.J. and Garcia-Rojo, E., *Synthesis of Na zeolites from fly ash in a pilot plant scale. Examples of potential environmental applications*. Fuel, **80**: 857–865 (2001).

13 Querol, X., Moreno, N., Alastuey, A., Juan, R., Andrés, J.M., López-Soler, A., Ayora, C., Medinaceli, A. and Valero, A., *Synthesis of high ion exchange zeolites from coal fly ash*. Geologica Acta, 5(1): 47–55 2007.

14 Singer, A. and Berkgaut, V., *Cation exchange properties of hydrothermally treated coal fly ash*. Environ. Sci. Technol., **29**: 1748–1753 (1995).

15 Amrhein, Ch., Haghnia, G.H., Kim, T.S., Mosher, P.A., Gagajena, R.C., Amanios, T. and de la Torre, L., *Synthesis and properties of zeolites from coal fly ash*. Environ. Sci. Technol., **30**: 735–742 (1996).

16 Moreno, N., *Valorización de cenizas volantes para la síntesis de zeolitas mediante extracción de sílices y conversion directa. Aplicaciones ambientales*. Ph.D. thesis dissertation, Universitat Politècnica de Catalunya (2002).

17 Rayalu, S., Meshram, S.U. and Hasan, M.Z. *Highly crystalline faujasitic zeolites from fly ash*. J. Hazard. Mater. 77: 123–131 (2000).

18 Querol, X., Alastuey, A., López-Soler, A., Plana, F., Andrés, J.M., Juan, R., Ferrer, P. and Ruiz, C.R., *A fast method for recycling fly ash: Microwave-assisted zeolite synthesis*. Environ. Sci. Technol., **31**: 2527–2533 (1997).

19 Park, M., Choi, C.L., Lim, W.T., Kim, M.C., Choi, J. and Heo, N.H., *Molten-salt method for the synthesis of zeolitic materials. I. Zeolite formation in alkaline molten-salt system*, Microporous and Mesoporous Materials, **37**: 81–89 (2000).

20 Park, M., Choi, C.L., Lim, W.T., Kim, M.C., Choi, J. and Heo, N.H., *Molten-salt method for the synthesis of zeolitic materials. II. Characterization of zeolitic materials*, Microporous and Mesoporous Materials, **37**: 91–98 (2000).

21 Moreno, N., Querol, X. and Ayora, C., *Utilisation of zeolites synthesized from coal fly ash for the purification of acid mine waters*. Environ. Sci. Technol., **35**: 3526–3534 (2001).

22 International Soil Reference and Information Centre, *Procedures for soil analysis. Technical paper 9*. ISRIC, FAO-UN pp. 9.1–9.13 (1995).

23 Hollman, G.G., Steenbruggen, G. and Janssen-Jurkovičová, M., *A two-step process for the synthesis of zeolites from coal fly ash*. Fuel, **78**: 1225–1230 (1999).

24 Querol, X., Moreno, N., Andres, J.M., Janssen, M., Towler, M., Stanton, K., Nugteren, H.W. and Cioffi, F., *SILEX, Recovery of major elements from coal fly ashes*. Final report, BRITE–EURAM Program BRPR-CT98-0801, Commission of the European Communities (2002).

25 Mason, B., *Principles of Geochemistry*. John Wiley & Sons, Inc.: New York (1952).

26 Rimstidt, J.D. and Barnes, H.L., *The kinetics of silica–water reactions*. Geochim. Comochim. Acta, **44**: 1683–1699 (1980).

27 Moreno, N., Querol, X., Andrés, J.M., López-Soler, A., Janssen-Jurkovičová, M., Nugteren, H., Towler, M. and Stanton, K., *Determining suitability of a fly ash for silica extraction and zeolitisation*. J. Chem. Tech. Biotechnol., **79**: 1009–1018 (2004).

28 Klug, P.H. and Alexander, E.L., *X-ray diffraction procedures: for polycrystalline and amorphous materials,* John Wiley & Sons, Inc.: New York (1974).

29 Catalfamo, P., Corigliano, F., Patrizia, P. and Di Pasquale, S., *Study of the pre-crystallisation stage of hydrothermally treated amorphous aluminosilicates through the composition of the aqueous phase*. J. Chem. Soc., **89**: 171–175 (1993).

30 Berkgaut, V. and Singer, A., *High capacity cation exchanger by hydrothermal zeolitization of coal fly ash*. Applied Clay Science, **10**: 369–378 (1996).

31 Lin, C.F. and His, H.C., *Resource recovery of waste fly ash: synthesis of zeolite-like materials*. Environ. Sci. Technol., **29**: 1109–1117 (1995).

32 Querol, X., Moreno, N., Umaña, J.C., Juan, R., Hernández, S., Fernández, C., Ayora, C., Janssen, M., García, J., Linares, A. and Cazorla, D., *Application of zeolitic material synthesised from fly ash to the decontamination of waste water and flue gas*. Journal of Chemical Technology and Biotechnology, **77**: 292–298 (2002).

33 Suyama, Y., Katayama, K. and Meguro, M., *NH_4^+-adsorption characteristics of zeolite synthesized from fly ash*. Nippon Kagaku Kaishi, **2**: 136–140 (1996).

34 Patane, G., Di Pascuale, S. and Corigliano, F., *Use of zeolitized waste materials in the removal of copper (II) and zinc (II) from wastewater*. Ann. Chim., **86**: 87–98 (1996).

35 Patane, G., Mavillia, L. and Corigliano, F., *Chromium removal from wastewater by zeolitized waste materials*. Mater. Eng., **7**: 509–519 (1996).

36 Lin, C.F., Lo, S.S., Lin, H.-Y. and Lee, Y.J., *Stabilization of cadmium contaminated soils using synthesized zeolite*. Hazard. Mater., **60**: 217–226 (1998).

37 Querol, X., Alastuey, A., Moreno, N., Álvarez-Ayuso, E., García-Sánchez, A., Cama, J., Ayora, C. and Simón, M., *Immobilization of heavy metals in polluted soils by*

the addition of zeolitic material synthesized from coal fly ash. Chemosphere, **62**: 171–180 (2006).

38 Mimura, H., Yokota, K., Akiba, K. and Onodera, Y., *Alkali hydrothermal synthesis of zeolites from coal fly ash and their uptake properties of cesium ion.* J. Nucl. Sci. Technol., **38**: 766–772 (2001).

39 Querol, X., Plana, F., Umaña, J., Alastuey, A., Andrés, J.M., Juan, R. and López-Soler, A., *Industrial applications of coal combustion wastes: zeolite synthesis and ceramic utilisation.* European Coal and Steel Community Contract 7220/ED/079. Final report, (1999).

40 Srinivasan, A. and Grutzeck, M.W., *The adsorption of SO_2 by zeolites synthesized from fly ash.* Environ. Sci. Technol., **33**: 1464–1469 (1999).

41 Breck, D.W., *Ion exchange reactions in zeolites.* In *Zeolite Molecular Sieves, Structure, Chemistry, and Use* (Chapter 7). Robert E. Krieger: Malabar, Fla, TIC:245213 (1984).

42 Juan, R., Hernández, S., Andrés, J.M., Querol, X. and Moreno, N., *Zeolites synthesised from fly ash: use as cationic exchangers.* Journal of Chemical Technology and Biotechnology, **77**: 299–304 (2002).

43 Jeong, H.-L., Dong, S.-K., Sung, O.-L. and Bang, S.-S., *Treatment of municipal landfill leachates using artificial zeolites* (in Korean). Chawon Risaikring, **5**: 34–41 (1996).

44 Moreno, N., Querol, X., Ayora, C., Alastuey, A., Fernández-Pereira, C., and Janssen, M., *Potential environmental applications of pure zeolitic material synthesised from fly ash.* J. Environ. Eng., **127**: 994–1002 (2001).

45 Cama, J., Ayora, C., Querol, X. and Moreno, N., *Metal adsorption on illite from a pyrite contaminated soil.* J. Environ. Eng., **131**: 1052–1056 (2005).

4.5   Reinforcing Materials: Fibres Containing Fly Ash

Flavio Cioffi

Contento Trade srl, Terenzano (UD), Italy

Types of combustion residues involved

- Municipal solid waste incinerator fly ash (MSWA)
- Coal fly ash (CFA)

For both these ashes no strict specifications are given or required. However, the batches should be as homogeneous as possible, and the overall chemical compositions of the two ashes should be known for proper formulation of the final material mixtures. Low sulfate content is considered an advantage.

State of development
The presented research has been carried out at a laboratory scale using a maximum sample size of 1 kg. Pilot plant testing is planned and the believed time to markets is in the order of 3–5 years.

4.5.1   Introduction

Glass fibres are extensively used for the production of reinforced plastic components, bituminous sheaths, technical textiles and nonwoven materials. Such fibres are classified on the basis of their physical appearance – that is, continuous or discontinuous – and on the type of glass from which they were produced. Continuous fibres are produced by forcing the molten feed material through spinnerets with regular holes, while discontinuous fibres are produced when the melted mass is thrown on a rotating disc at high speed. Most conventional reinforcing fibres are at present produced by spinning E glass, a high-quality boron-containing glass produced from rather expensive natural raw materials, at high temperatures.

The commercial fibres are usually used as reinforcement material in composites in a polymeric matrix (E glass fibre), bituminous matrix (C glass fibre) or cement matrix (AR, Alkali Resistant fibre). Continuous fibres have a defined diameter and homogeneous mechanical properties, and are used in high added value applications – for example, fibro-reinforced plastic materials – whereas discontinuous fibres have variable diameter and variable mechanical properties and are used in the insulating sector.

Substitution of such natural raw materials by secondary raw materials is of interest economically in decreasing the cost as well as environmentally in the conservation of natural resources. The possibility of converting a mixture of different combustion residues and other industrial residues into a glass from which continuous glass fibres can be produced has been explored by a consortium working in an EC-funded project (Life 98 ENV IT 00132 'WBRM'), where WBRM is an acronym for 'Waste Based Reinforcing Materials' and the resulting production process is further referred to as the WBRM process. Combustion residues included in this process are municipal solid waste incinerator fly ash (MSWA) and coal fly ash (CFA).

The study presented here shows that with this process it is possible to produce continuous fibres with chemical, physical and mechanical properties similar to those of conventional E glass fibres. Moreover, there is an overall reduction in energy consumption and only a minimal amount of waste is generated.

4.5.2 Raw Materials

The bulk of raw materials used for the production of the WBRM fibres consist of industrial residues such as fly ash, used foundry sand and scrap glass. The amount of virgin raw materials such as lime, alkali oxides, boric acid and other additives has been limited to less than 17 wt %. Formulation of an appropriate mixture of these materials to obtain a spinnable glass similar in composition to a commercial E glass was one of the main tasks of the project. Satisfying results have been obtained when using mixtures containing about 50 % MSWA and 10 % CFA (by weight). A typical WBRM mixture is given in Table 4.11, along with the

Table 4.11 A comparison of raw materials used for the production of E glass fibres using conventional technology and the WRBM process.

Conventional	Wt%	WBRM fibres	Wt%
Sand	31.1	MSWI ash	50
Fired china clay	29.6	Foundry sand A	27
Lime	14.0	Coal fly ash	10
Dolomite	9.4	Lime	11
Colemanite	15.6	Boric acid	2
Fluorspar	0.37		

ingredients of conventional E glass fibres. Knowing the compositions of the MSWA and CFA, optimal formulations can be obtained by slightly adjusting the ratios of the different materials.

4.5.3   WBRM Production Process

The process consists basically of three steps: pre-treatment of the MSWI fly ash by removing soluble salts; producing a homogenised mixture of the components; and converting this into a glass that is finally spun to obtain fibres. A simplified flow sheet of the WBRM process is given in Figure 4.18.

MSWI Ash Pre-treatment

The high salt content in most MSWI ashes (20–40 % in some samples) requires that a preliminary washing step is carried out, as the main anions in these salts, chlorides, sulfates and phosphates are not suitable for incorporation into the glass.

The ash-washing process is designed to eliminate the salts almost completely and at the same time produce a salt concentrate that can be recycled (see Figure 4.18). This washing is performed in a three stage counter-current process with water, using a liquid : solid ratio of 0.8. After the last vacuum filtration, the washed ash is dried in a drum dryer. High metal concentrations are removed from the concentrated salt solution by precipitation with specific organic complexing agents.

A mixture of salts suitable for use in de-icing roads can be obtained from this purified brine by evaporation under controlled conditions. The high-metal-containing slurry from the brine purification is the only waste product from this pre-treatment operation. Technically, this slurry may be mixed with the washed ashes before drying, as it has no major effect on the quality of the final products as long as the sulfate content is below 30 %. However, whether this is an environmentally sound solution depends on the application of the final products. If not, this fraction, which is always less than 1 % of the treated ashes, should be treated as a hazardous waste.

Glass Production

The dry washed MSWI fly ash is mixed in batches with coal fly ash, foundry sand and the other ingredients. The highly variable composition

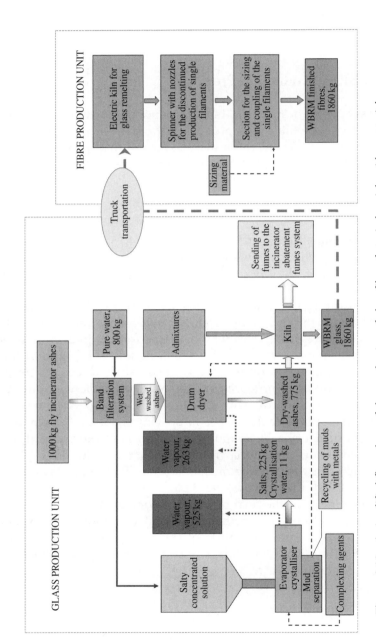

Figure 4.18 A simplified flow sheet of the production of glass fibres from industrial residue materials.

of the MSWI ash requires that the appropriate mixing ratios must be calculated for each batch from the chemical analyses of the components. The mixtures are then homogenised and melting takes place in a kiln at about 1350–1400 °C. Near the end of the melting process, the temperature is raised to 1450–1550 °C in a so-called 'conditioning chamber' to force gaseous inclusions out of the molten mass and increase its homogeneity. The melt is subsequently quenched to form a glass, as is shown in Figure 4.19.

The fumes generated from the complete combustion of the organic fraction in the fly ashes and volatilisation of the remaining heavy metals require subsequent gas cleaning treatment. If the melting process takes place at the incinerator plant, these fumes can be sent to the incinerator gas cleaning installation, where it will only constitute a small additional fraction of the gases to be cleaned. Thus the location of the melting process in or near to the incinerator plant will reduce the environmental impact and overall costs by avoiding transportation and handling of hazardous waste.

The melting of WBRM mixtures is faster and takes place at a lower temperature than the melting of mixtures for the production of conventional E glass. This results in a significant reduction in energy consumption.

Figure 4.19   The intermediate glass product (right) and the final WBRM fibre (left).

Fibre Production

Fibre production takes place in a similar way to that typically used in plants for discontinuous production of glass fibres. The glass is re-melted in an electric kiln and a multi-nozzle spinner draws the melted glass to spin the fibres at a temperature of 1200–1330 °C and 2400 rpm. The production of the fibres is controlled to ensure that nearly uniform diameter fibres are produced. Furthermore, the fibre diameter is maintained above 8 μm to make sure that no inhalable and dangerous particles are produced. Fibres may be further treated by coating or finished by coupling or cutting operations depending on their application. Fibres produced in this way are shown in Figure 4.19.

4.5.4 Products Obtained

The physical and mechanical properties of the fibres obtained from the WBRM process vary depending on the processing conditions. However, these properties are very similar to those for conventional E glass fibres, as shown below. Moreover, WBRM fibres exhibit better resistance to alkali attack when compared with conventional E glass fibres, although they are not as stable as the zirconium oxide based fibres. The process conditions guarantee complete thermal destruction of organic pollutants and immobilisation of any remaining heavy metals. Availability tests according to the Dutch standard NEN 7341 confirm that WBRM fibres are excellent from the environmental point of view.

In addition to the fibres, the WBRM process also produces a mixture of purified salts – for example, calcium, sodium and potassium chlorides – from the pre-washing process. This material, which may make up 20–40 % of the treated ash, is free of heavy metals, and therefore forms an excellent alternative for the calcium and sodium chlorides used for de-icing salts on roads.

4.5.5 The Market for Fibres

In Europe, about 700 000 tonnes of glassy reinforcement materials, excluding alkali-resistant fibres, were produced in 2000, which represents 28 % of the world production. E glass fibres represent 96–98 % of the world market for reinforcing fibres, and from 1994 to 2000 this market showed an average annual growth of 6.5 % in Europe (5.7 %

world wide). In 1994, the size of the European glass reinforcement market was about 400 000 tonnes, with a production of 1 300 000 tonnes of composite materials comprising 1 200 000 tonnes of plastic materials and 100 000 tonnes of textiles.

Four possible applications for WBRM fibres have so far been tested and these are described below:

- nonwoven fabrics;
- reinforced plastics;
- reinforced bitumen products; and
- reinforced cement products.

Nonwoven Fabrics

The nonwoven fabric sector uses large quantities of glass fibres to create composite materials, thermal insulating materials, industrial mufflers, filters and so on. Coupled glass fibres 60 mm long are usually applied in the production of nonwoven fabrics. The manufacturing process of nonwoven fabrics includes the following steps:

(1) opening the fibres, to avoid agglomerates and packets;
(2) carding and cross-lapping – typical textile processing operations necessary to orient and cross the fibres in layers;
(3) needling of the web to link the different fibre layers to increase cohesion.

Table 4.12 shows a comparison of laboratory measurements of the properties of nonwoven fabrics with conventional E glass fibres and WBRM

Table 4.12 A comparison of the properties of nonwoven fabrics from conventional and WBRM fibres.

Sample	Thickness, e (mm)	Mass $(g\,m^{-2})$	Thermal resistance, R $(m^2\,K\,W^{-1})$[a]	R/e[a]
1 Conventional E glass fibres	5.9	230	0.173	0.029
2 Conventional E glass fibres	10.6	460	0.286	0.027
3 WBRM glass fibres	5.8	230	0.151	0.026

[a] Measured following standard NF EN 31092.

fibres. These results show that, for the same thickness, the thermal properties of the WBRM fibres are similar to those for conventional E glass fibres.

Reinforced Plastics

For this product, needled WBRM felts were impregnated with thermosetting resins, using a high-pressure moulding technique widely used in the industry. Samples (15×15 cm) of various thicknesses have been produced, both with WBRM and E glass felts, under the following processing conditions:

- resin – unsaturated orthophthalic polyester resin in solution with styrene;
- catalyst – methylethylketone peroxide;
- curing conditions – 2 h at 70 °C.

The plastic samples produced were analysed to determine their optimal thickness, the reinforcement rate and their mechanical properties (Table 4.13). By comparing different samples, it can be seen that the reinforcement rate of conventional E glass is slightly lower than that of WBRM products and this means that the WBRM fibres can be satisfactorily impregnated by thermosetting resins.

Table 4.13 A comparison of properties of reinforced-plastic with E glass and WRBM fibres.

Sample	Thickness (mm)	Fibre reinforcement weight (g)	Sample weight (g)	Percentage reinforcement (%)
WBRM 1	1.3	22.0	49.6	44.3
WBRM 2	2.4	41.9	89.4	46.9
WBRM 3	3.3	51.3	119.8	42.8
WBRM 4	3.0	49.0	100.4	48.8
E 1 GLASS	3.7	47.1	112.1	42.0
E 2 GLASS	2.9	40.2	101.9	39.4

Flexural experiments according to the EN 63 standard for glass-reinforced plastics were performed and the results obtained reported in Table 4.14. It is clear that the mechanical properties of the WBRM fibro-reinforced plastics are very similar to those obtained with conventional

Table 4.14 Flexural properties of reinforced plastics.

Sample	Stress (MPa)	Bending (mm)	Modulus (MPa)
Un-reinforced polyester resin	95		4100
Resin + E glass fibres	170	3.0	6200
Resin + WBRM fibres	160	3.1	6600

E glass fibres. This confirms that the properties of WBRM fibres are very attractive and easily used in this product.

Reinforced Bitumen Products

Another interesting application for WBRM fibres is in bituminous conglomerate reinforcement for road applications. The recent introduction into the market of high-resistance modified bitumen for the production of porous wear layers stimulated research into high-performance fibro-reinforced composite materials. The fibres conventionally used in this field are mainly cellulosic, and a comparison of the effects of the loading of cellolosic and WBRM fibres in bituminous conglomerate on the Marshall stability is reported in Table 4.15 (the Marshall stability tests are used to determine the performance of asphalt under loads).

The quality : price ratio in this sector is crucial and therefore WBRM fibres compare very favourably for this application. WBRM fibres seem to have a high potential for application in bituminous conglomerates when applied with high compaction, although further study is required to understand their effects.

Table 4.15 The Marshall stability of fibre-reinforced bitumen specimens (the Marshall stability tests are used to determine the performance of asphalt under load).

Kind of fibre	Quantity of fibre (%)	Medium compaction (kN)	Heavy compaction (kN)
Cellulose	0.0	7.50	9.93
Cellulose	0.3	6.55	9.20
Cellulose	0.5	5.09	9.81
Cellulose	1.0	3.64	9.50
WBRM	0.0	7.50	9.93
WBRM	0.2	5.83	11.64
WBRM	0.3	6.29	11.97
WBRM	0.5	3.81	12.88
WBRM	1.0	3.68	3.53

Reinforced Cement Products

The tensile behaviour, and hence the deformation properties, of ordinary concrete is limited, but may be modified by adding discontinuous fibres – for example, steel, glass or plastic fibres – to the cement matrix during mixing. The fibres act as bridges between fissure borders, increasing the tenacity of the material. Varying the length of the fibres allows adherence to the matrix to be exploited, limiting the aperture of the fissures and promoting a residual post-fissure resistance. The fibres show great efficacy in the control of hydrothermal shrinkage and its mechanical behaviour.

Fibro-reinforced concrete is widely used in the USA, where 10 % of the ready-mix concrete is fibro-reinforced. There are many possible applications of this product as, for example: flags and pavements; refractory concrete; sprayed concrete (shotcrete) used for provisional or definitive finishing of tunnels, slopes, mines, pools, channels, bridges, pits and so on; and concrete pipes, joints, beams, guard rails, electric cabins, cisterns and so on.

Glass fibres used in this sector can be divided into two groups: alkali-resistant fibres for structural use and 'crack-stop' fibres, used to avoid cracking caused by shrinkage. The latter generally have a low chemical resistance and are normally conventional E glass fibres. The production of alkali-resistant fibres for structural use in Europe is about 2500 tonnes per year, but the amount of 'crack-stop' fibres used in Europe is not known.

WBRM fibres were initially tested in fibre-reinforced concrete (GRC), a high-quality application of alkali-resistant fibres such as the zirconia-rich CEM-FIL fibres. The fibre-reinforced cementitious product specimens were made using the premix process in accordance with UNI EN 1169. This process involves the blending of the slurry and chopped strands of fibre in a mixer prior to casting. To produce a premix of the correct quality, it is necessary to mix in two steps. After casting, the material is compacted by vibration. Composition of fibre-reinforced cement mixes are given in Table 4.16.

Experimental Results   The following evaluations can be made from the experimental results:

- using both 'APS' and 'OS' WBRM fibres, there is a significant loss of workability on fresh mortar;
- WBRM fibres do not disperse inside the cementitious matrix;

Table 4.16 The composition of fibre-reinforced cement mixes for measuring bending strength.

Mixes	1	2	3	4	5[a]	6[b]
Number of specimens	8	8	8	8	4	4
White PC 54.5 (g)	2500	2500	2500	2500	1250	1250
Sand (0.3–0.9 mm) (g)	1650	1650	1650	1650	825	825
Water (g)	940	940	740	740	470	370
Super-plasticiser[c] (g)	98	98	98	98	—	16
W/C ratio	0.40	0.40	0.32	0.32	0.38	0.31
Polymer admixture (g)	5	5	5	5	—	2.5
WBRM APS (g)	180	—	—	—	—	—
WBRM OS (g)	—	180	—	—	—	—
CEM-FIL 60/2 (g)	—	—	180	—	—	—
CEM-FIL 62/2 (g)	—	—	—	180	—	—

[a] Reference plain mix for the WBRM fibres.
[b] Reference plain mix for the CEM-FIL fibres.
[c] 30 % of the super-plasticiser was the dry compound.

- it was impossible to release the panel from the mould after curing for one day, probably due to a retardant effect caused by WBRM fibres;
- using both '60–2' and '62–2' types of CEM-FIL commercial fibres, good workability on fresh mortar has been verified using a lower water content;
- AR fibres disperse themselves inside the cementitious matrix quite well during the premixing phase, resulting in an homogeneous panel surface;
- It was possible to release the panel from the mould after one day's curing.

Technical Characterisation of Fibre-reinforced Cement Products The following tests were carried out on specimens from the fibre-reinforced cement samples:

- bending tests after 28 days aging;
- load–deflection curves to study ductility.

The results obtained are presented in Figure 4.20. Samples labelled 'O' were parallel to the long edge of the panel and those labelled 'V' were taken across the panel. The results obtained show that there is only a little improvement in bending strength in products using WBRM fibres instead of plain concrete, and that properties of panels made

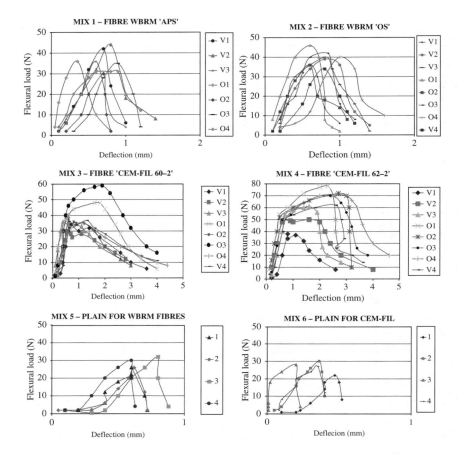

Figure 4.20   Load–deflection tests for samples of fibre-reinforced cement.

with standard CEN-FIL alkali-resistant fibres are considerably higher, especially with increasing deflection capacity.

In spite of these results, the properties of WBRM fibres could be exploited in lower-quality concrete applications, where a little improvement in tensile strength of the conglomerate can be really useful; that is, as a crack stopper.

4.5.6   Cost Estimations

On the basis of the production on a small scale (about $1 \, kg \, hr^{-1}$), preliminary cost estimations have been carried out for an industrial plant producing 12 000 tonnes per year of WBRM fibres. The amounts of raw materials required for this production, as well as some key figures for streams to be handled (compare with the flow sheet in Figure 4.18), are given in Table 4.16. Assuming an investment of €4.5 million for the construction of a WBRM plant to be depreciated in five years (working 24 hours a day for 300 days a year and a maintenance incidence of 2 %), and taking current costs for energy, water and raw materials, a cost estimation as presented in Table 4.17 results. For this purpose, the cost of the industrial residue raw materials was taken as an arbitrary €3 per tonne.

Table 4.17 The production cost for WBRM fibres.

Summary of costs	Thousands of euros per year	Euros per tonne of fibres
Depreciation	898	74.83
Raw materials	941	78.40
Electric power	1 925	160.45
Methane	438	36.51
Manpower	375	31.25
Maintenance	90	7.48
Total	4 667	388.92

For comparison, the present market price for E-glass fibres is between €2000 and €2500 per tonne, so the production cost of these fibres must be considerably higher than for WBRM fibres. The raw material costs for E glass production already reach €160 per tonne of E glass. Thus, WBRM fibres will be serious competitors in this market, as raw materials are much less expensive and additional important savings are made during the production stage. Moreover, the simultaneous production of de-icing salts adds to the revenues of the process.

4.5.7 Future Outlook

The results presented here have been obtained from tests carried out in a small pilot installation with a production capacity of about 1 kg of fibres per hour. However, this allowed comparative tests to be carried out on potential applications and evaluation of engineering design aspects for scaling up as well as preliminary cost calculations. The next step will be scaling up to a larger pilot plant production unit with a capacity of

2000 kg per day of fibres. This will allow extensive application testing of WBRM fibres at an industrial level under real conditions. Provided that positive results are obtained, the products may be on the market within a few years.

4.5.8 Conclusions

To summarise it can be concluded that by using mainly industrial residue raw materials treated with the WBRM process, it is possible to produce glass fibres with physical, mechanical and chemical properties similar to those of conventional E glass fibres. The reduction in the use of primary raw materials as well as energy savings during production makes the WBRM fibres competitive with the traditional fibres. WBRM fibres are sustainable products because of the use and immobilisation of industrial residues, the saving of primary raw materials and energy and the fact that virtually no waste is generated. Furthermore, the waste that is produced during pre-treatment of the MSWI ash provides mineral salts that can be used for road de-icing.

Bibliography

BRITE EURAM Recbuild Project Proposal n. 5536, *Recbuild: Advanced Building Materials Obtained by Recycling Toxic Wastes and Fly Ashes*. Final report, April 1992.

Contento, M.P. and Cioffi, F., *Procedimento per la produzione di materiale vetroso ad elevata resistenza agli alcali, ottenuto a partire da miscele di reflui industriali*. Italian Patent M192/A IT MI 92 A 002115.

Environment 'ASHREC', 'Ash Recycling Technologies'. Contract EVV-CT92-0196, Final report, December 1991.

Hreglich, S., Mancini, A. and Mancini, R., *Possibilità dell'uso delle frazioni fini agglomerate di sabbia nella produzione del vetro: una ricerca preliminare su scala di laboratorio*. Riv. Staz. Sper. Vetro, 15(5): 219–221 (1985).

Hreglich, S., Scandellari, M. and Verità, M., *Impiego delle scorie d'altroforno come materia prima nella produzione del vetro*. Riv. Staz. Sper. Vetro, 9(5): 205–214 (1979).

LIFE 98 ENV IT 00132 *WBRM: Waste Based Reinforcing Materials*. Final report, December 2001.

Locardi B and Zambon A, *Possibilità di impiego di idrossido di calcio di recupero nelle miscele vitrificabili*. Riv. Staz. Sper. Vetro 9(5): 193–198 (1979).

Locardi, B. and Hreglich, S., *Caratterizzazione chimico fisica delle scorie calciosilicatiche e loro prospettive di appliccazione nel settore ceramico*. Ceramica Inform., 221: 474–475 (1984).

Locardi, B., Barbon, F., Zambon, A. and Sorarù, G., *Indagini preliminari per lo sviluppo in laboratorio di fibre di tipo vetroceramico*. Riv. Staz. Sper. Vetro, **12**(3): 101–108, (1982).

Scarinci, G., Festa, D., Maddalena, A., Locardi, B. and Meriani, S., *Resistenza all'attacco alcalino di alcuni tipi di fibre vetrose*. Riv. Staz. Sper. Vetro, **14**(1): 5–10, (1984).

Scarinci, G., Festa, D., Sorarù, G.D., Locardi, B., Guadagnino, E. and Meriani, S., *Alkali resistance of some Zn–Al silicate glass fibres modified with Fe and Mn or Ti oxides*. J. Non-Crystall. Solids, **80**: 351–359, (1986).

Scarinci, G., Festa, D., Sorarù, G.D., Locardo, B., Guadagnino, E. and Meriani, S., *Resistenza alcalina di alcune fibre vetrose Zn–Al-silicatiche modificate con ossidi di Fe e Mn o Ti*. Riv. Staz. Sper. Vetro, **15**(4): 169–173, (1985).

4.6 Glass Polyalkenoate Cements

Ann Sullivan and Robert Hill
University of Limerick, Ireland; Imperial College, London, UK

Types of combustion residues involved
Coal gasifier slag has to be a calcium aluminosilicate glass. For optimum reaction and thus the best product, the Si : Al ratio should be close to unity and the Ca : Al ratio must be greater than or equal to 1 : 2. The silica mole fraction should ideally be below 0.5.

State of development
The presented research has been carried out at a laboratory scale, using a maximum sample size of approximately 1–10 g. The test specimens used were in accordance with BS 6039:1981 *Specifications for Dental Glass Polyalkenoate Cements*. Due to the differences in chemistry and physics between the gasifier slag based polyalkenoate cements and current ordinary Portland cement based systems, new standards to evaluate their properties for pilot plant studies would have to be developed for the intended building application.

4.6.1 Gasifier Slag

Gasifier slag, a by-product produced during the manufacture of gas from coal, forms a calcium aluminosilicate glass when cooled.[1] The inorganic residue from the gasification of coal forms as molten slag, which is tapped from the bottom of the gasifier into a quench chamber where it is rapidly cooled by water to form glassy granules. The viscosity of the liquid slag may be controlled by the addition of fluxes such as limestone. The physical form and chemical composition of the slag depends on processing conditions in the gasifier.[2] Slag composition also depends on the type and source of coal used and also the type of gasifier in operation (Chapter 1.8).

Gasifier slag consists of three major components: silica and alumina from the minerals in coal, and calcium oxide, mainly added as a flux in the gasification process. The range of typical gasifier slag compositions can be represented by a ternary phase diagram of the three main oxides: CaO, Al_2O_3 and SiO_2 (Figure 4.21). The relationship of gasifier slag to other by-products – namely, granulated blast furnace slag (GBFS) and pulverised fuel ash (PFA) – is also included in this figure.

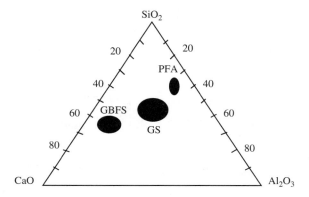

Figure 4.21 The ternary system $CaO–Al_2O_3–SiO_2$ (GS, typical gasifier slag compositions; GBFS, typical granulated blast furnace slag compositions; PFA, typical fly ash compositions).

There are currently two commercial gasifiers in Europe. One is operated by DEMKOLEC in the Netherlands; the other by ELCOGAS in Spain.[3-6] The DEMKOLEC gasifier produces 60 000 tonnes of slag per year, plus a further 20 000 tonnes of a fly ash with the same chemical composition. It is estimated that millions of tonnes of gasifier slag will be generated each year in the early 21st century and that disposal will become a major problem. The challenge facing the industry is to develop efficient and economical ways to recycle this slag. One possibility is the formation of glass polyalkenoate cements.

4.6.2 Glass Polyalkenoate Cements

Historical Background

Glass Polyalkenoate Cements (GPC) were originally developed by Wilson and Kent[7] as a semi-translucent adhesive restorative dental material. More recently, these cements have found use in a range of biomedical applications.[8] Literature on the development, materials science and clinical application of GPC is extensive.[9-12]

Setting Reaction

Glass polyalkenoate cements are formed by reacting acid-degradable calcium aluminosilicate glasses with aqueous poly(acrylic acid) (PAA).[13]

The PAA degrades part of the glass network, releasing metal cations, which ionically crosslink the PAA chains, resulting in a hard cement.[14] This setting reaction is shown schematically in Figure 4.22.

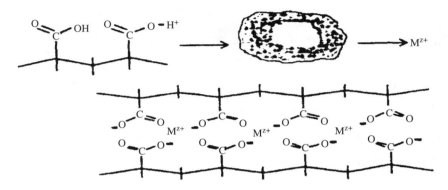

Figure 4.22   The glass polyalkenoate cement setting reaction.

These cements are currently used for dental and biomedical applications, and for such uses the glass component is specially synthesised and therefore too expensive for large-tonnage applications. However, gasifier slag has a similar chemical composition to the simple model calcium aluminosilicate glasses studied by Wilson et al.[15] Therefore it is possible to produce 'inexpensive' glass polyalkenoate cements by replacing the expensive glass component used in dental materials with low-cost gasifier slag.

Advantages of Glass Polyalkenoate Cements and Potential for the Building Industry

During the past 20 years, there has been considerable interest in the so-called 'macro defect free cements' (MDF) based on ordinary Portland cement (OPC) and water-soluble polymers. These materials are characterised by their extremely low porosity and superior mechanical properties compared to OPC produced conventionally. The major problem with MDF cements, however, is their hydrolytic stability. Rodgers et al.[16] highlighted similarities between MDF cements based on polyacrylamide and glass polyalkenoate cements. The setting reaction of MDF cements can also be described as an acid–base reaction, as shown schematically in Figure 4.23.

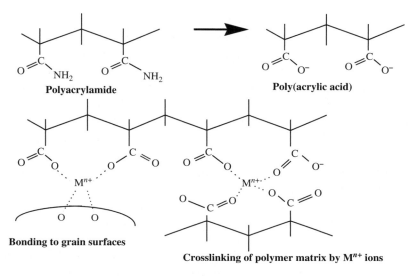

Figure 4.23   The MDF cement setting reaction.

However, unlike MDF cements, glass polyalkenoate cements are hydrolytically stable and do not lose their strength in water.[17] Inexpensive GPCs have many attractive properties that would be useful to the building industry, including:

- high flexural and compressive strengths;
- low porosity and excellent surface finish;
- the potential for resisting freeze–thaw;
- fast setting characteristics;
- the ability to bond chemically to wood and Portland cement;
- the ability to be rapidly processed to give complex mouldings, using polymer production methods;
- suitability as a matrix for fibre reinforcement, using conventional glass fibres; and
- high strength and hydrolytic stability on exposure to water.

With these advantages, inexpensive slag–glass polyalkenoate cements could potentially be used in the applications described below.

Rapid Repair Material Currently, various rapid-hardening cements are available for concrete repair, with considerable interest shown in magnesia-phosphate based systems.[18–20] However, with commercial quick-setting cements, inadequate surface preparation has always been the main cause of failure in any repair work. It is essential during preparation to prepare the area to a square or rectangular shape, extending beyond all defective concrete. It is also necessary to break out the concrete within the repair area to a clean, sound surface, and where reinforcement is present to clean the steel until it is rust free.

Potentially, glass polyalkenoate cements could be used advantageously as a rapid repair material. The available ranges of compressive strength (120–200 MPa) and tensile strength (4–20 MPa) are quite broad and arise from the variety of formulations on the market for various applications. However, the mechanical properties of glass polyalkenoate cements more than meet the requirements of a potential quick-setting cement. They are dimensionally stable and chemically adhere to many materials such as Portland cement, wood and stainless steel. Thus delays due to surface preparation of the repair area would be drastically reduced or entirely eliminated. Also, preliminary results show that glass polyalkenoate cements have good freeze–thaw resistance and acid resistance.[21]

Matrix for Glass Fibre Reinforcement Rapid expansion in the use of glass fibre reinforced cements (GRC) has not been achieved, owing to concern about their long-term durability, particularly in wet environments. For example, in continuously wet and UK weather conditions, an OPC matrix GRC becomes brittle and much weaker with time.[22] This reduction in GRC properties is often attributed to chemical corrosion of the fibres. The silicon–oxygen–silicon network structure, which forms the skeleton of all conventional silicate and borosilicate glass fibres, is severely attacked in highly alkaline solutions.[23] All hydraulic cements provide a highly alkaline environment both during the curing/hydration period and thereafter when moist. Portland cement is the most commonly used material and is the most alkaline.

Glass polyalkenoate cements would also have potential as a matrix for glass fibre reinforcement, as they are slightly acidic and, as shown by infrared spectroscopy, contain free carboxylic acid groups when fully hardened and at long-term ageing times.[24] Therefore, the pH in the pore solution of the matrix phase is on the acid side of neutral, thus producing a less aggressive environment for the fibres.

Foamed Cements and Concrete Foamed concrete is a lightweight material formed by entrapping or generating small bubbles of air or other gases in a Portland cement paste of mortar by mechanical or chemical processes. For example, one method for making a porous, inorganic construction material is to mix a gas-generating agent such as finely divided aluminium metal with a hydraulic cement and then cause foaming by adding water to the mixture.[25] This method has many processing problems; for example, uniform mixing of metal powder with mortar is difficult; the fine metal powder is likely to pollute the environment and endanger health; and the foam distribution in the construction material lacks uniformity. Foamed concretes are lightweight, durable and frost resistant, and have qualities as both acoustic and thermal insulators. Although the strength range is limited, it is quite adequate for a large range of applications.

Glass polyalkenoate cement could potentially be used to form a lightweight structural material, as shown by Drake et al.[26] They discovered that a 'foamed' glass polyalkenoate cement could be produced by mixing an acid-degradable aluminosilicate glass with poly(acrylic acid) in the presence of a blowing agent. The blowing agent may be selected from air, nitrogen, carbon dioxide or a volatile halohydrocarbon. Carbon dioxide can be formed in situ most easily by adding a metal carbonate – for example, calcium carbonate – to the glass/polyacid powder mixture and on addition of water, carbon dioxide is produced, causing the cement to foam.

4.6.3 Glass Polyalkenoate Cements Based on Waste Gasifier Slags

The Glass Component: Selection Criteria for Slag-based Glass Polyalkenoate Cements

Wilson et al.[15] and Hill et al.[27] have reviewed a wide range of simple calcium aluminosilicate glasses and related their chemical compositions to the acid degradability and the mechanical properties of the resulting polyalkenoate cements. These studies provided strong evidence that the aluminium to silicon ratio is a dominant factor determining cement properties in simple $CaO-Al_2O_3-SiO_2$ glasses. Hill et al.[27] specified that the requirements for a suitable glass are as follows:

* An Al : Si ratio close to unity, since the first step in the setting reaction is the hydrolysis of the Al–O–Si bonds of the glass network by PAA.

- The silica mole fraction should ideally be below 0.5.
- A minimum Ca:Al ratio of $1:2$ to allow sufficient Ca^{2+} ions to maintain the Al^{3+} ions in a fourfold coordination within the glass network (Figure 4.24); sufficient nonbridging oxygen atoms (NBO) to allow for a well disrupted glass network. For glass compositions with a Ca:Al ratio greater than $1:2$, the Ca^{2+} ions not required for charge balancing the Al^{3+} have the ability to disrupt the glass structure, creating nonbridging oxygen atoms (NBO) and consequently increasing glass reactivity (Figure 4.25).

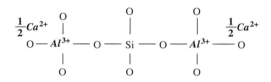

Figure 4.24   The structure of an aluminosilicate glass.

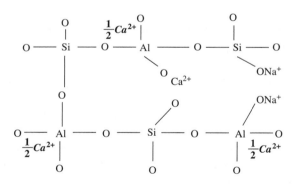

Figure 4.25   A 3-D representation of the glass structure, with the bridging oxygen atoms disrupted by the calcium and sodium modifying ions.

Gasifier slags, the composition of which are given in Table 4.18, meet these criteria, or come close to them, and therefore should be suitable for forming polyalkenoate cements.

Table 4.18   Gasifier slag compositions in mole fraction.

	Drayton	Newland	Cerrejon	British Gas
SiO_2	0.50	0.38	0.56	0.45
Al_2O_3	0.16	0.15	0.10	0.15
Fe_2O_3	0.04	0.04	0.04	0.06
CaO	0.30	0.43	0.29	0.33
Si : Al	1.56	1.26	2.80	1.50

Note: 'Drayton', 'Newland' and 'Cerrejon' were coals used in the Buggenum Gasifier and 'British Gas' was a pilot scale unit operated by British Gas

Poly(Acrylic Acid) Component and Cement Formulation

The liquid component of the original glass polyalkenoate cements developed by Wilson and Kent[7] was a 50 % aqueous solution of poly(acrylic acid). The molar mass and distribution of the poly(acrylic acid) together with its concentration influence the viscosity of the liquid. A low molar mass was required to achieve a high concentration without gelation. Poly(acrylic acid)s with higher molar masses can be incorporated into the cements by the preparation of vacuum dried poly(acrylic acid), which is added in powder form with the glass powder. Cements of this type have an almost unlimited shelf life if kept dry, as reaction does not occur until mixed with distilled water. The incorporation of powdered vacuum dried poly(acrylic acid) within the cement powder does not affect the overall setting reaction or the structure of the set cement. In general, different dental and biomedical applications require different physical and mechanical properties; thus cement properties may be tailored by varying cement formulation to the intended application. For example, for a dental filling material, cement samples would be formed by mixing the glass powder with poly(acrylic acid) in a weight ratio of 5 : 1 and adding this mixture to distilled water in a weight ratio of 4 : 1. This represents a glass powder to poly(acrylic acid) solution ratio of 2 : 1 with an acid concentration of 40 %, with the number average molar mass of the poly(acrylic acid) solution ranging from 3.25×10^4 to 1.08×10^5.

Mechanical Properties

Several compositional factors affect the mechanical properties of glass polyalkenoate cements, including variation in glass powder, variation in the powder : liquid ratio, and the molecular mass and concentration of the poly(acrylic acid). The mechanical properties are influenced by the conditions of testing, including storage conditions and duration and loading/strain rate. The effect of each of these factors is treated extensively in the literature.[9,10,12,28,29]

Poly(Acrylic) Acid Molar Mass and Concentration   Hill has shown that dental GPC cements can be treated as polymer composites where fracture takes place almost exclusively in the polymer matrix and mechanical properties are dominated by the polymer component.[30] Using a Linear Elastic Fracture Mechanics (LEFM) approach, Hill *et al.*[31] determined that the physics of the failure process of the dental glass polyalkenoate cements were dominated by poly(acrylic acid) molar mass and concentration as a function of time. Consequently, the mechanical properties could be tailored for specific biomedical applications by changing these parameters.

Drayton Slag Based Glass Polyalkenoate Cements

Gasifier Slag Cement Formation   Cement samples were formed by hand mixing < 53 μm gasifier slag powder with four different poly(acrylic acid)s in a weight ratio of 5 : 1 and then adding this mixture to water containing 30 % tartaric acid in a weight ratio of 4 : 1. This represents a slag powder to poly(acrylic acid) solution ratio of 2 : 1 with a poly(acrylic acid) of 40 % m/m. Cements were allowed to set in a compression mould (4 mm × 6 mm cylinders) for one hour at 37 °C, then removed from the mould and stored in distilled water at 37 ± 2 °C prior to testing. Compression tests were carried out after 1, 7 and 28 days. The specimen preparation techniques are based on Specifications and Standards for Dental Glass Polyalkenoate Cements (Table 4.19).

Table 4.19   Specifications and standards for dental polyalkenoate cements.

Property	Test method
Compressive strength, 24 hours (MPa)	ISO 7489[33]/BS 6039[34]
Diametral tensile strength, 24 hours (MPa)	Smith[35]
Flexural strength (MPa)	ASTM D 790–71[36]
Fracture toughness	Lloyd[37]
Toughness	Goldman[38]
Young's modulus	Hill *et al.*[27]
Dimensional stability	ISO 7489[33]/BS 6039[34]
Acid erosion	Wilson *et al.*[16]
Water leachability	ISO 7489/BS 6039

In terms of larger-scale industrial production of gasifier slag based cement, different preparation and test methods could potentially be developed depending on the intended application. For example, their high poly(acrylic acid)/polymer content allows glass polyalkenoate cements to be processed rapidly using established polymer production methods to give complex mouldings such as roof tiles. Thermal curing above the glass transition temperature of the medical-grade dental cements enables the flexural strength to be doubled.

Sullivan, Hill and Waters[32] carried out a preliminary study of the cement-forming ability of four different gasifier slags, termed Drayton, British Gas, Newlands and El Cerrejon. This study has shown that the Si : Al ratio of the glass is an important factor in determining the acid degradability and subsequent reactivity of the gasifier slag. Drayton slag was found to be the most suitable slag, and this slag gave a cement with a compressive strength > 50 MPa at 24 hours and > 100 MPa after six weeks. Immersion in water one hour after mixing had no significant influence on the compressive strength. Flexural strength after immersion in water was in the range 5–10 MPa. These results indicate that it is feasible to produce polyalkenoate cements from gasifier slags.

A Linear Elastic Fracture Mechanics approach, based on dental cement standards (Table 4.19), was used to evaluate cement mechanical properties so that the physics and chemistry of existing dental materials and slag based cements could be compared. The results of this study by Sullivan and Hill[39] demonstrated that like the dental materials, the molar mass and concentration of poly(acrylic acid) have a pronounced influence on the toughness and strength of slag-based cements. This insight into the failure modes of this novel cement allows cement formulations to be manipulated for specific applications in the building industry. Further work is required on 'inexpensive' glass polyalkenoate cements to evaluate their suitability for replacing existing rapid repair or concrete roof tile materials.

However, these properties may not necessarily be evaluated using the standards and specifications of current rapid repair/concrete roof tile applications.[40,41] It is envisaged that due to the difference in chemistry and physics between 'inexpensive' glass polyalkenoate cements and rapid repair/concrete roof tile material, new standards to evaluate their potential use in these applications will have to be developed.

4.6.4   Future Outlook

Coal gasification is an environmentally attractive route for power generation.[42] The process has much greater efficiency than conventional coal combustion, resulting in zero levels of SO_2, NO_x and reduced levels of CO_2 emissions, along with benign solid by-products. The ultimate success of the IGCC technology will depend on the economics of the process relative to alternative technologies for electric power generation, particularly the pulverised coal and natural gas combined cycle. The principal factors affecting the economics of IGCC include capital costs, operating and maintenance costs for the plant, fuel costs and the efficiency of conversion. The main disadvantages of IGCC plants are economic. However, substantial cost reductions have been achieved over recent years and coal-fired IGCC plants are becoming more competitive with mature technologies. IGCC technology advances are aimed at achieving higher efficiency of gasification, greater throughput, lower capital costs and further reductions of the emissions, making it the technology of the 21st century. Therefore, it is expected that future availability of slag for glass polyalkenoate cement formation would not be of concern. The formation of glass polyalkenoate cements from a by-product of coal gasification would also benefit the IGCC plant. The gasifier slag would be recycled to form a more useful product, reducing disposal problems but, even more importantly, adding value to the slag. Gasifier slag, unlike granulated blast furnace slag (GBFS) or pulverised fuel ash (PFA), cannot be used as an aggregate for concrete and is disposed of to landfill, incurring disposal costs.[43] If the gasifier slag could be recycled to form a viable cement for building applications and expanding the uses for the material, then it would not be unrealistic for the gasifier slag to command a price similar to that of other residues. In the UK, GBFS is used as an aggregate in concrete and is valued at approx €70 per tonne.[44] The value of the slag will depend on many factors, including the properties of the cements formed from it and the price of poly(acrylic acid) used in the cement, which will depend indirectly on the price of oil. The key to developing a successful glass polyalkenoate cement for use in the building industry is to reduce cost. The cost of glass polyalkenoate cement is drastically reduced by the incorporation of a gasifier slag as the glass component, and therefore the polyacid component determines the cost of the cement. Poly(acrylic acid) is an inexpensive polymer, the cost of which does not increase with molar mass, but only with the amount used in cement formation. Therefore, the mechanical properties of the cement can be improved by increasing the polyacid molar mass at a lower concentration at no added cost. An estimate of €360 per tonne is calculated as a typical cost for a polyalkenoate cement based on a gasifier slag. This is much more expensive than conventional Portland cement, at approximately €80 per tonne, but the cost would be offset by the improved properties.

References

1 Nixon, P.J., Osborne, G.J. and Shepperd, C.N., *Cementitious properties of slags from the British Gas/Lurgi slagging gasifier*. Silic. Ind., **12**: 253–262 (1983).

2 Shepperd, C., *Slag from coal gasification: its production, properties and possible uses*. In Proc. 2nd Int. Conf. Ash. Tech. & Mktg., London (1984).

3 Zuideveld, P.L. and Postuma, A., *Overview of experience with the Shell coal gasification process*. Mater. High Temp. **11**: 19–23 (1993).

4 Pastoors, J.W.T., *Materials test exposures in the demo KV-STEG in Buggenum*. Mater. High Temp., **11**: 139–143 (1993).

5 Mendez-Vigo, I., Chamberlain, J. and Pisa, J., *ELCOGAS IGCC plant in Puertollano, Spain*. Mater. High Temp., **14**: 81–86 (1997).

6 Schellberg, G.W. and Kuske, E., *Experience from the PRENFLO plant*. Mater. High Temp. **11**: 15–19 (1993).

7 Wilson, A.D. and Kent, B.E., *The glass ionomer cement: a new translucent cement for dentistry*. J. Appl. Chem. Biotechnol., **2**: 313–318 (1971).

8 Brook, I.M. and Hatton, P.V., *Glass ionomers: bioactive implant materials*. Biomaterials, **19**: 565–571 (1998).

9 Smith, D.C., *Development of glass ionomer cement systems,* Biomaterials, **19**: 467–478 (1998).

10 Pearson, G.J. and Atkinson, A.S., *Long term flexural strength of glass ionomer cements*. Biomaterials, **12**: 658–660 (1991).

11 Wilson, A.D., *Developments in glass ionomer cements*. Int. J. Prosthodont., **2**: 438–446 (1989).

12 Wilson, A.D., *Glass ionomer cements – origins, development and future*. Clin. Mater., **7**: 275–282 (1991).

13 Wilson, A.D., *The development of glass ionomer cements*. Dental Update, 401–412 (1977).

14 Wilson, A.D. and Kent, B.E., *A new translucent cement for dentistry: the glass ionomer cement*. Brit. Dent. J., **132**: 133–135 (1972).

15 Wilson, A.D., Crisp, S., Prosser, H.J., Lewis, B.G. and Merson, S.A., *Aluminosilicate glasses for polyelectrolyte cements*. Ind. & Eng. Chem. Prod. Res. & Dev., **19**: 263–270 (1980).

16 Rodgers, S.A., Brooks, S.A., Sinclair, W., Groves, G.W. and Double, D.D., *High strength cement pastes: Part 2 – reactions during setting*. J. Mater. Sci., **20**: 2853–2860 (1985).

17 Cattani-Lorente, M.-A., Godin, G.G. and Meyer, J.-M., *Mechanical behaviour of glass ionomer cements affected by long term storage in water*. Dent. Mater., **10**: 37–44 (1994).

18 Pera, J. and Ambroise, J., *Fiber reinforced magnesia phosphate concrete compositions*. Cement Concrete Compos., **20**: 31–39 (1998).

19 Mangat, P.S. and Limbachyia, M.C., *Repair material properties for effective structural application*. Cement and Concrete Res., **27**: 601–617 (1997).

20 Sarkar Asok, K., *Phosphate cement-based fast-setting binders*. Ceramic Bulletin, **69**(2): 234–238 (1990).

21 Sullivan, A., *Inexpensive glass polyalkenoate cements based on waste gasifier slags*. Ph.D. thesis, University of Limerick (2000).

22 Purnell, P., Short, N.R., Page, C.L., Majumdar, A.J. and Walton, P.L., *Accelerated ageing characteristics of glass fibre reinforced cement made with new cementitious matrices*. Composites Part A, **30**: 1073–1080 (1999).

23 Proctor, B.A., *Alkali resistant glass fibres for reinforcement of cement*. In Proc. NATO Adv. Study Inst. on Glass Current Issues, A.F. Wright and J. Dupuy (eds), Nijhoff: Dordrecht, Lancaster, pp. 555–572 (1985).

24 Crisp, S., Pringuer, M.A., Wardleworth, D. and Wilson, A.D., *Reactions in glass ionomer cements, II: an infrared spectroscopic study*. J. Dent. Res., **53**: 1414–1419 (1974).

25 Kuramote, N., Satio, H. and Yamamoto, W., *Foam containing slurry to make porous, inorganic construction material*. US Patent 3,963,507 (1976).

26 Drake, P.H., Humphries, C.J. and Preedy, J.E., *Structural foams from polymer ceramics*. British Patent 1,559,002 (1980).

27 Hill, R.G. and Wilson, A.D., *Some structural aspects of glasses used in glass ionomer cements*. Glass Technol., **29**: 150–188 (1988).

28 Cook, W.D., *Dental polyelectrolyte cements: III. Effect of powder : liquid ratio on their rheology*. Biomaterials, **4**: 21–24 (1983).

29 Hill, R.G., Wilson, A.D. and Warrens, C.P., *The influence of poly(acrylic acid) molecular weight on the fracture toughness of glass ionomer cements*. J. Mater. Sci., **24**: 363–371 (1989).

30 Hill, R.G., *Relaxation spectroscopy of polyalkenoate cements*. J. Mater. Sci. Lett., **8**: 1043–1047 (1989).

31 Hill, R.G. and Labok, S.A., *The influence of polyacrylic acid molecular weight on the fracture of zinc polycarboxylate cements*. J. Mater. Sci., **26**: 67–74 (1991).

32 Sullivan, A., Hill, R.G. and Waters, K., *A preliminary investigation of glass polyalkenoate cements based on waste gasifier slags*. J. Mater. Sci. Lett., **19**: 323–335 (2000).

33 International Organisation for Standardisation, *ISO 7489 Dental glass polyalkenoate cements* (1986).

34 British Standards Institute, *BS 6039 Specifications for dental glass ionomer cements*. BSI: London (1981).

35 Smith, D.C., *A new dental cement*. Brit. Dent. J., **125**: 381–384 (1968).

36 American Society of Testing and Materials, *ASTM D790-1, Standard methods of tests for the flexural properties of plastics*. ASTM International, West Conshohocken, PA (1971).

37 Lloyd, C.H. and Mitchell, L., *The fracture toughness of tooth coloured restorative materials*. J. Oral Rehabil., **11**: 257–272 (1984).

38 Goldman, M., *Fracture properties of composite and glass ionomer dental restorative materials*. J. Biomed. Mater. Res., **19**: 771–783 (1985).

39 Sullivan, A. and Hill, R., *Influence of poly(acrylic acid) molar mass on the fracture properties of glass polyalkenoate cements based on waste gasifier slags*. J. Mater. Sci., **335**: 1125–1134 (2000).

40   British Standards Institute, *BS 6319, Testing of resin compositions for use in construction.* BSI: London (1971).
41   American Society of Testing and Materials, *ASTM C1492, Standard specification for concrete roof tile.* ASTM International, West Conshohocken, PA (2001).
42   Medha, M. and Sunggyu, L., *Integrated gasification combined cycle – a review of IGCC technology.* Energy Sources, **18**: 537–568 (1996).
43   Berg, J. van der, Vliegasunie BV, Groningenhaven 7, 3433 PE Nieuwegein, The Netherlands, personal communication, April 2004.
44   Hill, R.G., Imperial College, London, personal communication, April 2004.

4.7 Fire-resistant Materials

Constantino Fernández-Pereira and Luis Vilches Arenas
Universidad de Sevilla, Spain

Types of combustion residues involved
Combustion residues of all kinds have been used for the very different applications treated in this section. No specific requirements can be given in general terms. However, due to high unburned material content and high alkali content in biomass and coal–biomass mixture residues, fire resistance properties for materials produced from such residues are better compared to coal fly ash based products.

State of development
Most of the available literature is found in patents. It is not always clear whether and to what extent such patents are being exploited. Research from Spain is published in the scientific literature and is in the stage between research and pilot production.

4.7.1 Introduction

Fire is defined as a destructive burning manifested by any or all of the following effects: light, flame, heat and smoke.[1] For fire to occur, a combustible product, an oxidant and an ignition source are all required. Fire resistance is the property of a material or structure to withstand fire or provide protection from it.[1] The term 'fire resistance' as applied to elements of buildings is characterised by the ability to confine a fire or to continue to perform a given structural function, or both. Thermal insulation materials are very important for the construction of fire-safe environments.

A thermal insulation material is a material that reduces heat transfer through the structure against which, or within which, it is installed. Thermal conductivity increases with the ordering of molecules, and for this reason amorphous and vitreous solids show lower thermal conductivity compared to crystalline materials. The structure and composition of solid materials have the most influence on thermal transport, with dense solid structures having a high thermal conductivity. Thermal insulating materials preferably should have a microporous structure in which the size of the pores is of the same size, or smaller than, the mean free path of the gas molecules occupying the pores, so that gas movement is hindered and the thermal conductivity decreases.

Numerous references can be found that mention the use of different ash and combustion by-products as components of fire-resistant materials, fireproof products or fire- (flame-) retardant formulations. However, most of these are patents, from which it is not always easy to deduce the role played by the combustion residues within a more or less complex formulation. In most of the cases, the products are composite materials based on an inorganic or organic matrix or ceramic materials.

The use of ash and other combustion residues in this field is determined by their physical and chemical characteristics. The applications can be subdivided into two broad groups:

(a) Direct use, where the combustion residue is used directly as a (functional) filler in a more or less complex mixture, which is then processed without provoking major changes to the ash. There are several reasons for the introduction of combustion residues:

   (i) the beneficial bulk properties of a low-cost inert material, as in many composite materials used in fire protection;
   (ii) certain beneficial physical properties – for example, products containing cenospheres or very fine ash that improves thermal and insulating properties;
   (iii) beneficial chemical properties – for example, alumina and silica phase content that conveys favourable fire-resistant and fire-retardant properties.

(b) Processed use, where the combustion residue is either physically or chemically converted into an end product with improved fire-resistant characteristics. This processing can be done in several ways:

   (i) using its pozzolanic properties to form a structural product such as concrete (pozzolanic concrete), boards or plates (calcium silicate plates);
   (ii) processing with other materials (e.g. magnesium oxychloride cement or gypsum) to give a product with better thermal and/or mechanical properties than the base materials;
   (iii) promoting a valuable component (mullite or xonotlite) through a thermal process (sintering or hydrothermal).

For practical reasons, in the following sections the use of combustion residues in different products showing fire resistance properties is subdivided according to the manufacturing process (sintering and hydrothermal processing), the type of material (concrete and mortar), its composition (calcium silicate) or the shape of the product and the application type (boards and panels, fire-doors and coatings).

4.7.2 Sintering Processes

Ceramic products containing coal ashes, aluminium phosphates, water and hardening agents are obtained by shaping-drying or drying-shaping, followed by sintering. The process comprises casting of slurries and heating at 250–600 °C and also includes drying the compositions to give pre-ceramic compositions and hot pressing. Shaping may be performed by injection moulding or extrusion moulding of the pre-ceramic compositions. The resulting ceramic products have high strength and fire resistance.[2]

Ceramic mouldings with bulk densities of 1.0–2.0 g cm^{-3} and water absorption between 0.01 and 0.8 vol % are manufactured from a mixture of coal ash, inorganic sintering material and inorganic thickener by kneading with water (allowing degassing), extruding, drying and heating at 1000–1300 °C. Ceramic mouldings are useful for lightweight aggregate and fire- and heat-resistant building materials.[3]

The thermal behaviour of eight Spanish fly ashes was investigated[4] by subjecting them to temperatures up to 1200 °C. The mineral composition of fired specimens, determined by XRD, for some ashes showed the formation of large amounts of refractory Al-minerals, such as mullite ($Al_6Si_2O_{13}$). This seems a favourable property for the application of these ashes as ceramics.

4.7.3 Hydrothermal Processing: Calcium Silicate Products

The effect of pore size, as explained in the introduction, makes the finest ashes suitable for the production of microporous thermal insulating materials. Such material may have pore sizes of about 10 nm, resulting in a thermal conductivity lower than that of air.[5]

Calcium silicate plates are widely used for fire protection in industry and as thermal insulating materials in both industrial and civil construction. These materials are among the best as far as density and thermal conductivity are concerned, and in addition offer other beneficial properties not possessed by other thermal insulating materials – as, such for example, their favourable hardness and vapour permeability. The maximum applicable temperature (1050 °C) is higher than for most other insulating materials; however, workability is as good as most thermal insulating materials.

Calcium silicate thermal insulating (CSTI) products are highly porous materials, formed by hardening the reactants in an autoclave. Under the hydrothermal conditions in an autoclave, the reactants $Ca(OH)_2$ (C), SiO_2 (S) and H_2O (H) react to form the hard calcium–silicate–hydrate (CSH) phases of the CSTI products. Major minerals in CSTI products are tobermorite, $Ca_5Si_6O_{16}(OH)_2 \cdot 4H_2O$, and xonotlite, $Ca_6Si_6O_{17}(OH)_2$. Different pozzolanic materials, such as coal fly ashes or biomass ashes (rice hull or bagasse) along with a calcium oxide material, are used to provide hydrothermally cured pozzolanic cements capable of forming strong and durable concrete and building materials by processing at temperatures as low as 20 °C. In addition, for the formation of lightweight blocks and tiles, expanded fillers can be added, such as hollow glass cenospheres from fly ash. Such products are cheap, and possess high strength, acid resistance and fire resistance. These cements can also be closed-cured at elevated temperatures in an autoclave, if desired.[6]

Fly ash and lime are used as raw materials for the hydrothermal production of a lightweight insulating material, based on xonotlite, which is used for the production of material with a high limiting temperature (1100 °C). The process is carried out in an aqueous suspension at a temperature of 190–240 °C.[7]

In a hydrothermal process to decompose and remove dioxins from refuse incineration ashes, a calcium silicate porous by-product is synthesised that may be used as a heat-resistant, sound-absorbing and fire-prevention material.[8] Rice husk ash is particularly rich in silica and has been successfully used for the manufacturing of calcium silicate fire-resistant insulation materials, with excellent thermal durability up to 1000 °C. Well-grown xonotlite crystals, which provide the body of the insulation material, were formed in reaction with lime.[9]

The manufacture of a calcium silicate thermal insulating fire-resistant material comprising xonotlite, wollastonite ($CaSiO_3$), and reinforcing fibres has been proposed using combustion residues as the raw material.[10] Xonotlite is synthesised by heating a mixture of amorphous silica micro-powder (such as silica ash or rice husk ash) and a Ca compound at 80–100 °C for 1–3 hours. Wollastonite and 2–8 % fibres are added, and the material is shaped in the required form and allowed to react for 12–24 hours at 190–220 °C, after which it is cooled and dried at 100–140 °C.

A cement-free moulded product can be prepared from a mixture of ground CaO- and SiO_2-containing components (fly ash) and hollow silicate microspheres. The mixture is steam cured in an autoclave at 200 °C and 8–21 bar for 1–5 hours. The resulting articles are suitable as sound and thermal insulators, and fireproofing materials.[11]

The manufacture of similar thermal insulation and fire-resistant materials has been proposed from coal fly ash or bottom ash with spent lime from sugar beet processing,[12] and from fly ash, CaO, mica and organic fibres.[13]

Timtherm is a lightweight insulation material that uses fly ash as the raw silicate material. Processing takes place under high pressure in an autoclave, resulting in a slurry that may be further processed with additives to form calcium silicate plates in a filter press.[14] Under this registered name, hot insulated pipe supports designed for high-temperature (up to 1000 °C) piping systems are commercialised by the Dutch company Powerpiping.

4.7.4 The Fire Resistance of Pozzolanic Concrete and Mortar

The effects of fire on the properties of concrete are mainly the effects of exposure to elevated temperatures. Many investigations have been carried out on the residual mechanical properties of concrete subjected to elevated temperatures. However, only limited research has been carried out on concretes containing fly ash, where damage is not only due to a combination of chemical and mechanical changes of the constituents of the concrete, but is greatly influenced by factors such as the properties of the fly ash, its dosage and the curing regime.[15,16]

The strength and durability performance of normal- and high-strength pozzolanic concretes incorporating silica fume, fly ash and blast furnace slag was compared at 800 °C by Poon et al.[15] Fly ash improved the performance of concrete at elevated temperatures compared to concretes containing silica fume or pure OPC. However, this improvement was more significant at temperatures below 600 °C. The residual compressive strength and hence the fire resistance of concrete improved when fly ash was added.[16] This may be ascribed to the pozzolanic reactivity of the fly ash. Nasser and Marzouk[17] found an increase in strength of concrete containing 25 % of lignite fly ash in the temperature range of 121–149 °C that was as high as 152 % of the original strength, due to the formation of tobermorite. So, in general, fly ash containing concretes show better performance at elevated temperatures than pure OPC concretes, mainly due to the pozzolanic reaction of the ash, reducing the free lime content and enhancing strength and durability.

Apart from its presence in pozzolanic concrete as a binder material, different ashes, mainly coal fly ashes, are used as synthetic aggregate in fire-resistant concrete. Coal fly ash sintered agglomerates with an average size of 0.5–12.0 mm along with polypropylene fibres (m.p. 160–165 °C) were added to a concrete mixture to increase its fire resistance. In the case of fire, the fibres melt or char to give a finely divided capillary net, allowing the steam formed to escape from the concrete and so preventing spalling. The samples reached a temperature of 1350 °C after 60 minutes and remained undamaged and free from spalling after 120 minutes.[18] Concretes containing aggregates made from municipal solid waste incinerator ashes have been reported to show high mechanical strength and fire resistance.[19] *Haydite*, an expanded shale lightweight aggregate, has high strength, high fire and corrosion resistance, and good thermal insulation properties. A *Haydite*-like material may be manufactured by mixing 65–90 parts fly ash, 10–25 parts of a binder and 3–5 parts additives, adding water, stirring, drying at 300–500 °C for 10–25 minutes, and finally calcining at 800–1000 °C for 10–25 minutes and at 1100–1200 °C for 5–10 minutes.[20]

Concrete structures, such as wall surfaces, natural groundwork, linings of tunnels and so on, may require fireproof coatings. Metallic structures used in construction have the disadvantage of being vulnerable to fire, and therefore also require fire-resistant coatings. Spray type mortars and concrete materials are most appropriate for such purposes. Addition of coal fly ash to such pastes has been reported to improve adhesive strength during spraying, thereby reducing remarkably the rate of bounce and dust generation.[21]

A waterproof insulating mortar prepared from coal fly ash comprises a mixture of expanded perlite (rhyolitic volcanic glass) and sepiolite ($Mg_4Si_6O_{15}\cdot6H_2O$) and an impervious organosilicon waterproofing agent. Its advantages include simple preparation, low cost, wide application range, high adhesion and fire-retarding properties.[22]

Cementitious binders comprising $CaSO_4$ hemihydrate ($CaSO_4 \cdot \frac{1}{2}H_2O$), Portland cement optionally mixed with fly ash and/or ground blast slag, and either silica fume or rice husk ash, have been used in fireproofing sprays applied to metal surfaces.[23] A material containing a high proportion of fly ash (minimum 70 wt%), has been sprayed on metallic structures and has shown acceptable physical and mechanical properties.[24] The fire resistance of this product is comparable to that of current commercial products and therefore offers future prospects. Details of this material developed by AICIA in Spain are further described below in Section 4.7.8.

4.7.5 Boards, Panels and Other Moulded Building Materials

Modern prefabricated building components require characteristics such as light weight, good thermal and sound insulation, simple construction, decreasing intensity of labour and cheap building costs. Fire-resistant products possessing such properties include partition wallboards, building boards, fireproofing plates, panels and blocks, and other moulded and hardened products. Most references concerning such products, in which combustion residues constitute a significant part of the raw materials, are patents. An exception is the work of the AICIA[25–27] group in Spain, considered below in Section 4.7.8.

First, a brief overview of patented fire-resistant moulded building materials is presented, excluding calcium silicate insulating materials, which are treated separately.

Reinforced Panels

Panels are normally reinforced using an internal metallic mesh sheet or frame, most often made from stainless steel. These are manufactured by introducing the reinforcement metal into a slurry of fly ash/binder mixture, followed by drying, shaping and sintering.[28,29] Alternatively, using the steel frame as a mould, boards with external reinforcement are produced.[30]

Fly ash composite materials and fibres form the raw materials for the integral manufacturing of reinforced panels that consist of a sheet and structural ribs on the inner surface. It is reported that such panels are stronger than concrete and have good fire-resisting performance.[31]

Sandwich and Multilayer Panels

Sandwiching of fire-resistant materials between two facing sheets is a common technique, where the cavity between the two surface materials may be filled with ash-containing hydraulic cements,[32] silicate foams[33] or aerated mortar.[34] When such sheets alternate in a multi-layer arrangement, a surface plate and a filler plate are distinguished. The surface plate can be made from a composite containing unsaturated polyester resin and the filler plate from lightweight coal ash-based raw materials.[35] Insulator sheets made of ashes and water-glass may be bonded together as panels[36] or laminated with a foamed layer.[37] A frame made of glass-fibre cloth can also be used between the layers.[30,38]

Composite Boards

Fire-resistant composite materials containing coal fly ash and urea-formaldehyde resin have been proposed as decorative board[39] and as a mould-pressed composite wall.[40] In such common composite materials an organic (a thermosetting resin) or inorganic matrix is reinforced with fibres and/or thermally stable particles (for example, combustion residues).[41] The binder in those composites may be sodium (or potassium) silicate alone[28,33,36,42] or in combination with other binders; for example, water-glass and polyvinyl acetate (PVAC).[43] The use of cement together with a polycondensation adhesive and a PVAC dispersion agent has also been proposed.[44]

Magnesium-containing Boards

Magnesium oxychloride cement (*Sorel* cement), a lightweight aerogel material with good strength, rapid hardening, fire resistance and easy production, has great potential for prefabricated fire-resistant products. However, due to decomposition or carbonatisation of the main crystal phase ($5Mg(OH)_2 \cdot MgCl_2 \cdot 8H_2O$) in water or humid air, the durability or water resistance of this material is poor. Resistance to deformation is also poor and therefore in the past its use has been limited. However, by adding a high percentage of fly ash, such properties of this material are greatly improved. The more fly ash that is added to the matrix, the better is the water resistance of the product.

The presence of fly ash in magnesium oxychloride cement has the advantage of increasing the specific surface, especially when cenospheres are used. This reduces the reaction speed and provokes a more homogeneous and complete reaction. Moreover, fly ash reacts with the cement to form new phases, thereby stabilising the matrix structure and improving water resistance. Lightweight and fireproof elements manufactured using magnesium oxychloride cement and fly ash have been described in the literature.[45,46]

Magnesium hydroxide decomposes endothermically with the formation of water and therefore is used as a flame retardant in polymeric products. A blend containing alumina cement, fly ash, perlite, $Mg(OH)_2$, and methylcellulose has been patented as strong inorganic fire-resistant sheet material.[47] A blend containing MgO and fly ash has been used as a lightweight partition wallboard.[48]

Gypsum Boards

Fly ash may be one of the components, along with vermiculite, fibreglass and other materials, in gypsum boards. A cementitious binder comprising $CaSO_4$ hemihydrate and Portland cement mixed with fly ash and/or rice husk ash has been used to prepare fireproofing sprays or fire-stopping boards.[23] Phospho-gypsum has also been used with fly ash for manufacturing thermally insulating materials.[49] A mixture containing fly ash, gypsum powder and glass fibre was patented for use in fireproof decoration boards.[50] Boards manufactured from perlite, cement, gypsum, fly ash, adhesive and glass fibre cloth have high impact strength, thermal and sound insulation properties and fire resistance.[30]

Cellulose Panels

As shown above, the introduction of cellulosic materials as fibre reinforcement can contribute to the creation of a porous structure facilitating the elimination of vapour in the event of fire. With combustion residues as one of the components, wallboard containing plant fibre powder,[47] gypsum-like boards including fibrous cellulose material from paper mills,[12] bricks and panels containing 15–30 wt% of cellulose-based material,[51] composite plates composed of wood fibre (60–70 wt%),[40] wastepaper de-inking sludge[52] and sediments from paper manufacture[43] have all been used in building materials.

Other Ash Panels

Apart from the most commonly used coal fly ashes and solid waste incineration ashes, biomass ashes, such as wood ashes,[36,43] rice husk ashes[23,53] and sewage sludge incineration ashes, have also been used in the manufacturing of plates.[54]

Cenospheres from coal fly ash, the hollow microspheres or floating beads with densities less than 1.0 are of particular value in filler applications. Since cenospheres give improved thermal and electrical properties, they are also used to improve the fire resistance property of boards and walls, especially in sandwich panels,[33,38] but also in composite plates for floors and walls.[40]

4.7.6 Fire-doors

Fly ash pellets can be used as a lightweight aggregate for manufacturing frames for fire-resistant doors.[55,56] Other ingredients are cement and/or gypsum and a fibrous aggregate to prevent cracking. A fireproof core can be made of foamed perlite, gypsum, pulverised coal ash, cement and fibreglass. This lightweight core exhibits excellent fire resistance properties, and together with a frame and two panels can be made into a fireproof door.[57] A similar construction for a fire-resistant door has been proposed by Shitiwatt and Berl,[58] where the core consists of perlite, gypsum, vermiculite, fly ash, cement and glass fibre. Both sides of the core are equipped with decorative veneer, and formed into an integrated lightweight and nondeforming fire-resisting door. High percentages of fly ash can be incorporated in the low-density gypsum core of a fire-resistant door to provide improved flexural strength and residual hardness on exposure to fire compared to an unmodified gypsum plaster door.[59]

4.7.7 Coatings and Foam Materials

According to a large number of patents, ashes and other combustion residues are used in the manufacturing of coatings, paint or lining materials in building structures, ducts, material cables, furniture, textile, cardboards and so on, to convey fire-retardant or fire-resistant properties. The materials are often applied as a foam to benefit from their fire retardance and thermal insulation properties.

Foam

An alkali silicate is most frequently the main compound of such foams. A liquid glass binder and a refractory filler, in the form of hollow ash micro-spheres (30–40 wt%), may be used for manufacturing foamed thermal insulation material.[60] Tanabe and Matsuura[61] used a ceramic silicate foam with water-glass and powder components for heat-resistant coatings on high-strength concrete and steel frames. A paste containing among other ingredients fly ashes, Al and water-glass has been proposed for making fireproof porous inorganic rigid foam products with densities in the range 40–1000 kg m^{-3} and pore diameter 0.1–10 mm.[62] These products have a foam ratio up to 10 : 1 and a fire resistance greater than 180 minutes, according to DIN 4102, Part 9. A foam composed of a silicate solution produced by caustic digestion of biomass ash combined with a polymer forming agent shows improved thermal insulation and fire retardant properties.[63] Fujimasu[64] proposes a fire-resistant foam composed of thermoplastic or thermosetting resins mixed with Na silicate as flame retardant that includes fly ash as a filler. A foam made from fly ash has been developed using phenolic resin binders at low levels. The fly ash consists of hollow micro-spheres and needs to be treated to remove contaminants.[65] Initial testing for fire resistance has indicated very encouraging results. A low-cost foam material made of magnesium oxide, magnesium chloride and a high proportion (20 : 80) of powdered coal ash used as filler has been patented by Zhu et al.[66] A lightweight, high-strength, fireproofing material has also been prepared from light burnt magnesium, magnesium chloride, polyacrylamide, powdered coal ash, fibre and a foaming agent. The product is made by foaming, moulding, solidifying and curing the slurry at room temperature.[67]

Plastic Fire Retardants

Chen[68] has developed a fire-retarding and smoke inhibiting polyvinyl chloride formulation, which uses 10–70 weight portions fly ash to 100 portions of PVC as the smoke-inhibiting agent. This product is suitable for use as construction material, cable material and so on. Blount[69] also uses fly ash as filler for the production of a flame retardant coating for an open-celled porous organic mass. A flame-resistant coating for electric wire made of fly ash and self-hardened inorganic material has been described by Arai et al.[70] that, on burning, does not generate hazardous gases. Another fire-retardant composition that can be applied to a plastic material consists of a combustible substrate such as a lignocellulosic material or a thermoplastic resin in combination with wood fly ash.[71]

A MgO-containing sealing compound prepared by mixing two separate mixtures, one of MgCl$_2$ with 5–15 wt% fly ash and the other of fired magnesite (MgCO$_3$ → MgO) with 10–30 wt% fly ash has been shown to be especially suitable for electric cable channels and ventilation ducts.[72] Orimulsion ash has been also used as a fire-retardant with a wide range of plastic construction materials and formed into beads, powders, films, fibres, plates, filaments or rods for use in thermal and electrical insulators, furniture, and road and roofing products.[73]

Paints and Coatings

A water-based fire-resistant coating, containing an alkali metal silicate, a reactive calcined filler (e.g. fly ash) and a latent acid catalyst is suitable for application to cardboard substrates.[74] Fire-resistant coatings for building, containing 20–33 wt% fly ash have been proposed as well.[75] A fireproof composition comprising liquid glass or silico-phosphate binder, kaolin, calcium carbonate, vermiculite and fly ash (10–15 wt%) has been used for concrete, metal and wood building structures.[76] A water-based fireproof insulation lining, in which ash and slag are the principal ingredients of the skeletal structure, was developed for industrial boiler stacks.[77]

A water-based fire-retardant coating containing paper sludge incineration ash, acrylic polymer powder and PTFE powder was applied on to polypropylene fabrics and subsequently dried to give a flexible sheet.[78] Another nonhalogen fire-retardant sheet was prepared by mixing fly ash (10–70 parts by weight), with a metal hydroxide (42–110 parts by weight) and polyolefin (100 parts by weight).[79] A flexible refractory insulating felt for pipelines and containers made from coal fly ash, magnesium silicate and silica fume by mixing, baking and demoulding has been patented.[80]

Various other coatings containing combustion residues as an integral or optional ingredient have been reported; based, for example, on Na silicate,[81] unsaturated polyester resin,[82] polyamides[83] or acrylic binders.[84] Such coatings are normally intended for very specific applications.

4.7.8   Case Study: Fireproof Products Using Fly Ash Together with Industrial Waste

Introduction

AICIA, the Andalusian research association connected to the University of Seville, Spain, has developed the process of mixing coal fly ash with different types of residue materials with the aim of producing fireproof and fire-retardant materials. With the collaboration of two industrial partners, a project, known under the acronym CEFYR, and sponsored by the European Coal and Steel Community Coal Programme, was carried out.

Among the most widely adopted solutions for fire protection is the compartmentalisation of the structure by applying fire-doors, ceilings and divisions with plates or panels and the protection of metallic structures with sprayed insulating mortars. Autoclaved cellular concrete containing fly ash is known for its low density and for its high heat resistance, arising from the low thermal conduction properties of the cenospheres in the ash. Most commercial materials used to provide passive protection against fire, such as calcium silicate, vermiculite or perlite, have chemical compositions and physical properties similar to those of inorganic mixtures of fly ash and other inorganic and organic components. Therefore, the main objective of the CEFYR project was to study new fireproof products, in the laboratory and at pilot-scale level, employing fly ash, industrial wastes and cellulose derivatives.[25] Such products would be applied as panels or sprayed (gunited) surfaces and would at the same time decrease the environmental impact from the disposal of wastes.

Fireproof Panels

Plates of different compositions and sizes (15×30 cm and 30×45 cm, and varying thickness) were moulded using fly ash from the Los Barrios power plant in Spain and other components, such as vermiculite, Portland cement, aluminous cement and wastes from titanium dioxide pigment manufacture, rice hulls and waste paper, these being selected from preliminary laboratory studies. Based on the results obtained with different pastes, two formulations were finally selected for further study. The first had fly ash (> 50 wt%) and cellulose as main components (Type 1)[25] and the second fly ash (60 wt%), Portland cement or lime, and vermiculite (Type 2).[26] As a major variation of the Type 2 formulation, a composition was used in which titanium waste (> 35 wt%) was the second most important ingredient, replacing the Portland cement and lime, and in which vermiculite was added in small quantities (< 10 wt%).[27] The titanium waste (RTi) is from the initial sulfuric acid attack on ilmenite in the production of titanium dioxide. In the presence of fly ash at high temperatures, it may generate different thermally stable titanates, and as such improve fire resistance. The exfoliated vermiculite was added to the mixture to increase its porosity and decrease its density.

Depending on the residues used, certain pre-treatment steps may be required so that mixing and blending can be carried out effectively to produce a smooth paste. The components were then mixed according to the formulation and put into a planetary mixer or a pug mill with sufficient water to provide the appropriate fluidity and the mixture blended to a smooth paste. Simple moulding or compaction, as well as vacuum filtration with pressure compaction were used as moulding techniques. Curing was carried out simply at ambient conditions or after special thermal treatment at temperatures up to 1000 °C, following a predetermined thermal ramp (firing).

Fireproof Sprayed Surfaces

The performance of the fireproof compositions was tested with a unit designed for wet and dry guniting. Spraying was restricted to Type 2 formulations. Mixing was carried out in a 50 l blender fitted with 10 blades attached at 45° along a shaft attached to a 1.5 kW variable-speed geared motor (0–50 rpm). For wet guniting, a positive-displacement pump (flow 0–150 l h^{-1}) was used. For dry guniting, a cone-shaped injector (Venturi) was designed, with air supplied at 120 m^3 h^{-1} at 3 bar pressure to a variable-diameter injector, which guarantees speeds of around 40 m s^{-1} inside the injector. Such flows can handle air/paste ratios of around five. Two types of spray nozzle were designed for injecting air or water: with outlet diameters between 10 and 20 mm spraying rates of 20–90 m s^{-1} (appropriate for this type of material) could be tested. The same nozzles can be used for either wet and dry guniting. The paste was sprayed on an adaptable, concave surface with a total spraying area of 1 m^2, from which 30 × 20 cm test pieces could be obtained.

Production

In the Type 1 samples, the amount of paper in the mixture was optimised to show the best thermal behaviour. At the same time, different thermal treatments for improving the insulating properties of the samples were tested. Type 2 samples had as major constituents fly ash, RTi and vermiculite, to which, optionally, Portland cement or lime or aluminous cement was added as a major additional ingredient. Different methods of moulding and curing were tested and the influence of vermiculite particle size on the insulation properties of the mixtures was studied. Figure 4.26 shows some of the test pieces.

Figure 4.26   Fireproof plates, 33 mm thick.

Once suitable conditions for guniting the mortar were found, these were further optimised by studying the influence of the proportion of fly ash, the thickness of the spray and the nozzle type. To facilitate comparison with the fireproof plates, samples of the sprayed materials with similar thickness (20–35 mm) were obtained. Figure 4.27 shows the guniting process.

Thermal Properties

A special furnace (Figure 4.28) was constructed so that the moulded plates and the sprayed surfaces could be subjected to the Spanish standard fire resistance test (UNE 23-093-81). The furnace was equipped with two doors, one of which had a bracket for plates (or sprayed

Figure 4.27   Spraying (guniting) fireproof mortars.

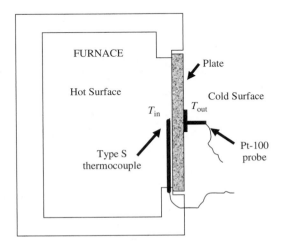

Figure 4.28   An insulating capacity test furnace.

surfaces), which would form the surface exposed to the normalised temperature–time heating curve. With the data provided by this furnace, the thermal behaviour of the products was studied. The resistance to fire, the so-called fire-rate, expressed by the time required for the unexposed side of the plate (the cold surface) to reach 180 °C (t_{180}), was measured. The regulation includes materials with fire-rates varying from 15 to 360 minutes. For materials sprayed on to metal, the standard considers the time taken for the metal surface to reach 500 °C (t_{500}), as this is a critical temperature at which a metallic structure starts to lose mechanical strength. This was measured by inserting a thermocouple into the sample, between the metal surface of the support and the sprayed material.

Figure 4.29 shows heating curves for Type 1 and Type 2 plates. The resulting t_{180} values are between 75 and 90 minutes, which is considered good compared to current commercial fire-retardant plates. There is a clear difference in shape of the curves, due to transformations in the Type 2 plate as vermiculite loses water. Thus Type 2 plates with lime replacing Portland cement showed better insulating behaviour than identical plates manufactured only with Portland cement, as a consequence of the water released by the thermal decomposition of calcium hydroxide. However, although the thermal properties were better, the mechanical strength of the plates containing lime was clearly less than that of the Portland cement.[26] Furthermore, based on the power supplied by the furnace,

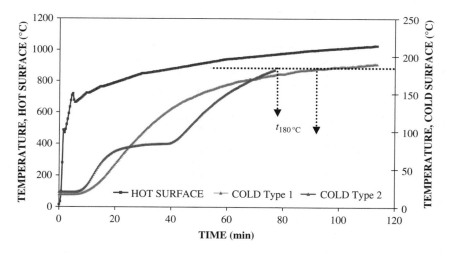

Figure 4.29 Heating curves for Type 1 (25 mm thick) and Type 2 (33 mm thick) plates.

the temperatures of the exposed and unexposed sides of the samples and the dimensions of the samples (exposed area and thickness), values for the thermal conductivity of the material, k, were estimated, from Fourier's Law, to be of the order of 0.15–$0.28 \, W \, m^{-1} K^{-1}$ at $800 \, ^\circ C$. As a comparison, values between 0.13 and $0.35 \, W \, m^{-1} K^{-1}$ for commercial fire retardants are reported in technical specifications.

Environmental Quality

In order to apply fireproof materials containing industrial residues in construction projects, in many countries environmental standards must be met. Therefore, the products were subjected to two of the most commonly used leaching tests, TCLP and DIN 38414-S4, and an ecotoxicological assessment of the leachates was carried out. In all cases, the concentrations of metals are far below the limits generally in force. Also, none of the samples exceeded the Spanish regulations for hazardous waste in the *P. phosphoreum* luminiscence test and the *D. magna* inhibition test.[25,27]

Conclusions

The results from this project showed that it is possible to produce fireproof products competitive with other commercial fire-resistant products, using fly ash and residues as main components. In summary, this project has achieved the following:

• the manufacture of new fireproof products with properties similar to or better than those of existing commercial products;
• a proposal for a simplified methodology for the evaluation of fireproof products;
• the design and construction of a versatile gunite unit to test insulating mortars;
• the manufacture of fireproof products using cost-effective methods.

New Developments

After the finalisation of the CEFYR project, AICIA has carried out studies in which it has been proven that different conventional fly ashes

(ASTM Class F) from different power plants, independently of the coal burned and the combustion conditions, may be used to manufacture insulating and fireproof plates, similar to those obtained from the ashes from the Los Barrios power station. Thus the effect of different ashes was studied using gypsum as binder.[85] The results indicate that all the mortars have a similar insulating capacity.

On the other hand, the recent proliferation of new thermal processes for energy production – for example, the combustion or gasification of biomass, or of coal–biomass mixtures – has generated new types of ash and slag that are different from those generated in coal combustion and that may have interesting properties for use as constituents of fire-resistant products.

Thus AICIA has studied the behaviour of such ashes from the combustion and gasification of waste biomass produced by the extraction of olive oil and from the co-combustion of coal and such biomass with the objective of creating new routes for the recycling of these residues, using their cementing and insulating properties. Thus, the influence of the type of ash on the insulating and fire-resistant capacity of mortars containing a high proportion of these ashes has been studied.[86,87]

In these studies, it has been shown that mortars with a high proportion of biomass ash show better insulating properties than the coal ash mortars. The reasons for this seem to be the properties induced by the high unburned material content (LOI) and the alkali content of the ashes. These characteristics are manifested in the mortars by a high water storage capacity that favours their fire resistance. Thus, it shows that some physical–chemical properties of the ashes that can invalidate their use in more classic applications, as in cement and concrete, can be very advantageous in regard to the insulating properties of the mortars and their application as fire-resistant products, opening up new prospects in this field.

4.7.9 State of Development

In general, the degree of development of the potential applications described in the bibliography and patents in this section is difficult to evaluate, due to the commercial secrecy used by the insulating materials manufacturers in relation to the complete description of the materials and the use of residues in their composition. Thus the lack of specific legislation in many countries defining the characteristics that the residues should meet in order to be recycled as fire-resistant materials in certain applications (e.g. as passive protection against fire in industrial buildings), can make the use and development of this type of product difficult.

The fact that many of the patents have been filed by the inventors themselves suggests a situation in which the majority of the patents referred to in this section are not yet being exploited. In some cases, the patent applicants are companies that actually market fire-resistant products, but they do not usually indicate which of their products contain combustion residues, or under what patent number their products are covered.

Some data concerning companies that commercialise fire-resistant products containing fly ash may be found on the Internet. For example, TG Advance Concrete Co. Ltd (Thailand) and Prime Structures Engineering Pte Ltd (Singapore) produce fire-resistant lightweight precast concrete panels (named Finewall) from Portland cement type I, fly ash, lime, recycled cellulose, chemical additives and water (more information may be found at www.sabyehome.com): see also the previously mentioned *Tintherm*® hot insulated pipe supports.

4.7.10 Conclusions

The large number of research groups that have generated bibliographical references in this field over recent years indicates the interest in the recycling of combustion residues in fire-resistant materials. In addition, the level of technological innovation of the research as shown by the numerous patent applications indexed in this section must be emphasised. This is firstly motivated by the good insulating characteristics of some of the constituents of the residues and, secondly, the availability and purchase price of these secondary materials makes them very competitive with virgin raw materials as constituents of fire-resistant materials. Consequently, the added value recycling these residues in such applications is very high.

On the other hand, as has been noted previously, new types of combustion or gasification residues may contribute favourably to the desired properties of fire-resistant materials produced from such residues. Thus, a promising future can be envisaged for this application, although still greater technological developments are required.

References

1   American Society for Testing and Materials, *Standard terminology of fire standards.* *ASTM E-176-89a.* Philadelphia (1989).
2   Kojima, A. and Shimada, K., *Ceramic compositions containing coal ashes, shaped ceramic products, and their manufacture.* JP Patent 09100153 A2 (1997).
3   Shiotani, T., *Manufacture of ceramic molding from coal ash.* JP Patent 2000086348 A2 (2000).
4   Querol, X., Umaña, J.C., Alastuey, A., Bertrana, C., López-Soler, A. and Plana, F., *Physico-chemical characterization of Spanish fly ashes.* Energy Resources, 21(10): 883–898 (1999).
5   Häussler, K. and Schlegel, E., *Calciumsilicat-Wärmedämmstoffe.* Freiberger-Forschungshefte. A834 Grundstoff-Verfahrenstechnik Silikattechnik. Technische Universität Bergakademie Freiberg (1995).
6   Mallow, W.A., *Recycled material.* In Proceedings of the 3rd Materials Engineering Conference, ASCE, NY, pp. 828–835 (1994).
7   Borst, B. and Krijgsman, P., *Hydrothermal synthesis of light-weight insulating material using fly-ash.* In Waste Materials in Construction, J.J.J.R. Goumans, H.A. van der Sloot and Th.G. Aalbers (eds). Elsevier: Amsterdam, the Netherlands, pp. 659–662 (1991).
8   Mie, K., *Manufacture of calcium silicate porous substance, used as heat resistant material, involves performing hydrothermal process of boiled and settled refuse incineration ashes in presence of oxidizing agent and alkali.* JP Patent 2001151506-A (2001).
9   Hara, N., Inoue, N., Noma, H. and Inoue, K., *Utilization of rice husk ash for calcium silicate products.* In Proc. Beijing Int. Symp. Cem. Concr., 4th, Z. Wu (ed.). International Academic Publishers: Beijing, Peoples Rep. China, 2: 466–472 (1998).
10 Sun, Y., Sun, X., Zhou, J. and Yu, X., *Manufacture of calcium silicate thermal insulating fire-resistant material.* CN Patent 1204678 A (1999).
11 Burtscher, W., *Manufacture of cement-free hydrothermal hardened molded articles and their application.* WO Patent 200035826 B1 (2000).
12 Strabala, W.M., *Fly ash and lime powder composition for production of strong lightweight building materials.* WO Patent 9709283-A (1997).
13 Yang, E., Sun, H., Ma, H. and Dong, W., *Manufacture of light weight wallboards from low cost raw materials.* CN Patent 1210831 A (1999).
14 Clarke, L.B., *Applications for coal-use residues.* IEACR/50. IEA Coal Research, London (1992).
15 Poon, C.S., Azhar, S., Anson, M. and Wong, Y.L., *Comparison of the strength and durability performance of normal- and high-strength pozzolanic concretes at elevated temperatures.* Cem. Con. Res., 31(9): 1291–1300 (2001).
16 Xu, Y., Wong, Y.L., Poon, C.S. and Anson, M., *Impact of high temperature on PFA concrete.* Cem. Con. Res., 31(7): 1065–1073 (2001).
17 Nasser, K.W. and Marzouk, H.M., *Properties of mass concrete containing fly ash at high temperatures.* J. Amer. Concr. Inst., 76(4): 537–551 (1979).
18 Frech, K. and Bos, J.W., *Polypropylene fiber containing concrete mixture with increased fire resistance.* EP Patent 1156022 A1 (2001).
19 Fujimasu, J., *Manufacture of lightweight aggregate using waste polyvinyl chloride resins and cement-based composition for high-strength lightweight concrete using the aggregate.* JP Patent 2001240441 A2 (2001).

20 Zhao, C., Li, R. and Zhang, Y., *Manufacture of light-weight fly ash haydite for construction.* CN Patent 1264688 A (2000).
21 Kunihija, S., Ishii, M. and Murai, H., *Concrete mixing material for spray type concrete for fireproof coating*, JP Patent 11180743-A (1999).
22 Tong, Z., *Energy-saving, imperious, water-proof and insulating mortar for building.* CN Patent 1313260-A (2001).
23 Stav, E., Burkard, E.A., Finkelstein, R.S., Winkowski, D.A., Metz, L.J., Mudd, P.J., *Cementitious binder – contains calcium sulphate hemihydrate, Portland cement, fly ash or slag, and silica fume or rice husk ash.* WO Patent 9852882-A; EP Patent 991606-A (1999).
24 Vilches, L.F., Fernández-Pereira, C., Leiva Fernández, C., Olivares del Valle, J. and Vale Parapar, J., *Use of fly ash in a sprayed mortar for the passive protection against fire of metallic structures.* In 2001 International Ash Utilization Symposium. 22–24 October, Lexington. KY (2001).
25 Vilches, L.F., Fernández-Pereira, C., Olivares del Valle, J., Rodríguez-Piñero, M. and Vale, J., *Development of new fire-proof products made from coal fly ash: the CEFYR project.* J. Chem. Tech. Biotechnol., 77: 361–366 (2002).
26 Vilches, L.F., Fernández-Pereira, C., Olivares del Valle, J. and Vale, J., *New insulating elements made of coal fly ash and lime to be applied in fire doors.* G159/2002, Wascon 2003, San Sebastián, Spain (2003).
27 Vilches, L.F., Fernández-Pereira, C., Olivares del Valle, J. and Vale, J., *Recycling potential of coal fly ash and titanium waste as new fire-proof products.* Chem. Eng. J., 95: 155–161 (2003).
28 Tanabe, S., *High strength thin ceramics and metallic mesh sheet manufacture*, JP Patent 5279136-A (1993).
29 Oda, K. and Oda, M., *High-strength ceramic composite boards and their manufacture from fly ashes.* Jpn. Kokai Tokkyo Koho JP 10025170 A2 (1998).
30 Han, X. and Chen, L., *Manufacture of light weight cement board with steel mesh frame for building construction.* CN Patent 1277946 A (2000).
31 Fraval, H.R. and Emens, G.W., *Prefabricated structural panel for use in buildings.* WO Patent 200181267-A2 (2002).
32 Shiu, H., *Light-weight mineral fireproof complex plate and the cement material fill for the layered partition wall – with the harmless ingredient.* TW Patent 411374-A (2001).
33 Bykova, E.V., Korshunova, G.K., Dorofeev, A.A. and Laricheva, N.F., *Production of multi-layer thermal insulation panels by end-to-end assembly of individual blocks formed by outer layers and filler.* WO Patent 200116442 A (2001).
34 Shishu, R.C., Jagadeshwariah, S. and Pattabhi, V., *Prefabricated light-weight construction blocks and panels.* AU Patent 9714731-A (1998).
35 Li, J., *One-step formation tempered fire-resistant colour board.* CN Patent 1198418-A (1999).
36 Heise, M., *Thermal insulation, useful at low and high temperature for technical plant, e.g. in refinery, power station food or paper industry or agriculture, consists of ash and water glass.* DE Patent 19930653-A1 (2001).
37 Nozaki, Z. and Yamamoto, M., *Inorganic laminates having high durability and fire resistance for building materials.* JP Patent 07304124 A2 (1995).
38 Niu, T. and Shi, D., *Fly ash baking-free non-steaming light partition board.* CN Patent 1137596-A (1998).

39 Lu, L., Zhou, C. and Lu, Y., *Manufacture of multi-function decoration board*. CN Patent 1128774 A (1996).
40 Liu, Y., Chen, M., Li, H., Liu, Z. and Shang, Z., *Mold-pressed composite wall and floor and their manufacture*. CN Patent 1218775 A (1999).
41 Nagatani, Y., Okamura, K., Fukushima, Y. and Yoshida, M., *Binder compositions for building materials with excellent durability and water resistance*. JP Patent 2000001659 A2 (2000).
42 Sakamoto, M., Kamiya, M. and Hayamizu, Y., *Inorganic hardened material for purifying atmosphere*. JP8196902-A (1996).
43 Mader, H.B., *Bricks and boards, and their manufacture*. DE Patent 4432019 A1 (1996).
44 Kehr, E. and Krug, D., *Fire-resistant constructional material from waste products*. DE Patent 19625251-A1 (1998).
45 Yu, F.C., *Fireproof board for use as building board*. EP Patent 867573-A (1998).
46 Han, M., Li, B. and Wang, G., *Research on production of lightweight building materials from fly ash*. In Proc. 15th Ann. Int. Pittsburgh Coal Conf., University of Pittsburgh, pp. 1093–1100 (1998).
47 Sanuki, I., *Extrusion moulding of fire resistant strong inorganic sheet material*. JP Patent 5318437-A (1994).
48 Tian, X., *Fire-resistance, lightweight partition for use in buildings*. CN Patent 1280962-A (2001).
49 Psyllides, A.M., *Fly ash-based mixtures for manufacturing of construction materials with predetermined properties*. GB Patent 2344341 A1 (2000).
50 Cai, J., *Process for synthesising fireproof decoration board*. CN Patent 1103830-A (1997).
51 Strabala, W.M., *Fire resistant, insulating structural products*. WO Patent 9622952-A (1996).
52 Gerischer, G.F.R., Van Wyk, W.J., Ysbrandy, R.E. and Crafford, J.G., *New ideas on the use of deinking residues*. PTS-Symp., PTS-SY 01/98 (PTS-CTP-Deinking-Symposium, 1998), 41/1–41/13 (1998).
53 Kamio, N., Kimura, K., Suzuki, M. and Kawamura, K., *Cement water-based slurry compositions for lightweight building materials and manufacture thereof*. JP Patent 09086995 A2 (1997).
54 Maeda, A. and Yoshikawa, A., *Hardening of sewage sludge incineration ashes and hardened body*. JP Patent 09155316 A2 (1997).
55 Lee, H.T.E., *Fire-resistant door or door frame*. GB Patent 2285470-A and 2285470-B (1995).
56 Lee, H.T.E., *Fire resistant frame for doors, etc*. SG Patent 41918-A1 (1998).
57 Blair, J.H. and Stewart, M., *Method for manufacturing the fireproof door where fireproof core is made of foam perlite, gypsum, pulverized coal ash, cement and fiber glass*. TW Patent 402658-A (2001).
58 Shitiwatt, M. and Berl, J.S., *Production method of safety fire-resisting door*. CN Patent 1263200-A (2002).
59 Wexler, J., *Method of producing fire-resistant panels used as a fire-resistant core for a fire-door leaf*, NZ Patent 333778-A (2000).
60 Bykova, E.V., Korshunova, G.K., Dorofeev, A.A. and Laricheva, N.F., *Silicate composition for manufacture of foamed thermal insulation material using liquid glass binder, refractory filler, hardener and modifier*. WO Patent 200073238-A (2001).

61 Tanabe, E. and Matsuura, K., *Compositions for ceramic foams for heat-resistant coatings on high-strength concrete and steel frames.* JP Patent 10025174 A2 (1998).

62 Krafft, A.P., *Foamable paste and multicomponent paste compositions for obtaining porous inorganic rigid foam products, the fireproofing porous rigid foam products obtained, and process for their formation.* WO Patent 9725291 A2 (1997).

63 Stephens, D.K., *Manufacture of fine-celled foam composition having improved thermal insulation and fire retardant properties.* WO Patent 2001085638 A1 (2001).

64 Fujimasu, J., *Fire-resistant resin compositions.* JP Patent 08048808 A2 (1996).

65 Argade, S.D., Shivakumar, K.N., Sadler, R.L., Sharpe, M.M., Dunn, L., Swaminathan, G. and Sorathia, U., *Mechanical and fire resistance properties of a core material.* In International SAMPE Symposium and Exhibition, 49(SAMPE 2004), pp. 281–292 (2004).

66 Zhu, H., Xu, Y. and An, B., *Inorganic heat insulating, sound-proof and fire-proof foam material is made from magnesium oxide, powdered coal ash, magnesium chloride solution, foamer, additive and reinforcing material.* CN Patent 1150988-A (2001).

67 Zhao, F., *Composite foamed material, used as insulating material in buildings.* CN Patent 1193615-A (1999).

68 Chen, W., *Fire-retardant and smoke-inhibiting type polyvinyl-chloride compositions and its preparation method.* CN Patent 1288020-A (2001).

69 Blount, D.H., *Production of flame retardant open-celled porous organic mass.* US Patent 5721281-A (1998).

70 Arai, T., Mochizuki, A., Sawada, H., Yoshina, A. and Mizuno, Y., *Fire resistance electric wire with improved fire resistance.* JP Patent 8335412 (1997).

71 Chow, S. and Casilla, R.C., *Fire retardant formulation for lignocellulosic and plastic materials.* CA Patent 2161315 AA (1997).

72 Rzechula, J., Werling, J.Z., Wrona, J., Czerwinski, Z., Marjanowski, J., Ostrowski, J., Dabrowski, A., Adamczyk, M., Kras, K. and Karski, E., *Sealing compound for use on fire-resistant partitions, especially electric cable and ventilation duct passages.* PL Patent 178360 B1 (2000).

73 Roth, T.J., Jones, D.R., Grzybowski, K.F. and Welliver, W.R., *Fire retardant plastic construction material composition.* US Patent 5658972-A (1997).

74 John, A.J. and Quist, P.I., *Aqueous fire-resistant compositions containing metal silicates, calcined reactive fillers, and latent acid catalysts.* WO Patent 2001044404 (2001).

75 Belikov, A.S., Krikunov, G.N. and Stankevich, S.N., *Fire-resistant coatings used in building operations.* SU Patent 1805118-A1 (1994).

76 Krivtsov, J.V. and Ladygina, I.R., *Fireproof composition for concrete, metal and wood.* RU Patent 2140400 (1999).

77 Yamamoto, K., *Fireproof insulation lining material for incinerator in chimney of industrial boiler.* JP Patent 8145323-A (1996).

78 Himeno, M., *Ash-containing curable compositions and their use for flexible dimensionally stable lightweight weather-and fire-resistant thermally-insulating building materials.* JP Patent 2001220518 A2 (2001).

79 Sengirik, K. and Azuma, Y., *Fire resistant sheet contains predefined amount of fly ash, metal hydroxide and polyolefin.* JP Patent 2000144581-A (2000).

80 Lu, P., Wang, Z. and Tian, F., *Fly-ash insulating fireproof felt and production and mixing method.* CN Patent 1170073 (1998).

81 Ion, M., Lencu, V. and Calota, S., *Fireproof silicate paints and their manufacture for wood.* RO Patent 108873 B1 (1994).
82 Tanaka, K., *Electroconductive, fire resistant resin compositions.* JP Patent 6128411-A (1994).
83 Wakamura, K. and Ohsawa, T., *Fire-resistant polyamide compositions and covers for segment bolt boxes therewith having high bending and compression strength.* JP Patent 11043601 A2 (1999).
84 Ahluwalia, Y., *Substrate coated with a low bleed composition having the same ionic charge.* WO Patent 9900338-A; EP Patent 991602-A (1999).
85 Leiva, C., Vilches, L.F., Vale, J., Olivares, J. and Fernández-Pereira, C., *Recycling of power station coal fly ashes as building elements to be used in passive fire protection.* In International RILEM Conference on the Use of Recycled Materials in Building and Structures, E. Vazquez, Ch.F. Hendriks and G.M.T. Janssen (eds). RILEM, Bagneux, France, pp. 881–889 (2004).
86 Leiva, C., Vilches, L.F., Vale, J. and Fernández-Pereira, C., *Influence of the type of ash on the fire resistance characteristics of ash–enriched mortars.* Fuel, **84**: 1433–1439 (2005).
87 Vilches, L.F., Leiva, C., Vale, J. and Fernández-Pereira, C., *Insulating capacity of fly ash pastes with a potential use for passive protection against fire.* Cement and Concrete Composites, **27**: 776–781 (2005).

4.8   Fly Ash as a Replacement for Mineral Fillers in the Polymer Industry*

Henk Nugteren and Richard Kruger
Delft University of Technology, The Netherlands; Ash Resources (Pty) Ltd, Republic of South Africa

Types of combustion residues involved

The polymer industry comprises a huge variety of processes and products, all with their specific properties and requirements. The range of fillers used is very broad and much depends on the functionality sought. Therefore, in theory all types of combustion residues may be potential fillers in polymer products. However, fly ashes containing predominantly spherical particles are much in favour, because no natural mineral filler may offer this functionality. High sphericity gives many advantages in workability and material properties. Another important factor influencing functionality is particle size. In general, the finer the particle size, the better for most applications, although concentrates of the larger cenospheres may be required for their low density. Besides the size itself, the size distribution is important. Well-graded fillers are often an advantage and those may be obtained from fly ash by size classification technologies. A third important criterion for some applications is the colour of the filler; in general, the lighter the better.

State of development

As reported below, there are some current industrial uses of fly ash fillers in paint. Much research work has been done and is under way, notably in South Africa, on the use of ash fillers in rubbers and plastics. The fact that very often this is done with poorly characterised fly ashes shows that we are still in the beginning of the process of understanding the influence that ash properties have on the properties of polymer products. Much testing is still required to achieve comprehensive formulation for different polymeric products using highly specified combustion residues as functional fillers. The potential is there and the future seems bright.

---

*Part of this section has been published as Nugteren, H.W. and Kruger, R.A., *Fly ash as a replacement for mineral fillers in the polymer industry*. In World of Coal Ash Conference (WOCA), Lexington, Kentucky, 11–15 April 2005 (Proceedings on CD, paper #146).

4.8.1   Introduction

Strictly speaking, polymers can be considered composite materials, since functional fillers are often incorporated in order to provide specific properties. Phenolic resins would never have been extensively utilised were it not for the incorporation of mineral fillers, which facilitated the required engineering properties and functionality.

While commodity resins such as polyvinyl chloride (PVC), polystyrene (PS), polyethylene (PE) and polypropylene (PP) are often sold as pure resins, price escalations and possible uncertainty in petroleum feedstocks, as well as increased performance criteria, have established a widespread market for functional fillers.

Filled polymer composites extend the available volume of resins, improve many of the physical properties and are generally cheaper. In certain instances, they facilitate faster production cycles and better dimensional stability.

4.8.2   Filler Characteristics

No filler can be considered as ideal and there are many inherent properties – for example, morphology, particle size distribution, density, colour, oil absorption, thermal and electrical characteristics, and so on – which need to be exploited in order to provide the required functionality.

The advent of spherical fillers recovered during the beneficiation and processing of fly ash opens up new and exciting opportunities in the polymer industry. Sphericity is an important feature that provides processing advantages whilst imparting specific characteristics to the polymer. The use of a sphere allows maximum volume to be filled at a minimum surface area. The contact surface between the organic matrix and the inorganic filler is thus minimised, and along with it the resin demand (oil absorption). Spheres aid processing since they promote flow, thus improving the rheology, and decreasing the viscosity during compounding and extrusion. Filler loadings higher than for conventional fillers can be achieved without sacrificing processability. Spherical fillers recovered from fly ash are considerably cheaper than both their artificially manufactured equivalents and the polymer matrix that they replace. The financial implications are obvious. Despite these significant advantages, the commercial use of fly ash based spherical fillers is not widespread, with the possible exception of South Africa.[1]

Apart from sphericity, there are a variety of other technical and commercial requirements. The spheres must be readily available in various ranges of particle size and their properties must be consistent. South African coals have a low energy value, necessitating precise and efficient combustion conditions for effective utilisation. As a result, fly ash with a very low carbon content ($< 0.5\,\%$) is produced in quantities of up to 6 million tonnes annually at individual power stations. Beneficiating the ash from this huge resource allows the extraction of only the highest-quality spheres, which are then used as fillers in the polymer industry.

Two grades of micro-spheres are recovered from fly ash by aerodynamic classification. The coarser material has a d_{50} of $11.5\,\mu$m, a d_{90} of $52\,\mu$m, and a d_{99} of $110\,\mu$m. The finer material had a d_{50} of $3.8\,\mu$m, a d_{90} of $8.5\,\mu$m, and a d_{99} of $19.5\,\mu$m. These materials are on the market in South Africa under the trade names *Plasfill 15*® and *Plasfill 5*®, respectively. They consist of glass spheres encapsulating a skeletal of mainly mullite, as well as traces of magnetite and quartz.

Changing to accept new fillers will meet with resistance from a polymer industry that is well versed in the performance, quality and benefits derived from conventional fillers. In the following sections, opportunities for the use of fly ash as a filler in a variety of polymer products are reviewed. A comprehensive introduction to the use of industrial minerals and fly ash as fillers is given by Ciullo,[2] as well as by Katz and Milewski.[3]

4.8.3 Paint

Worldwide, the average annual production of paint is estimated to be about 10^{10} litres,[2] consuming about 10 million tonnes of fillers. Paint formulations consist of the polymeric binder; a solvent; pigments, both organic and inorganic; additives; and functional fillers.

Functional fillers such as talc, mica and kaolinite both extend the polymeric film and, by virtue of their plate-like morphology, improve the durability and tensile properties. The rheology and gloss are enhanced by spherical fillers such as precipitated silica and/or barytes. All these functional fillers are extremely fine, generally in the sub-micron range, and predominantly white. In addition, calcium carbonate ($3-7\,\mu$m) is used as a bulk extender. The fineness of these fillers is crucial in maintaining gloss and brightness.[4]

It is thus obvious that formulating a paint is a complex procedure, requiring a fine balance between the components and the total amount of minerals used.[2] Exceeding the critical pigment volume concentration (CPVC) will reduce the gloss and surface characteristics of the paint (Figure 4.30).

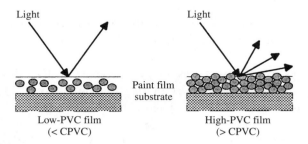

Figure 4.30   The effect of filler particles on gloss. If the pigment volume concentration (PVC) excesses a certain critical value (CPVC), filler particles are forced to extend outside the surface of the paint film. This will, of course, also happen if particles have a larger diameter than the thickness of the paint film (after Ciullo[2]). Reproduced by permission of William Andrew Inc.

While fly ash is often mentioned as a possible filler for paint, commercial applications are limited primarily due to the lack of whiteness, the relatively large particle size of fly ash and the high pH of fly ash, which could necessitate significant reformulation. Reluctance to change is thus understandable. Despite this, Pruett-Schaffer produces a range of waterborne and solvent-based coatings utilising both fly ash and cenospheres as fillers. Cenospheres are the lightweight fraction (RD = 0.6–0.7) of fly ash that float on ash disposal ponds. A paint containing cenospheres has been developed for aluminium rail cars. In this case, the large size of the cenospheres ($> 75\,\mu m$) is used to aid flow, improve drying and provide opacity.[5]

Smaller, higher-density fly ash spheres are also used as a pigment extender, especially for the darker paints, so that substantial reduction in the pigment loading, for example Fe_2O_3, is achieved. Fly ash also provides an improvement of up to 30 % wear resistance for road marking paints.[5] One of the more significant improvements is the development of a solvent-free coating that can be applied over damp surfaces and adheres to steel, galvanised metal and aluminium. As proof of their paints' performance, Pruett-Schaffer have painted several transmission towers for American Electric Power. The results are excellent.[5]

Pruett-Schaffer also report that fly ash fillers are used in high-temperature paints for automotive and industrial applications. Furthermore, they are used for deck coatings on aircraft carriers and automobile undercoatings. In South Africa, cenosphere-filled underbody coatings are used for motor cars. In this case, the cenospheres aid application to the substrate, and by virtue of the fact that they encapsulate a partial vacuum, provide excellent sound-insulating characteristics.[1]

Valspar use fly ash in high-build coatings for transmission towers, epoxy mastics, floor coatings, high-build urethanes and some primers.[6] Test work has shown that after 1000 hours in a salt spray chamber (ASTM B117), fly ash filled coatings outperform the same product with conventional fillers. After 3500 hours of exterior exposure (ASTM D1014), fly ash filled coatings remain unaffected.

Because of their sphericity, particle size distribution and flowability, fly ash allows improved packing density and subsequently a reduced requirement for the most expensive ingredient – the resin. In addition to their technical advantages, paints containing fly ash are thus also cheaper to formulate.[7,8]

In a study commissioned by KEMA, SKIM INDIS investigated the possibility of using washed fly ash. It was found that black paints manufactured for the underwater parts of ships benefits from fly ash as a filler.[9]

A new type of elastic waterproof emulsion coating was prepared using fly ash as filler.[10] The material could be applied at low temperature (0–5 °C) and showed good film forming performance, high tensile strength and good extensibility.

4.8.4   Rubber (Elastomer)

Elastomer is the generic name for a class of materials (rubber) that exhibit elastic recovery after significant deformation. Their additional properties of impermeability to gases, excellent chemical resistance and tenacity make elastomers an indispensable part of modern industry. There are a wide variety of rubbers, both natural and synthetic,[11] with a total world consumption of currently about 19.5 million tonnes per year.

Natural rubber (NR) is obtained from plants containing latex. The most important of these is *Hevea brasiliensis*. Other sources are Guayule, Gutta Percha and Balata. All NR exhibits different processing and vulcanisation properties, depending on both the climate and soil upon which the plants have been grown as well as the methods of recovery. NR is classified according to the Standard Malaysian Rubber (SMR) system.

Synthetic rubber (SR) is more consistent in its properties and is differentiated into general-purpose rubbers such as styrene–butadiene rubber (SBR), butadiene rubber (BR) and isoprene rubber (IR). In addition, there is a large variety of speciality rubbers engineered to provide specific properties. Included in these are chloroprene rubber (CR), ethylene propylene rubber (EPDM), and so on.

Both NR and SR utilise a wide variety of additives and functional fillers to produce the desired engineering properties, aid compounding, improve processing, counteract degradation or reduce cost.

Formulating rubber is a complex process, since some or all of the following are required: activators, accelerators, protective agents, vulcanising aids, processing aids, peptisers, retarders, tackifiers, extenders and reinforcing functional fillers such as carbon black and ultra-fine silica.

Approximately 2 million tonnes of functional fillers are consumed annually in the rubber industry. Of these, 70 % is carbon black, 15 % kaolinite, 8 % $CaCO_3$ and 4 % precipitated silica. Furthermore, other minerals such as talc, barytes and diatomites are also occasionally used. The term 'filler' is rather misleading, as it suggests inexpensive inert diluents that occupy space of the more costly primary material, the elastomer. This might have once been the case, but a 100 % pure unfilled elastomer is of little use. So, nowadays, all fillers perform a function. In rubber, this is mainly to improve processability and the properties of the end-product. Important filler properties for rubber are as follows:

(a) *Particle size.* In general, filler particles in rubber are below $10 \mu m$ and it is often considered that the smaller the particles, the better the rubber. Figure 4.31 shows the particle ranges of the different fillers in rubber. The main reinforcing fillers are indeed the aggregated nano-particles of carbon black (on the right, denoted by their quality codes, N220 to N990) and precipitated and fumed silica. Larger particles are less reinforcing but have other specialised function, such as, for example, the barytes in increasing the density.

(b) *Surface area.* Preferably up to $20–200 \, m^2 \, g^{-1}$, the larger the better – an anisometric particle shape is favoured.

(c) *Structure.* Fractal-like aggregates of small particles are preferred rather than single primary particles, thus giving the filler a high structural index.

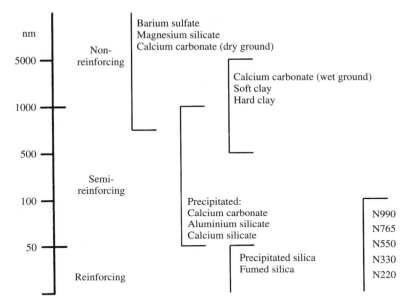

Figure 4.31 The particle size range of fillers in rubber (after Ciullo[2]). Reproduced by permission of William Andrew Inc.

(d) *Surface activity.* The surface area must be chemically active so that the elastomer can attach to it. Carbon black has functional active groups; other fillers may have some or none at all. However, the surfaces of such fillers can be treated by silanisation to provide activity for adhesion. Figure 4.32 illustrates how the OH-groups of a mica particle interact with the silane molecule and how this facilitates the bridging to and adhesion on to the polymer.

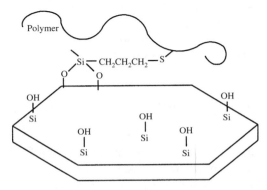

Figure 4.32 Silane coupling: the surface treatment of a mica particle by silanisation to make it available for polymer attachment (after Ciullo[2]). Reproduced by permission of William Andrew Inc.

The main properties to be considered when fly ash is used as a filler in rubber are the particle size distribution, particle morphology, surface properties and colour. Most fly ashes will probably be too coarse. A beneficiation process is thus required to separate and recover the finer fractions of the ash. Thus in South Africa, *Plasfill 5* and *Plasfill 15* are recovered from fly ash for use as a filler. The spherical morphology of the fly ash particles imply a low surface-to-volume ratio, and it is therefore expected that fly ash filler will probably be nonreinforcing. Colour is less important than in paint and paper, but becomes an issue for light-coloured products where carbon black cannot be used.

Garde *et al.* tried to modify the surface of fly ash particles by treatment with acids and bases, but found only limited reinforcement and little bonding between the polyisoprene rubber and the ash particles.[12] The fly ash filler they used had a large surface area ($146 \, m^2 \, g^{-1}$), but even after treatment with bis(triethoxy-silylpropyl)tetrasulfane (TESPT) only little bound rubber formation was found, possibly because of the lack of hydroxy groups on the surface of the fly ash. Therefore, it is not realistic to aim for replacement of carbon black by fly ash. However, comparing the performance of fly ash filler in rubber with other nonreinforcing fillers such as $CaCO_3$ and kaolinite would be of more interest.

Kruger investigated this by examining the effects on processing and properties of the final product while filling natural rubber with $CaCO_3$, kaolinite and fly ash.[13,14] Rubbers were prepared using the *Plasfill* powders, $CaCO_3$, kaolinite and 'whiting' at loadings of 50, 100 and 150 parts per hundred rubber (phr). Due to their low surface area, these fly ashes show relatively low oil absorption (17–18 %) compared to the fillers to be replaced (20–35 %). Kaolinite ($2.60 \, g \, cm^{-3}$) and $CaCO_3$ ($2.75 \, g \, cm^{-3}$) have a higher relative density compared to *Plasfill* powders (2.15–$2.25 \, g \, cm^{-3}$), so that it results in a lower-density rubber. Thus, at equivalent mass loadings the volume of the rubber will increase. This can be a significant economic factor in extending the substantially more expensive polymer.

Compared to the other fillers, the compounding time – that is, the time required for the homogenous incorporation of the filler into the rubber matrix – is significantly shorter (30–70 %) when using fly ash. The ease of processing NR depends mainly on the flow behaviour of the filled rubber compound and is measured by the Wallace rapid plasticity test at $100 \, ^\circ C$. The lubricating effect of the fine fly ash spheres (acting as slip planes between polymer chains) improves the flow properties compared to that of the unfilled natural rubber (Wallace plasticity number 16–18 versus 20). Moreover, when filling with *Plasfill*, the plasticity seems independent of the filler loading. This is in contrast to the other fillers which lose their plasticity at higher loadings (Wallace plasticity number up to 30 for kaolinite loadings of 150 phr). This means that, contrary to conventional fillers, the fly ash may be used at high loadings without affecting the processability. The use of peptisers (organo-metallic complexes), often introduced to reduce viscosity, may be decreased when using fly ash fillers.

The behaviour of rubber during curing or vulcanisation is important, since it determines the process of cross-linking the polymer chains and therefore the physical properties of the final product. Curing characteristics were determined using an oscillating disc rheometer (Monsanto) and results are presented in Table 4.20. The minimum torque value is indicative of the plasticity. With *Plasfill* the values are lower and therefore confirm its positive effect on processability. The maximum torque is a measure of the stiffness of the rubber. Kaolinite has little influence, whereas $CaCO_3$ increases the stiffness and fly ash only does so at high loadings. Measurements of Shore hardness confirm the findings indicated by the maximum torque as presented in Table 4.20. *Plasfill* had no effect on either curing or induction time of natural rubber, whereas $CaCO_3$ reduces the time and kaolinite increases it. This can be a major advantage for fly ash, as it removes a variable from the already complex process of manufacturing a rubber with specific physical properties.

Table 4.20   Rheometer results for the various natural rubber formulations with different fillers at 150 °C. Reproduced by permission of Ash Resources Pty Ltd.[13]

Filler		Minimum torque (inch pounds)	Maximum torque (inch pounds)	Curing time (minutes)	Induction time (minutes)
Type	Level (phr[a])				
Plasfill 15	50	5	46	5.0	2.2
	100	5	46	5.0	2.5
	150	3	62	4.9	2.2
Kulu 15	50	7	48	4.6	2.3
	100	5	56	3.5	1.4
	150	7	67	3.2	1.2
Kaolin	50	6	43	6.3	3.2
	100	5	41	5.8	2.8
	150	6	44	6.4	2.1
Natural rubber	0	10.5	39	5.1	2.1

[a] phr = Parts per hundred rubber.

The physical properties of the rubber produced with the fly ash filler look promising. All fillers reduce the tensile strength of natural rubber, with $CaCO_3$ reducing it the most. The finer fly ash (*Plasfill 5*) performs as well as kaolinite, especially at higher loadings (Figure 4.33). Tear strength is also reduced by all the fillers except for kaolinite, which may even restore tear strength at high loadings. The ability of rubber to stretch before breaking is a very important property for most rubber applications. Compared to natural rubber, all fillers reduce the elongation at break. The least reduction in elongation at break occurs with *Plasfill* (Figure 4.34). The greatest increase in the modulus of elasticity is achieved by using kaolinite as a filler (Figure 4.35). Given the effects on the two last-mentioned properties, it would be interesting to use fly ash and kaolinite together. The fly ash filler would thus have the function of promoting the compounding efficiency and improving the elongation at break, whereas the kaolinite would be used to engineer specific physical properties.

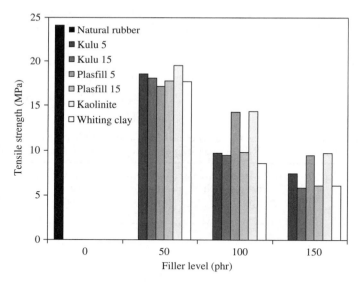

Figure 4.33 The effect of filler loading and filler type on the tensile strength of natural rubber (Reproduced by permission of Ash Resources Pty Ltd[13]).

In a follow-up investigation, Kruger found that when using a mixture of *Plasfill 5* and kaolin, the tear strength increased and the incorporation time decreased in direct proportion to the kaolin content. Kruger also

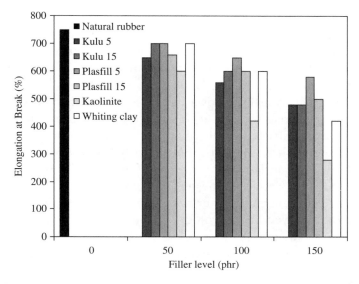

Figure 4.34 The effect of filler loading and filler type on the elongation at break of natural rubber (Reproduced by permission of Ash Resources Pty Ltd[13]).

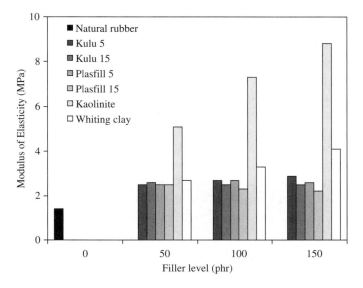

Figure 4.35 The effect of filler loading and filler type on the modulus of elasticity of natural rubber (Reproduced by permission of Ash Resources Pty Ltd[13]).

confirmed that an ultra-fine fly ash recovered from a source in Europe performs similarly to *Plasfill 5* in NR.[1]

Kruger also investigated the use of fly ash in SBR and confirmed that it aids processing.[1] The influence of particle size on processing time was also measured. By virtue of their larger surface area, finer fillers take longer to incorporate. At a filler loading of 100 phr, the increase in time for the finer fly ash filler over its coarser counterpart was marginal (30 seconds), whereas for $CaCO_3$ it was well in excess of 3 minutes.

The effect of particle size distribution was reflected in the plasticity values of the rubber. The ultra-fine fly ash (*Plasfill 5*) filled rubber always flowed easier than SBR using any of the other fillers. The coarser grade (*Plasfill 15*) showed equivalent flow to $CaCO_3$ at low filler levels (50 phr) and it is only at 150 phr that the effect of the spherical filler is reflected in improved flow rheology.

The fly ash only marginally accelerated the cure of SBR, whereas all the other fillers had a much greater effect. This result correlates with that of NR. Rather surprisingly, the ultra-fine fly ash (*Plasfill 5*) reinforced the SBR and at 150 phr reached values equivalent to those of kaolin. These results indicate that although the behaviour of fly ash as a filler in NR and SBR are broadly similar, there are some significant differences. In order to quantify the effect of fly ash, testing and comparison of the results for various formulations are therefore necessary.

In further unpublished work, Kruger showed that when used in conjunction with 20 phr of kaolin, *Plasfill 5* increased the extrusion rate of EPDM rubber by 12–20 %.[1] The swell was reduced along with the increase in *Plasfill* from 20 to 100 phr.

Vishwakarma *et al.* used a completely different approach.[15] Alkali digestion of fly ash followed by acid treatment was used to generate silica and alumina. The alumina was used for water treatment, whereas the precipitated silica was intended for use as filler in rubber. Properties such as purity, bulk density, water and oil absorption capacity, particle size distribution, surface area and morphology indicate that this silica is suitable as a filler in rubber formulations.

4.8.5 Polyvinyl Chloride (PVC)

PVC is one of the most versatile of all plastics and can be used in rigid (high-pressure pipe), flexible (hosing, electrical cable) and highly flowable form (plastisol coatings for tarpaulins).

During the processing of PVC, heat stabilisers, lubricants, impact modifiers, flame retardants, fillers, pigments and, if required, plasticisers are used. The degree of plasticity as well as the physical attributes are determined by the type and amount of plasticiser used. The ingredients are mixed in a high-speed blender to produce a powder, which can then either be processed (compounded) directly or used to produce a granulate for conversion into commodities at a later stage.

PVC is the largest consumer of industrial minerals in the polymer industry. Its general availability as a powder is an advantage, since this facilitates the ease with which fillers can be introduced. Formulation is, however, complex since the array of ingredients can each potentially influence the processing and properties of PVC.[16]

PVC is widely used in the bedding and sheathing of electric cables. In a cable formulation, when 28 phr of calcium carbonate filler is replaced with the same mass of *Plasfill 5*, the maximum torque during compounding was reduced from 136 to 50 mg. This was ascribed to the fluidity of the dry blend brought about by the sphericity of the filler. As a result of the lower energy input, gelation time was increased from 30 to 48 seconds. It was necessary to reduce the plasticiser content by 12 % to restore the 30 second gelation time. The modified formula met all technical criteria. In another instance, where the $CaCO_3$ was only partially replaced, 30 % less plasticiser was required and the properties of the PVC cable were well within the specifications.

Replacement of $CaCO_3$ with *Plasfill* in a rigid PVC formulation influences the processing characteristics and increases the gelation time. While the formulation could be modified by altering both the type of stabiliser and lubricant, a more practical approach was followed during commercialisation. The original formula was left unaltered and, depending on the type of pipe, between 2 to 6 phr of ultra-fine fly ash (*Plasfill 5*) was added as a processing aid. As expected, the extrusion rate was increased.[16]

Dip-coated PVC gloves are widely used in industry. The introduction of *Plasfill* into the dip-coating formulation bulked the product without affecting the rheology. The quality of the coating on the fabric improved along with abrasion resistance, thus making the products more durable.

The commercial implementation of spherical fillers in the South African PVC industry is widespread and growing. In order to gain maximum benefit from the introduction of spheres, it is necessary to take cognisance of the accompanying changes in rheology. The levels of processing aids, stabilisers and lubricants can be altered to modify processing parameters and ensure that sufficient energy is introduced to guarantee complete gelation. This is being successfully done in South Africa.

The wear resistance of polymers depends on the fillers used. PVC commonly is filled with $CaCO_3$, but this gives rise to a very low wear resistance.[17] SiC and Al_2O_3 as fillers improve the wear resistance of PVC significantly and also up to 10 % fly ash may be used to enhance wear resistance.

4.8.6 Polyethylene and Polypropylene

Together, polyethylene (PE) and polypropylene (PP) are by far the most widely used polymers and globally account for the usage of more than 50 million tonnes annually (PE about 35 and PP about 16 million tonnes).[18] While PE is generally processed without fillers, the potential for incorporation of such products exists. On the other hand, the properties of PP can be extensively modified by the inclusion of mineral fillers. Talc, mica and carbonates are used to increase stiffness and dimensional stability, important properties when PP is used in commodities that operate at higher temperatures as in electrical appliances and dishwashers. In automotive applications (bumpers), however, the increase in density, concomitant to the use of fillers, limits application.

The most extensive investigations into the potential of fly ash based fillers in PE and PP was carried out by Davies.[19,20] As expected from a spherical filler, it was found that dispersion into PP was easier and up to 60 % mass could be incorporated. The shrinkage of moulded articles was reduced, while the Vicat softening temperature remained unaltered at 152–155 °C. Compared to the virgin material, the density was significantly increased, while the melt flow index decreased by about 30 %. The tensile and impact properties declined with increased filler loading. The values achieved were, however, similar to those obtained with talc. The use of silane coupling agents considerably improved the physical properties.

Kruger investigated the effect of using either *Plasfill 5* or $CaCO_3$ on the physical properties of both HDPE and PP.[21] Blends of up to 60 % (mass) were prepared with a twin screw intermeshing extruder. Strands of compound were drawn, cooled and granulated. Test specimens were injection-moulded using these granules. In Table 4.21, the flexural and tensile properties of the two polymers are given for both *Plasfill* and $CaCO_3$ at different filler levels. Perusal of these results indicate that at

Table 4.21 A comparison of the physical properties of HDPE and PP filled with either $CaCO_3$ or *Plasfill* (both fillers have a mean particle size of 5μm). Reproduced by permission of Ash Resources Pty Ltd.[36]

Filler level (mass %)	HDPE				PP			
	Tensile yield (MPa)		Flexural yield (MPa)		Tensile yield (MPa)		Flexural yield (MPa)	
	Plasfill	$CaCO_3$	*Plasfill*	$CaCO_3$	*Plasfill*	$CaCO_3$	*Plasfill*	$CaCO_3$
0	27.5	27.5	18.0	18.0	34.2	34.2	31.4	31.4
20	24.3	24.1	19.5	18.6	30.3	30.6	34.9	34.4
40	19.9	23.2	19.7	21.6	27.2	28.6	33.0	36.0
60	16.1	19.1	20.2	21.7	18.8	18.4	27.7	30.6

20 % (mass) filler loading, there is no difference in the physical properties of the respective polymers. Sole also reached similar conclusions
when he compared the results of PP filled with $CaCO_3$ and *Plasfill*.[22]

Kruger showed that the effect of the two fillers on the rheology was,
however, totally different.[21] At 60 % mass loading of *Plasfill 5*, the melt
flow index for PP at 230 °C and with a 2.16 kg load was 3.1 g per 10 min.
In the case of $CaCO_3$ this was reduced to 2.1 g per 10 min. Extrusion
tests carried out on filled PP showed a significant difference in the power
required to produce extrudate at a specific production rate (Table 4.22).
When compared to $CaCO_3$, the spherical nature of *Plasfill* reduced the
shear stress required for a constant production rate. These results indicate
that faster extrusions are possible if $CaCO_3$ is substituted with *Plasfill*.

Table 4.22   The power required to produce a 40 %
(mass) filled PP extrudate at a specific rate. Reproduced
by permission of Ash Resources Pty Ltd.[36]

Extrudate	Production rate $(kg\,h^{-1})$	Current required (amp)
1 (40 % $CaCO_3$)	20	15
2 (40 % $CaCO_3$)	50	42
3 (40 % *Plasfill*)	20	13
4 (40 % *Plasfill*)	50	34

Over the past decade, the Institute of Materials Processing of the
Michigan Technological University, USA, has extensively worked on the
development of fly ash fillers for plastics.[23,24] It was found that fly ash
could replace commercial fillers in PVC, polypropylene, polyethylene
and nylon without loss of mechanical properties. A survey of plastic
manufacturers found that there is a reluctance to use fly ash as a filler
because of its low brightness, inappropriate pH and the presence of
broken cenospheres. Therefore, beneficiation of the ash was carried out
to improve the physical characteristics, such as to resemble those of
the widely used commercial mineral fillers. The patented MTU Fly Ash
Recycling and Processing System[25,26] first removes cenospheres from
a slurry by gravitation and subsequently separates unburned carbon
from the remaining slurry by froth flotation. The resulting processed
fly ash was then evaluated as a filler, but although there was some
improvement, results were not yet completely satisfactory.[23] Thus, in
addition to cenospheres and carbon, magnetic particles were removed as
well and a fine fraction was obtained by air classification ($d_{50} = 4\,\mu m$,

accounting for 17 % of the total ash).[24] This material performed much better. The particles were then coated with a silane coupling agent and used to replace $CaCO_3$ filler in polypropylene and polyethylene. The mechanical properties of the samples were equivalent to those using $CaCO_3$; however, the colour and brightness problem remained. Efforts to improve the brightness were unsuccessful.

Chamberlain and Hamm found that by replacing commercial fillers by fly ash in medium-density polyethylene, the flexural properties improved and the viscosity increased, but the tensile properties dropped and the impact strength dropped dramatically.[27] This means that high loadings of fly ash cannot be used if the product requires high impact strength or high tensile strength. These results were confirmed by other studies.[28] Biagini *et al.* determined the viability of coal fly ash to be used as a substitute filler in plastic resins.[29]

4.8.7 Polyester Resins, Epoxy Resins and Polyurethanes

Polyester Resins

Annual global demand for polyester is reportedly about 2.1 million tonnes.[18] Although this is lower than that for the traditional thermoplastic resins (PP, PE, PVC and PS), the high filler loading in polyesters and other thermoset polymers make them important consumers of fillers.

Unsaturated polyester resins are generally used in conjunction with glass fibre reinforcement. The glass fibre provides the mechanical strength required in many of the applications. In certain instances where flexural and tensile properties are less important (shower trays), nonreinforcing fillers are utilised.

Manufacturing methods used for polyesters include forming techniques such as casting, compression and sheet, dough and injection moulding, as well as hand lay-up and spray-up processes. All of these techniques require excellent flow properties. Products typically include automotive parts, electrical distribution components, countertops, appliance housings, tank linings, streetlight poles and recreational equipment such as swimming pools and boats.

The flow properties of cast polyester are particularly important in ensuring good surface finish and retention of mould detail. In order to assess the effect of flow, *Plasfill* replaced the fine mineral filler used as part of the existing formulation for the casting of statues and railings. The influence on rheology and compressive strength of two grades of

Table 4.23   The flow rate (g per 20 s) of a polyester resin at various mass loadings of *Plasfill*. Reproduced by permission of Ash Resources Pty Ltd.[36]

% Filler level	*Plasfill 15*	*Plasfill 5/45*
0	32	32
10	40	43
20	51	62
30	53	74
40	52	83
50	28	51

Plasfill was investigated. As can be seen in Table 4.23, the incorporation of spherical fillers increased the flow particularly when, as in the case of *Plasfill 5/45*, a specially produced fly ash in which the particle size range is essentially between 5 and 45 μm, there is a dearth of fines. The conventional formula containing 30 % calcium filler carbonate flowed at 38 g per 20 s.

At 40 % mass loading, the compressive strength of the filled polyester was 72 and 68 MPa for the *Plasfill 15* and *5/45* products, respectively. This exceeds the specification of 50 MPa. In both cases, the flexural strength was 19 MPa.

Upon commercialisation, it was established that, due to the higher volume loadings achieved, the new formulation resulted in a resin saving of 6 %. Furthermore, both the off-mould surface finish and dimensional tolerance of the manufactured articles was improved.

In some other studies, up to 25 wt % fly ash filler was used in an unsaturated polyester resin and compared to $CaCO_3$-filled unsaturated polyester resin.[30,31] It was found, however, that the fly ash filled resin was inferior to the $CaCO_3$-filled resin with respect to tensile and flexural strength. The introduction of fly ash increased the tensile and flexural modulus. Rebeiz *et al.* reported that 25 wt % addition of fly ash results in an increase in tensile modulus of the material by approximately 80 % and a decrease in its tensile strength and ultimate strain by 40 and 75 %, respectively.[31] Silanisation of the fly ash filler gave a significant increase in tensile strength. The most beneficial effect was reached with approximately 15 % addition of silanised fly ash.[32] These authors used a fly ash for which only was reported '*ca* 0.1 μm and 17 % hollow particles'. The influence on cure time was low, but viscosity increased to levels at which moulding becomes difficult (at 25 % fly ash). Silanisation helps here as well and also gives an improved dispersion of the fillers, which is important for many physical properties. These properties may be further improved using a hybrid filler composed of spherical fly ash particles and plate-like mica particles. The addition of 25 parts by weight of mica and 15 parts of fly ash to the polyester resin seems to be the optimum filler.[32]

Significant improvement and cost savings were achieved when a dough moulding formulation was adapted to include lightweight hollow glass spheres (cenospheres) recovered from fly ash.[1] The synergism between fillers with different morphologies was used to reduce the cost of a glass fibre reinforced electrical distribution box. In this case, the formula was amended to contain 40 phr cenospheres, with 57 phr $CaCO_3$, and 62 phr of glass fibre. As a result, the density was reduced by 8 % compared to the original formulation. The flow of the mix in the heated mould was substantially improved and fewer imperfections were noted in the finished article. As a result, the moulding temperature could be reduced from 160 to 148 °C without compromising quality. The physical (tensile strength 26 MPa, flexural strength 52 MPa), electrical and fire-resistant properties exceeded the required specifications.

Epoxy Resins

Kishore *et al.* used class C fly ash at up to 30 % as a filler in a medium-viscosity DGEBA epoxy resin.[33] The fly ash containing both solid and hollow spheres has a bimodal particle size distribution, the parameters of which are not given. Testing of samples show that fly ash filled epoxy composites display lower impact strengths with increase in filler content. However, with increasing ash content the material becomes more ductile and the fracture area decreases. This is because fly ash particles influence the crack paths during the initiation phase and the propagation phase, as well as in the termination phase. The filled systems require less energy of initiation compared to neat epoxy, but show higher energies of propagation of the crack. The study did not include any comparison with the behaviour of epoxy resins using other commercial fillers.

Polyurethane

High loadings of fly ash can be used for the preparation of polyurethane carpet backings without affecting the manufacturing process or physical properties.[34]

4.8.8   Specifications

In the majority of studies referred to above, the focus has been on the properties of fly ash filled polymer compared to the unfilled polymer, or to the polymer utilising commercial available fillers. This is of course important, but another aspect, the properties of the combustion residue itself, has received very little attention. Most authors provide limited general information about the residues used, such as 'C class fly ash, *ca.* 0.1 μm with 17 % hollow particles',[35] but others do not give any data at all.[27] The work at MTU[24] and that by Kruger[13,14,16] are the exception, not only because they better document the properties of the fly ash, but also because they study the influence of ash properties on the properties of the final product when different fractions of beneficiated ash are used.

The long history of use of ash in cement and concrete has resulted in clear specifications for the ash to be used, mainly because all relevant aspects have been studied and the relationships between ash properties and product quality are known. This is not yet the case for fly ash as a filler in polymers. Fillers in the polymer industry are sold in various grades, each conforming to a prescribed quality criterion. The same will necessarily have to apply to fly ash fillers in polymers. Certified fly ash batches based on the cement or concrete industry are not appropriate for filler applications, as they probably do not meet the properties relevant to fillers. So, new specifications will have to be set up for each particular product. In order to draw up such specifications, careful testing as set out in Section 4.1 is required.

4.8.9   Summary and Conclusion

Mineral fillers are no longer only being utilised to reduce the cost of polymers but, more significantly, to engineer products with specific properties. The diversity of features such as shape, size, density, composition and surface activity, is utilised to modify the optical, thermal, mechanical and electrical properties of polymers.

Predominant amongst these features is probably the particle size and shape. In this regard, the acicular nature of glass fibre, the plate-like shape of kaolin and the tabular form of feldspar are all utilised. However, can the same be said of spheres?

The inherent advantages of spherical fillers have probably been under-exploited, mainly due to the lack of availability of cost-effective products and a dearth of technical information on their performance in commodity polymers. New manufacturing technology has, however, enabled the cost-effective production of alumina silicate spheres.

As discussed in some of the examples, many manufacturers are exploiting the advantages offered by spherical fillers, not only to save costs but also to improve the quality of the finished products. Among the reasons cited for using *Plasfill*® are: lower oil absorption, resulting in a saving on resin; the ability to utilise higher loadings without sacrificing physical properties; improved polymer rheology, allowing thinner walled products to be manufactured; lower density; less differential shrinkage; and easier processing.

The use of spherical fillers in commodity plastics and rubber is still in its infancy. Much needs to be done and many challenges must be met. Not the least of these as far as *Plasfill* is concerned is the perceived negative effect of its light grey colour. In some cases, surface pigment interaction has been manipulated to limit this effect. The availability of a pale cream version will ameliorate the situation even further. Encouraging results are being achieved when *Plasfill*-to-polymer interfacial bonding is modified by using coupling agents.

In most of the examples cited, data is presented for the application of a single type of filler only. In many cases some of the properties of the polymer are enhanced, while others are sacrificed. This begs the following question: In pursuit of the ideal filler, can the unique properties of spheres and other minerals be synergistically utilised? Therein may lie the opportunity to widen the scope for polymers.

References

1   Kruger, R.A., Personal communication (2004).
2   Ciullo, P.A., *Industrial Minerals and Their Uses: a Handbook and Formulary*. Noyes Publications: Westwood, NJ (1996).
3   Katz, H.S. and Milewski, J.V., *Handbook of Fillers for Plastics*. Van Nostrand Reinhold: New York (1987).
4   Ledger, W.A., *Paints*. In *Plastics and Resin Compositions*, W.G. Simpson (ed.). Royal Society of Chemistry: London (1995).
5   Mainieri, J.F., *Problem solving forum*. Journal of Protective Coatings and Linings, 10(10): 15 (1993).
6   Delaney, J.P., *Problem solving forum*. Journal of Protective Coatings and Linings, 10(10): 21 (1993).
7   Growall, J., *Problem solving forum*. Journal of Protective Coatings and Linings, 10(10): 16–20 (1993).
8   Tarricone, P., *Fly ash for hire*. Civil Engineering, 61(10): 46–49 (1991).

9 SKIM INDIS Marktonderzoek, *Mogelijkheden voor toepassing van gewassen vliegas*. Project 5251. Uithoorn, the Netherlands (1996).

10 Lu, P., Yang, W., Yu, S. and Chi, P., *Development of elastic waterproofing F-CS emulsion coatings*. Huaxue Jiancai, **17**: 30–32 (in Chinese) (2001).

11 Schwartz, O., *Kunststofkunde*. Vogel-Buchverlag: Wurtzburg, Germany (1987).

12 Garde, K., McGill, W.J. and Woolard, C.D., *Surface modification of fly ash – characterisation and evaluation as reinforcing filler in polyisoprene*. Plast. Rubber Compos., **28**: 1–10 (1999).

13 Kruger, R.A., Hovy, M. and Wardle, D. *The use of fly ash fillers in rubber*. In Proc. 1999 International Ash Utilization Symposium, Lexington, Kentucky, pp. 509–517 (1999).

14 Kruger, R.A., Hovy, M. and Wardle, D., *The use of fly ash fillers in rubber*. In Proc. PROGRES Workshop on Novel Products from Combustion Residues, H.W. Nugteren (ed.), Morella, Spain, pp. 295–303 (2001).

15 Vishwakarma, M., Bandopadhyay, A.K. and Maitra, S., *Development of rubber filler grade precipitated silica and water treatment grade alumina from fly ash*. In Proc. Intern. Conf., Fly Ash Disposal & Utilisation, New Delhi, India, C.V.J. Varma (ed.), Central Board of Irrigation and Power, New Delhi, India, **1**: 2/20–2/25 (2000).

16 Kruger, R.A., Jubileuszowa Miedzynarodowa Konferecja Popioly z Energetyki, Warszawa, 14–17 pazdziernika 2003.

17 Yang, F. and Hlavacek, V., *Improvement of PVC wearability by addition of additives*. Powder Technol., **103**: 182–188 (1999).

18 Baker, R.A., *Minerals and polymers – high performance, high value*. Industrial Minerals, **369**: 73 (1998).

19 Davies, L.C.B., Hodd, K.A. and Sothern, G.R., *Pulverized fly ash, its use as a filler for polyolefins*. Plastics and Rubber Processing and Applications, **3**: 163–168 (1983).

20 Davies, L.C.B., Sothern, G.R. and Hodd, K.A., *Pulverized fly ash, its use as a filler for polyolefins. Part 2: Coupling agents and a comparison with Ballotini*. Plastics and Rubber Processing and Applications, **5**: 9–14 (1985).

21 Kruger, A., *Comparison of Plasfill® and $CaCO_3$ as fillers in plastics*. C.S.I.R., Pretoria, South Africa (1997) (Confidential Report).

22 Sole, D.M., *The effect of fillers on the mechanical properties of polypropylene*, B.Sc. (Hons) Report, University of Cape Town, South Africa (1993).

23 Huang, X., Hwang, J.Y. and Tieder, R., *Clean ash as fillers in plastics*. In Proc. 11th Int. Symp. on use and management of coal combustion by-products (CCBs), American Coal Ash Association, Orlando, Fla, USA, **1**: 33 (1995).

24 Huang, X., Hwang, J.Y. and Gillis, J.M., *Processed low NO_x fly ash as a filler in plastics*. In Proc. 12th Int. Symp. on Use and Management of Coal Combustion by-products (CCBs), American Coal Ash Association, Orlando, Fla (1997).

25 Hwang, J.-Y., *Wet process for fly ash beneficiation*. US Patent 5,047,145 (1991).

26 Hwang, J.-Y., *Wet process for fly ash beneficiation*. US Patent 5,227,047 (1993).

27 Chamberlain, A. and Hamm, B., *The viability of using fly ash as a polymer filler*. In 56th Annual Techn. Conf. of Soc. of Plastics Engineers, **3**: 3415–3417 (1998).

28 Chand, N. and Vashishtha, S.R., *Rheological studies of some fly ash filled PP/LDPE blends*. In Proc. IUPAC Int. Symp. Adv. Polym. Sci. Technol., **2**: 765–767 (1998).

29 Biagini, M., Paris, A., McDaniel, J. and Kovac, V., *Utilisation of fly ash as a filler in plastics*. In 58th Annual Techn. Conf. of Soc. of Plastic Engineers, **3**: 2848–2850 (2000).

30 Devi, M.S., Murugesan, V., Rengaraj, K. and Anand, P., *Utilization of fly ash as filler for unsaturated polyester resin.* J. Appl. Polym. Sci., **69**: 1385–1391 (1998).

31 Rebeiz, K., Banko, A.S. and Craft, A.P., *Temperature properties of polyester mortar using fly ash waste.* Pract. Period. Hazard, Toxic, Radioact. Waste Manage., **3**: 107–111 (1999).

32 Şen, S. and Nugay, N., *Uncured and cured state properties of fly ash filled unsaturated polyester composites.* J. Appl. Polym. Sci., **77**: 1128–1136 (2000).

33 Kishore, S.M., Kulkarni, S., Sharathchandra, S. and Sunil, D., *On the use of an instrumented set-up to characterize the impact behaviour of an epoxy system containing varying fly ash content.* Polymer Testing, **21**: 763–771 (2002).

34 Jenkines, R.C., *Carpet backing precoats, laminate coats, and foam coats prepared from polyurethane formulations including fly ash.* Patent Int. Appl. WO 9808893 A1 (1998).

35 Şen, S. and Nugay, N., *Tuning of final performances of unsaturated polyester composites with inorganic microsphere/platelet hybrid reinforces.* European Polymer Journal, **37**: 2047–2053 (2001).

36 Nugteren, H.W. and Kruger, R.A., *Fly ash as a replacement for mineral fillers in the polymer industry.* In World of Coal Ash Conference (WOCA), Lexington, Kentucky, 11–15 April 2005 (Proceedings on CD, paper #146).

4.9   Geopolymers

Henk Nugteren

Delft University of Technology, The Netherlands

Types of combustion residues involved

Geopolymers can be synthesised from materials containing soluble Si and Al through alkaline activation. Therefore, combustion residues that contain significant amounts of soluble Si and Al at the surface of a large proportion of their particles can react and develop strength as the reaction proceeds. Class F coal fly ashes are most commonly used, but class C coal fly ash may also be suitable. Lignite ashes and MWSI ashes seem less suitable for this application. Also, particle size and particle size distribution are important, fine and or well-graded materials having advantages. Therefore fly ashes have greater potential than bottom ashes.

State of development

Geopolymers in general represent a niche market in the building industry and waste-based geopolymers in particular still have to find their way into the market. Development is mainly at the state of research and pilot demonstrations, but the potential in the area of building materials and waste immobilisation technologies seems very high. Before fully exploring this potential, the reactions and the influence of the chemistry and reaction conditions must be better understood, so as to be able to tailor the material properties in an optimum way for specific applications. This applies especially for waste-based geopolymers, since the source material already has an inherent variability.

4.9.1   Introduction

During the past decade, new inorganic cementitious compounds have been commercially introduced into the US and European markets. These compounds are the results of developments in the field of inorganic aluminosilicate polymers or geopolymers, mainly inspired by the pioneering work of the French researcher Joseph Davidovits. This development started in 1972 with a patent on the polycondensation of kaolinite with NaOH.[1] Since then, the knowledge of such inorganic polymers has increased and different materials such as resins, foams and fibres have been developed based on geopolymers.[2] It was known that in ancient times cements based on geopolymers were used by the Egyptians and Romans,[3,4] so the present development may be seen as a revival of old technology.

Despite the improved knowledge on this presumed old technology, several questions still remain unsolved. In particular, the roles of some parameters during the geopolymerisation reaction still need to be defined, and the crystal chemical status of several cations that can be involved in a geopolymer lattice still need to be clarified.

4.9.2 The Chemistry of Geopolymeric Systems

Aluminosilicate binders are termed inorganic geopolymeric compounds, since the geopolymeric cement obtained is the result of an inorganic polycondensation reaction, so-called geopolymerisation. Such reactions yield three-dimensional tecto-aluminosilicate frameworks with the general empirical formula

$$M_n[-(SiO_2)_z - AlO_2]_n.wH_2O,$$

where M is a cation (K, Na, Ca), n is the degree of polycondensation, and z is 1, 2, 3 or $\gg 3$. Such frameworks are called polysialates, where 'sialate' stands for the silicon-oxo-aluminate building unit. The sialate network consists of SiO_4 and AlO_4 tetrahedra linked by sharing all oxygen atoms. Positive ions (Na^+, K^+, Ca^{2+} etc.) must be present to balance the negative charge of Al in fourfold coordination. Chains and rings may be formed and crosslinked together, always through a sialate Si–O–Al bridge (Figure 4.36).[2]

The amorphous to semi-crystalline three-dimensional geopolymeric silico-aluminate structures typically consist of:

poly(sialate)	$z = 1$	(Si–O–Al–O–)
poly(sialate-siloxo)	$z = 2$	(Si–O–Al–O–Si–O–)
poly(sialate-disiloxo)	$z = 3$	(Si–O–Al–O–Si–O–Si–O–)

For Si : Al $\gg 3$, the polymeric structure results from the crosslinking of polysilicate chains, sheets or networks with a sialate link (Figure 4.36).

This terminology suggests that only integer values of z are possible. This is the case in fully crystalline materials such as zeolites, but doubtful for the flexible structure of geopolymers.[5]

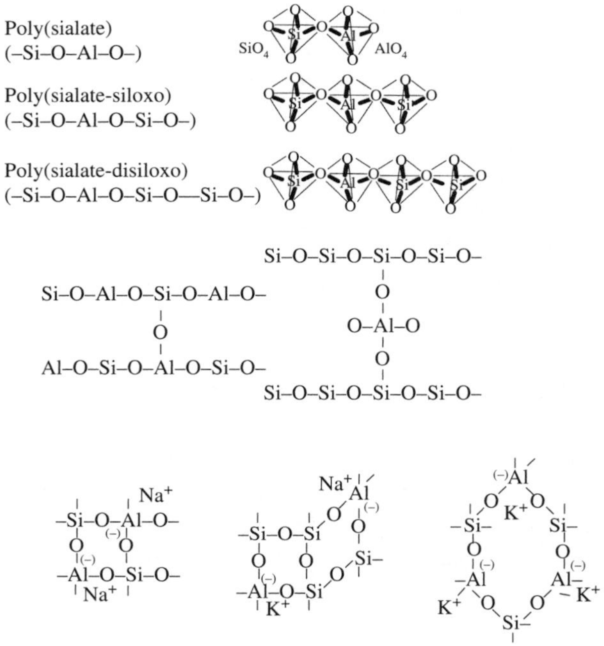

Poly(sialate)
(–Si–O–Al–O–)
SiO_4 AlO_4

Poly(sialate-siloxo)
(–Si–O–Al–O–Si–O–)

Poly(sialate-disiloxo)
(–Si–O–Al–O–Si–O—Si–O–)

$$Si–O–Si–O–Si–O–Si–O–$$

Si–O–Al–O–Si–O–Al–O–

O

Al–O–Si–O–Al–O–Si–O–

Si–O–Si–O–Si–O–Si–O–
|
O
|
O–Al–O
|
O
|
Si–O–Si–O–Si–O–Si–O–

Figure 4.36 Examples of sialate links and sialate rings. Reproduced by permission of Joseph Davidovits.[2]

The (potassium, calcium)–poly(sialate-siloxo)– (K,Ca)(Si–O–Al–O–Si–O), cements result from the geopolymeric reaction

$$2(Si_2O_5, Al_2O_2) + K_2(H_3SiO_4)_2 + Ca(H_2SiO_4)_2$$
$$\Rightarrow (K_2O, CaO)(8SiO_2, 2Al_2O_3, nH_2O)$$

According to this terminology, common silicate minerals are strictly termed 'polycondensated sialates'. Thus, using the terminology, the majority of the Earth's crust is composed of siloxo-sialates and sialates. Thus it can easily be shown that the common feldspar series albite–anorthite ($NaAlSi_3O_8$–$CaAl_2Si_2O_8$) can be described as from poly(cyclosialatedisiloxo) for albite to poly(cyclodisialate) for anorthite.

4.9.3 Synthesis of Geopolymers

Some natural aluminosilicates can easily be used to produce sialate-type ions that may subsequently be polycondensed to yield different polysialates. This is well known from the synthesis of zeolites from clay minerals (Section 4.4).

Reacting a 50/50 (wt ratio) kaolinite/quartz blend with NaOH by cold-pressing followed by hot-pressing (130–200 °C), Davidovits formed a so-called Na–polysialate–quartz block-composite, which he termed SILIFACE Q. This composite, with Si : Al = 1, shows the major lattice d-values of both quartz and sodalite but those of kaolinite have disappeared, suggesting complete transformation.[6]

By using aluminosilicate oxides with Al^{3+} in fourfold coordination together with alkali polysilicates, polymeric Si–O–Al bonds can be made with Si : Al = 2.[2] Such reactions will proceed at atmospheric pressure and can be carried out by calcining aluminosilicate hydroxides, or by condensation from the vapour phase. The calcined kaolinite normally used is called KANDOXI (acronym for KAolinite, Nacrite, Dickite, OXIde). The actual geopolymerisation is carried out with the mixture at 85 °C. The reactions are exothermic and the heat generated is proportional to the conversion rate, so that the yields may be monitored by simple thermography. Another possibility is to analyse the ^{27}Al MAS–NMR spectrum and control the ratios of Al(IV), Al(V) and Al(VI) present in the product. XRD characterisation is difficult, because the products are X-ray amorphous and usually show some strong reflections from the accessory minerals present in the raw materials.

By adding Ca polysilicates to the mixture, geopolymers setting at room temperature can be made. Such geopolymers are cement-like materials, but do not rely on limestone calcination and are acid resistant. For detailed properties, see Section 4.9.4.

Synthesis of geopolymers with Si : Al = 3 and Si : Al ≫ 3 is believed to lead to speciality products, and at present there is no reason for fly ash to be a suitable precursor for such geopolymers.

NMR studies suggest that potassium is preferentially incorporated into the geopolymeric matrix in mixed-alkali geopolymers, and that the amount of unreacted phase increases with the nominal Si : Al ratio and the Na : K ratio in the activating solution.[7] Furthermore, it seems that at relatively low Si : Al ratios (< 1.65) K-geopolymers have a more disordered structure compared to Na-geopolymers, and also have a higher content of Al–O–Al linkages,[8] which may have consequences for their mechanical properties.

X-Ray Diffraction (XRD), Differential Thermoanalysis (DTA), High Resolution Electron Microscopy (HRME, both SEM and TEM), Thermogravimetric Analysis (TGA) and (Magic Angle Spinning) Nuclear

Magnetic Resonance spectroscopy ((MAS)–NMR) are the main tools for the characterisation of the synthesised geopolymer phases.

4.9.4 Properties and Applications of Geopolymers

Special properties may be induced by following an appropriate synthetic route. Among the properties mentioned by Davidovits[2] are high-temperature resistance, water resistance, acid resistance, low thermal expansion and fire resistance. Some examples of such products involving fly ash in their synthesis are given below. Recently, fire-resistant geopolymers have been produced using blast furnace slag.[9]

In general, geopolymers with a Si : Al ratio of 1 or 2 will present a rigid 3D network and may be used for bricks, ceramics, cements and concretes, but also for fire protection and the immobilisation of nuclear and toxic waste. These are products and applications normally experienced in connection with cement and ash. The introduction of the geopolymerisation reaction in traditionally used processes may be beneficial for the enormous improvement that results with such products. The products are bulky and this is one more reason why fly ash should be thought of as an option for a geopolymer precursor. One of the most promising fields is the replacement of ordinary cements by geopolymers, discussed below. Geopolymers with Si : Al ratios of 3 and much higher display a real polymeric character, and are more appropriate for highly technological applications such as tooling for aeronautics and fire protection. The Si : Al ratios in fly ash and the presence of contaminants render fly ashes less suitable in their synthesis. Moreover, these markets are relatively small and therefore such geopolymers are not further considered here.

(K–Ca)-polysialates, or geopolymeric cements, are the most interesting geopolymers as far as fly ash is concerned. They offer a possible future alternative for bulk product cement. They are basically different from traditional hydraulic binders in which hardening is the result of the hydration of Ca-aluminates and Ca-silicates. They harden rapidly at room temperature and provide compressive strengths of the order of 20 MPa, after only 4 hours at 20 °C, when tested in accordance with the standards applied to hydraulic binder mortars. The final 28-day compression strength is in the range of 80–140 MPa.

Alkalis are a problem in ordinary cements, since they are known to cause deleterious alkali–aggregate reactions. However, alkali-activated aluminosilicate binders, as in an alkali-activated slag described by

Talling and Brandstetr,[10] do not generate any alkali–aggregate reaction. Davidovits[2] shows that geopolymeric cements with alkali contents as high as 9.2 % do not show any expansion during setting (Figure 4.37).

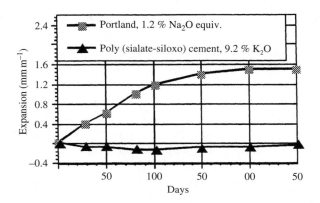

Figure 4.37 ASTM C227 bar expansion for (K–Ca)-poly(sialate-siloxo) cement and ordinary Portland cement. Reproduced by permission of Joseph Davidovits.[2]

In France, an SME (Géopolymère SA) connected to the Geopolymer Institute of Professor Davidovits (Cordi-Géopolymère) offers 1 kg samples of different geopolymer binders for testing on different applications. This company has a production capacity of 10 tonnes per month, and sells speciality products in air filtration, foundry and microwave technology, and for aeronautics, aerospace and racing cars.

4.9.5 Using Fly Ash as a Source Material for Geopolymers

Synthesis and Properties

From the use of fly ash as a precursor in the synthesis of zeolites (Section 4.4), it is only a small step to its use as a precursor for geopolymers. A mineralogist or geochemist may easily appreciate this when considering the nature of geopolymers described above. The average Si : Al ratio in coal fly ash seems appropriate for the synthesis of geopolymers with z in the order of 1–2. Proper mixing of ashes and/or the addition of small amounts of inexpensive industrial minerals will give the required mixes for synthesis.

The use of fly ash as the silica and alumina source for the synthesis of geopolymers was first proposed by Wastiels *et al.* in 1994.[11] This was further developed by Palomo and co-workers, who did not use the name geopolymer, but 'alkali-activated fly ash' instead.[12,13] Using fly ash alone, products with mechanical strengths in the 60 MPa range were obtained after curing at 85 °C for only 5 hours.

Although low-calcium fly ashes (Class F) are currently most commonly used in geopolymer synthesis, there seems to be no reason why Class C fly ash could not also be used, as the presence of calcium provides extra nucleation sites and may enhance some desired material properties.[14]

In general, compressive strength of fly ash based geopolymer concrete decreases with increasing $H_2O : Na_2O$ ratio (in the range 10–14) and increasing water : solids ratio (in the range 0.17–0.22).[15] However, the $Na_2O : SiO_2$ ratio in the operational range of 0.095–0.12 does not have any significant effect on the compressive strength. It must be noted here that some disagreement exists on suitable ratios to be applied. Davidovits[16] proposed much higher $Na_2O : SiO_2$ (0.20–0.28) and $H_2O : Na_2O$ (15–17.5) ratios than those investigated by Hardjito et al.[15] According to the latter, an $H_2O : Na_2O$ ratio above 14 provokes considerable segregation of mixture ingredients due to the presence of excess water. Barbosa et al.[17] found that the optimum composition occurred when the $Na_2O : SiO_2$ ratio was 0.25 (within Davidovits' range) and the $H_2O : Na_2O$ ratio was 10 (at the low end of Hardjito's range).

Although geopolymer synthesis is mostly reported to occur in the temperature range of 60–80 °C, ambient temperature synthesis has also been successful. In an optimisation study by Cioffi et al.[18] it was even shown that under some circumstances stronger products could be obtained with a long curing time at room temperature than at shorter curing times at higher temperatures. Furthermore, it was very clear that potassium-based geopolymers resulted in higher strengths than sodium-based geopolymers.

The performance of the geopolymers will be influenced by not only the reaction parameters but also the properties of the fly ash used. Since fly ash is highly variable in chemical and mineralogical composition, and physical properties, this seems to be a straightforward matter to investigate. As Si and Al present in the glass phase of the ash determine the concentrations of these elements in the interstitial fluid phase during geopolymerisation, a correlation should exist between Si and Al in glass and the developed compressive strength. However, in a study in which six different fly ashes were characterised and geopolymerised, no such clear correlation could be determined.[19] Since only a very thin skin of glassy spheres in the ash will dissolve and these spheres exhibit a high intra- and inter-particle inhomogeniety, it cannot be expected that the overall glass chemistry will determine the concentrations in the intermediate phases and thus the strength of the products.

Successful attempts using fly ash for geopolymer synthesis were reported at the international conferences Géopolymère'99[20] and Geopolymers 2002,[21] and research is believed to be continuing. Applications for the geopolymers produced from ash were in the fields of special cements, toxic waste management and low-CO_2 cements.

Microstructure

The key difference between zeolites and geopolymers is the lack of long-range order in the latter. Thus zeolites are crystalline, whereas geopolymers are amorphous to X-rays. However, geopolymers and zeolites are synthesised from the same reactants and under very similar conditions, with the main difference being the amount of water present. When synthesising zeolites from fly ash, the aim is to reach the highest possible yield. Therefore considerable amounts of water are first required to dissolve the main reactive components in the fly ash and have them reprecipitated in the form of a zeolitic powder. When synthesising geopolymers, the aim is to produce a strong monolithic material in which only the surfaces of the original grains react to bind them together. Compared to the zeolite case, the required amount of water is lower by a factor of between 10 and 100. Nevertheless, there is a possibility that the reaction products are similar; that is, that the geopolymeric network is actually an agglomeration of nanocrystalline zeolites.[5] However, an electron diffraction study of a sodium polysialate siloxo geopolymer prepared from metakaolinite confirms the essentially amorphous nature of the geopolymer matrix, even after heating at 1200 °C.[22]

Lloyd and Van Deventer[23] investigated the microstructure of a geopolymer, made from 75 % fly ash and 25 % ground granulated blast furnace slag activated with potassium silicate, by XRD, SEM and TEM. With high-resolution SEM, a geopolymeric reaction product consisting of at least two phases dendritically intergrown with each other could be seen. Since the dendritic phase is probably rich in Ca and is found near the slag remnants, it is suggested that this phase is the reaction product of the slag and the other phase the reaction product of the fly ash. It has also been shown that geopolymers from only fly ash appear to consist of a single phase. TEM microanalysis shows that the Ca-rich phase is amorphous and contains a significant amount of potassium, which indicates that it is a newly formed phase. The high Ca and Si content and the amorphous diffraction pattern indicate that this phase may be CSH (calcium silicate hydrate). This may have important implications, as it will affect the acid resistance of the material. Mullite needles about 300–500 nm long and 50 nm wide were found in the geopolymeric matrix, but it is not clear whether those are newly formed or not. Much smaller crystals of about 30–60 nm are probably a reaction product because of the significant potassium content, and could be anorthite or gehlenite, although the identification is ambiguous. The inability to detect this new phase by XRD can be explained by a combination of the small size and low abundance.

In a review article by Provis and co-workers,[5] the microstructure of geopolymers is discussed at length. The outstanding feature of all published diffractograms of geopolymers is that, regardless of the source material, activating solution and curing conditions, a broad hump centred at around a 2θ value of 27–29° is always present. From studies on zeolite synthesis, it is known that an aluminosilicate gel formed by mixing colloidal silica with sodium aluminate solution displays a broad peak centred at a 2θ of approximately 22°. During heating, this peak shifts towards $2\theta = 28°$, before transferring into the sharp peaks of the synthesised zeolite. The intermediate phase has been interpreted as 'precrystallisation', with the zeolite structure present on a scale of no more than four unit cells, or approximately 8–10 nm. By analogy with this, Provis et al.[5] suggest that the persistent hump at $2\theta = 27 - 29°$ in geopolymers may indicate that they contain a significant level of nanoscale crystallinity in the form of zeolitic nanocrystals. Further weight is given to this proposition by the observation of short-range ordered domains in HREM and TEM studies, Raman spectroscopic data, calorimetric data and mechanical properties. It is beyond the scope of this section to describe all this in detail: the interested reader is referred to the review article by Provis et al.,[5] in which a full discussion is given, together with all relevant references.

4.9.6   Some Specific Applications for Fly Ash Based Geopolymers

Special Cements and Concretes

The first commercial geopolymer was produced in the USA in 1997 by ZeoTech Corporation, under the name ZeoTech Concrete 100,[24] following research work at the Drexel University in Philadelphia.[25] This concrete is made of a blend of sodium silicate solution, alkaline acti-vator, Class F fly ash, silica-bearing aggregates and water. It is chemi-cally inert to chlorides, salts, most mineral and organic acids, alkalis and solvents. It further meets or exceeds standard requirements for structural concrete.

The material is not intended to replace standard Portland cement concrete, but offers an alternative to specialities such as polymer or admixture-heavy concretes used for protective coatings or linings on conventional-concrete surfaces and structures. The material enables precasting, and target applications include manholes, sumps, trenches, lift stations, waste containment, marine structures and speciality pipe. The first 20 tonnes of ZeoTech Concrete 100 were produced in 1998.

Heat-resistant geopolymeric cements have also been synthesised from mixtures containing class F fly ash.[26] Sodium silicate was used as the alkaline constituent and caustic soda as activator. It was found that the best performance occurred at an $SiO_2 : Al_2O_3$ ratio of 4 (adjusted by adding silica fume), using metakaolinite as the aluminosilicate source and fly ash as heat-resistant filler; or at an $SiO_2 : Al_2O_3$ ratio of 4–4.5, using fly ash as the aluminosilicate source and milled chamotte as heat-resistant filler. Such optimum compositions have setting times at $80\,°C$ of up to 4 hours. After firing at $800\,°C$, this gives a compressive strength of up to $88.7\,N\,mm^{-2}$, a residual strength up to 245 % and contraction of less than 4.2 %.

Geopolymers as Immobilisation Matrices

The first investigations into the use of geopolymers for immobilisation were undertaken from 1994 to 1997 in a European Union sponsored project (BRITE-EURAM BE-7355-93).[27] This GEOCISTEM project was aimed at the immobilisation of sludges contaminated with radionuclides, heavy metals and hydrocarbons at two uranium mining locations in the former Eastern Germany. The geopolymeric matrix was based on the CARBUNCULUS Cement™,[28] which did not contain fly ash and which is further described in the next subsection. In a pilot operation, 30 tonnes of low-level radioactive waste was treated to produce easily handleable geopolymeric monolithic products. It is reported that immo-bilisation using this technology requires approximately the same unit cost as conventional OPC technology, but in most aspects provides the same performance as vitrification.[27]

In a chemical modelling study, it was found that Al–O–Al bonds are significantly weaker than the Si–O–Al bonds in amorphous aluminosilicate materials.[29] The formation of Al–O–Al bonds may therefore be detrimental to the immobilisation efficiency of cationic forms of toxic and radioactive metals. Solid state NMR analyses of aluminosilicate glasses and geopolymers show that in geopolymers Al–O–Al bonds are less frequently formed than in glasses of similar composition. This suggests that in many instances geopolymerisation proves a more effective method of waste immobilisation than vitrification.[29]

Van Jaarsveld and co-workers also recognised that geopolymers are suitable candidate structures for hosting heavy metals and other toxic ions.[30] Since such applications must be cheap to be economically viable, it was recognised that the geopolymer itself should be synthesised from inexpensive raw materials. The proposed solution was to use coal fly ash for this purpose and the same authors have reviewed the potential for such new technologies.[31]

In a case study, Van Jaarsveld and Van Deventer[32,33] used coal fly ashes for the production of a geopolymeric matrix for the immobilisation of Pb and Cu from wastes. Two fly ashes were used in their experiments, one CaO-rich (8.2 %) with 50.1 % SiO_2 and 28.3 % Al_2O_3, and another CaO-poor (0.6 %) with 61.4 % SiO_2 and 33.0 % Al_2O_3. The matrices contained between 60 and 70 % of this fly ash, 14 and 16 % of kaolinite or metakaolinite and 3.7 and 6 % of NaOH or KOH, the remainder being water. The water : ash ratio was between 0.2 and 0.45. Cu and Pb ions were added to the matrices as solutions of their nitrates to give total concentrations of 0.1–0.2 %. These two metals were chosen because of their equal valence state but different ionic radii. The Pb-containing matrices gave stronger structures than those containing Cu, although the Pb-containing structures showed larger specific surfaces areas. This seems counterintuitive, but it is believed that the larger ionic radius gives more strength to the structure although the structural units get smaller. These effects are amplified with increasing concentrations of Pb and Cu.

The immobilisation of metals in the geopolymeric structure is not only physical encapsulation, but the metals also affect both the chemistry and the morphology of the structures. This is also shown by the resulting slightly different chemical compositions and d-spacing values. The effects are ionic radius (and valence) related, with the larger ions tending to be immobilised more efficiently. The results obtained with the fly ash derived geopolymers were as good as the previous results with more expensive commercial geopolymer binders.[32,33]

Concentrated salt solutions or brines often result from conventional waste-water cleaning operations. Sodium and potassium chlorides and sulfates are among the most common salts in such brines. These two cations are essential to the structure of a geopolymeric matrix and it is relatively expensive to supply them as pure reagents. Therefore, it is logical to try to supply the basic backbone structure of the geopolymer through fly ash and the balancing cations through a brine solution. In this way, fly ash would be beneficially reused and the brine would be immobilised. Swanepoel et al.[34] prepared geopolymers from a common alkaline fly ash (46 % SiO_2, 21 % Al_2O_3, 4.3 % Fe_2O_3 and 9.8 % CaO) and a brine water containing more than $100 \, g \, l^{-1}$ of combined Na and K chlorides and sulfates. Matrices were prepared with 50–60 % fly ash and 20 % brine, with the addition of between 10 and 20 % kaolin and 5 % NaOH. The samples were cured for 24 hours at 50 °C and showed early compressive strength around 1 MPa, increasing up to 4 MPa after 28 days curing at room temperature. In general, the compressive strength decreased with the inclusion of the salts compared to pure water. Leaching tests on the crushed samples obtained after seven days curing ($L/S = 10$ in distilled water) showed that 60–80 % of Na but only 20 % of K was immobilised. Equilibrium was reached after about 48 hours of leaching. Unfortunately, this study did not report on the fate of the chloride and sulfate.

Apart from immobilising hazardous waste in a fly ash based geopolymeric matrix, the contamination in fly ash itself may be immobilised by geopolymerisation. A good example is given by the immobilisation of wood ash contaminated by radionuclides (^{137}Cs and ^{90}Sr, specific activity $> 50\,000 \, Bq \, kg^{-1}$) in the Chernobyl region (Ukraine).[35] The backbone of the geopolymeric matrix was formed by the alkaline activation of calcined bentonite (at 550 °C) and was mixed with the ash. Wood ash and bentonite in a 1.25 : 1 mass ratio provided the required compressive strength for solidification and chemical capsulation of the radionuclides.

Brown coal fly ash (high in CaO and MgO and low in SiO_2 and Al_2O_3) was stabilised and immobilised in a similar way using kaolin and sodium silicate with sodium hydroxide as alkali activator.[36,37] Up to 60 % fly ash could be added to allow sufficient binding and decrease the leachability. However, some trace elements did not respond (Cu, Cr and Mo), whereas some even showed increased leaching (Mn, Ni, V and Zn).

Low-CO₂ Cement

The production of ordinary Portland cement is based on the calcination reaction of limestone. This takes place in a kiln at high temperatures, which are normally maintained by the combustion of carbon fuels: thus the production of 1 tonne of cement generates 0.4 tonnes of CO_2. This CO_2 is generally recognised and taxed in countries where such CO_2 taxes exist. However, calcination itself yields another 0.55 tonnes of chemically released CO_2, which often is neither considered nor taxed. This means that the production of 1 tonne of cement roughly emits about 1 tonne of CO_2. Looking at the world cement production in 1987 (approximately 1 billion tonnes), it can be calculated that this industry was responsible for 5 % of the world's total anthropogenic CO_2 emission.[38] Commercial development means building infrastructure and housing, which is directly related to cement and concrete production. With the rapidly developing Asian countries, cement production has increased exponentially, and even faster than the atmospheric CO_2 concentration (Figure 4.38). This means that cement production for the near future will provide a greater share of the CO_2 emissions. Davidovits[39] estimates that if the current trends continue, by the year 2015 the world cement production will be responsible for a CO_2 emission equal to the present anthropogenic CO_2 emission of Europe (estimated at 3500 million tonnes in 1998). With this perspective, it will be impossible to even approach the targets for CO_2 emissions set by international agreements. New revolutionary technologies are certainly required to solve this problem.

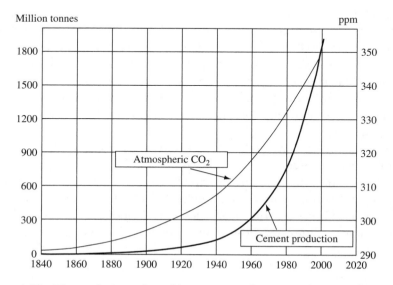

Figure 4.38 The evolution of world cement production and atmospheric CO_2 concentration. (from IPCC and Cembureau[39]). Reproduced by permission of Joseph Davidovits.

As explained above (Sections 4.9.2 and 4.9.3), the production of geopolymeric cement does not require any calcination of calcium carbonate, but instead requires the calcination of a few mineral ingredients such as kaolinite (KANDOXI). In the so-called CARBUNCULUS Cement™,[28] KANDOXI and volcanic tuff, together making up 70 %, are calcined at 800 °C. In addition, iron blast-furnace slag, up to 25 %, is used and this does not need any pre-treatment. Thus the calculated CO_2 emission for the production of 1 tonne of this geopolymeric cement is 0.184 tonnes, including the calcination of minerals and grinding and mixing energies. This is five times lower than for the production of Portland cement. Moreover, the geopolymer cement production is also more efficient in terms of reactive oxide use, as the mole ratio of CaO to SiO_4 (or AlO_4) groups in the structure of Portland cement is 3:1, whereas for the geopolymeric matrices the required amount of Na_2O or K_2O for each SiO_4 group is 1:6 or even lower.

Tuff is a common volcanic rock, defined as 'an indurated, pyroclastic deposit, predominantly consisting of fine-grained volcanic fragments, with maybe the presence of some sedimentary particles. The deposit may or may not be laid down in water, and it may be well sorted or heterogeneous.'[40] Thus, the term 'tuff' describes the appearance and origin of this rock, but not its composition. There are many different types of tuff, which are highly variable in composition. The requirement for geopolymers is that $K_2O + Na_2O$ should be high and CaO relatively low; thus the most appropriate materials are trachytic and phonolitic tuffs. These rocks have on average $K_2O + Na_2O = 10$–13% and CaO approximately 1–3 %. The only rocks with compositional implications named in Davidovits' work are andesitic and lamproitic lavas, which in general have lower alkali and higher CaO content.[41] By making a suitable choice of volcanic rock (even extremely alkali rocks – for example, sodic rhyolites – may be used,), its quantity may be reduced, the rest being replaced by an alkali-rich fly ash.

For the production of large quantities of geopolymers that may challenge Portland cement production, huge resources of suitable raw materials are required. In the research referred to here, volcanic tuff and blast furnace slag were used, but it is a small step to replace the latter by coal fly ash. This step would indeed be necessary, since blast-furnace slag production is insufficient and rapidly decreasing due to changes in metallurgical processes. Possibly, as proposed above, (part of) the volcanic tuff may be replaced by highly alkaline, low-calcium fly ashes. This would be an additional advantage, because suitable volcanic rocks are less well distributed in highly populated areas than power stations, so that considerable transport of bulky raw materials may be avoided. Moreover, combustion residues do not require thermal pre-treatment, yet another energetically attractive point for fly ash.

Davidovits,[39] in an optimistic view, states that by applying such new technologies that reduce the CO_2 emissions for cement production by 80–90 %, the future for coal combustion can be secured. Electricity utilities can then produce energy and low-CO_2 cement in the same plant. He foresees that by the year 2015, the required 3500 million tonnes of cement could be produced in this way.

References

1 Davidovits, J., *Sintered composite panels*. US Patent 3,950,470 (1972).
2 Davidovits, J., *Chemistry of geopolymeric systems, terminology*. In Proc. Geopolymer Int. Conf., Géopolymère'99, J. Davidovits, R. Davidovits and C. James (eds), 30 June – 2 July 1999, Saint-Quentin, France, pp. 9–39 (1999).
3 Davidovits, J., *La fabrication des vases de pierres au V et IV millenaires*. In 2nd Int. Congr. of Egyptology (abstracts), Grenoble, France (1979).
4 Davidovits, J., *Ancient and modern concretes: what is the real difference?* Concrete International: Des Constr., 9: 23–29 (1987).
5 Provis, J.L., Lukey, G.C. and Van Deventer, J.S.J., *Do geopolymers actually contain nanocrystalline zeolites? A re-examination of existing results*. Chem. Mater., 17: 3075–3085 (2005).
6 Davidovits, J., *Geopolymers of the first generation, Siliface process*. Proc. Geopolymer'88 1: 49–67 (1989).
7 Duxson, P., Lukey, G.C., Separovic, F. and Van Deventer, J.S.J., *Effect of alkali cations on aluminum incorporation in geopolymeric gels*. Ind. Eng. Chem. Res., 44: 832–839 (2005).
8 Duxson, P., Provis, J.L., Lukey, G.C., Separovic, F. and Van Deventer, J.S.J., *^{29}Si NMR study of structural ordering in aluminosilicate geopolymer gels*. Langmuir, 21: 3028–3036 (2005).
9 Cheng, T.W. and Chiu, J.P., *Fire-resistant geopolymer produced by granulated blast furnace slag*. Minerals Engineering, 16: 205–210 (2003).
10 Talling, B. and Brandstetr, J., *Present state and future of alkali-activated slag concretes*. In 3rd Inter. Conf. Fly Ash, Silica Fume, Slag and Natural Pozzolans in Concrete, Trondheim, Norway, 2: 1519–1545 (1989).
11 Wastiels, J., Wu, X., Faignet, S. and Patfoort, G., *Mineral polymer based on fly ash*. Journal of Resource Management and Technology, 22: 135–141 (1994).
12 Palomo, A., Grutzeck, M.W. and Blanco, M.T., *Alkali-activated fly ashes: a cement for the future*. Cem. Conc. Res., 29: 1323–1329 (1999).
13 Fernández-Jiménez, A. and Palomo, A., *Characterisation of fly ashes. Potential reactivity as alkaline cements*. Fuel, 82: 2259–2265 (2003).
14 Deventer, J.S.J. van, Lukey, G.C., Duxson, P. and Provis, J.L., *Reaction mechanisms in the geopolymeric conversion of inorganic waste to useful products*. In Proc. 1st

Int. Conf. on Engineering for Waste Treatment (WasteEng), 17–19 May 2005, Albi, France (on CD) (2005).

15 Hardjito, D., Wallah, S.E., Sumajouw, D.M.J. and Rangan, B.V., *Properties of geopolymer concrete with fly ash as source material: effect of mixture composition.* Paper to the Seventh CANMET/ACI International Conference on Recent Advances in Concrete Technology, 26–29 May 2004, Las Vegas (2004).

16 Davidovits, J., *Mineral polymers and methods of making them.* US Patent 4,349,386 (1982).

17 Barbosa, V.F.F., MacKenzie, K.J.D. and Thaumaturgo, C., *Synthesis and characterisation of materials based on inorganic polymers of alumina and silica: sodium polysialate polymers.* International Journal of Inorganic Materials, **2**: 309–317 (2000).

18 Cioffi, R., Maffucci, L. and Santoro, L., *Optimization of geopolymer synthesis by calcination and polycondensation of a kaolinitic residue.* Resour. Conserv. Recycl., **40**: 27–38 (2003).

19 Keyte, L.M., Lukey, G.C. and Van Deventer, J.S.J., *The effect of coal ash glass chemistry on the tailored design of waste-based geopolymeric products.* In Proceedings 1st Int. Conf. on Engineering for Waste Treatment (WasteEng), 17–19 May 2005, Albi, France (on CD) (2005).

20 *Proc. Geopolymer Int. Conf., Géopolymère'99*, Davidovits, J., Davidovits, R. and James, C. (eds), 30 June – 2 July 1999, Saint-Quentin, France, (on CD) (1999).

21 Lukey, G.C. (ed.), *Proc. Geopolymers 2002*, 28–29 October 2002, Melbourne, Australia (on CD: ISBN 0-9750242-0-5) (2002).

22 Schmücker, M. and MacKenzie, K.J.D., *Microstructure of sodium polysialate siloxo geopolymer.* Ceramics International, **31**: 433–437 (2005).

23 Lloyd, R.R. and Van Deventer, J.S.J., *The microstructure of geopolymers synthesised from industrial wastes.* In Proc. 1st Int. Conf. on Engineering for Waste Treatment (WasteEng), 17–19 May 2005, Albi, France (on CD) (2005).

24 Marsh, D., *Silicate-based mix technology elevates precast performance.* Concrete Products, 1 May 1998.

25 Silverstrim, T., Martin, J. and Rostami, H., *Geopolymeric fly ash cement.* In Proc. Geopolymer Int. Conf., Géopolymère'99. J. Davidovits, R. Davidovits, and C. James (eds), 30 June – 2 July 1999, Saint-Quentin, France, pp. 107–108 (1999).

26 Krivenko, P.V. and Kovalchuk, G.Y., *Heat-resistant fly ash based geocements.* In Proc. Geopolymers 2002, G.C. Lukey (ed.), 28–29 October 2002, Melbourne, Australia (on CD) (2002).

27 European R&D project BRITE–EURAM BE-7355-93: *Cost-effective geopolymeric cement for innocuous stabilization of toxic elements (GEOCISTEM).* Final Report, April 1997.

28 Davidovits, J., *Geopolymeric cement based on low cost geologic materials. Results from the European research project GEOCISTEM.* In Proc. Geopolymer Int. Conf., Géopolymère'99. J. Davidovits, R. Davidovits and C. James (eds), 30 June – 2 July 1999, Saint-Quentin, France, pp. 83–96 (1999).

29 Provis, J.L., Duxson, P., Lukey, G.C. and Van Deventer, J.S.J., *Modelling chemical ordering in aluminosilicates and subsequent prediction of cation immobilisation efficiency.* In Proc. 1st Int. Conf. on Engineering for Waste Treatment (WasteEng), 17–19 May 2005, Albi, France (on CD) (2005).

30   Van Jaarsveld, J.G.S., Van Deventer, J.S.J. and Lorenzen, L., *The potential use of geopolymeric materials to immobilize toxic metals: Part II. Material and leaching characteristics.* Mineral Engineering, **12**: 75–91 (1999).

31   Jaarsveld, J.G.S. van, Deventer, J.S.J. van and Lorenzen, L., *Factors affecting the immobilization of metals in geopolymerized fly ash.* Metallurgical and Materials Transactions B, **29**: 283–291 (1998).

32   Van Jaarsveld, J.G.S. and Van Deventer, J.S.J., *The effect of metal contaminants on the microstructure of fly-ash based geopolymers.* In Proc. Geopolymer Int. Conf., Géopolymère '99. J. Davidovits, R. Davidovits and C. James (eds), 30 June – 2 July 1999, Saint-Quentin, France, pp. 229–248 (1999).

33   Van Jaarsveld, J.G.S. and Van Deventer, J.S.J., *The effect of metal contaminants on the formation and properties of waste-based geopolymers.* Cem. Conc. Res., **29**: 1189–1200 (1999).

34   Swanepoel, J.C., Strydom, C.A. and Smit, J.P., *Safe disposal of brine water in fly-ash geopolymeric material.* Proc. Geopolymer Int. Conf., Géopolymère '99. J. Davidovits, R. Davidovits and C. James (eds), 30 June – 2 July 1999, Saint-Quentin, France, pp. 253–267 (1999).

35   Chervonnyi, A.D. and Chervonnaia, N.A., *Geosynthesis of immobilization matrix from ash of biomass and clay.* In Proc. Geopolymers 2002, G.C. Lukey (ed.), 28–29 October 2002, Melbourne, Australia (on CD) (2002).

36   Bankowski, P., Zou, L., Hodges, R., Singh, P.S. and Trigg, M., *Brown coal fly ash stabilisation by inorganic polymers.* In Proc. Geopolymers 2002, G.C. Lukey (ed.), 28–29 October 2002, Melbourne, Australia (on CD) (2002).

37   Bankowski, P., Zou, L. and Hodges, R., *Reduction of metal leaching in brown coal fly ash using geopolymers.* J. Hazard. Mater., **B114**: 59–67 (2004).

38   Davidovits, J., *Global warming impact on the cement and aggregates industries.* World Resource Review, **6**: 263–278 (1994).

39   Davidovits, J., *Geopolymeric reactions in the economic future of cements and concretes: world-wide mitigation of carbon dioxide emission.* In Proc. Geopolymer Int. Conf., Géopolymère '99. J. Davidovits, R. Davidovits and C. James (eds), 30 June – 2 July 1999, Saint-Quentin, France, pp. 111–121 (1999).

40   Visser, W.A. (ed.), *Geological Nomenclature.* Royal Geological and Mining Society of the Netherlands (1980).

41   Gimeno, D., Davidovits, J., Marini, C., Rocher, P., Tocco, S., Cara, S., Diaz, N., Segura, C. and Sistu, G., *Development of silicate-based cement from glassy alkaline volcanic rocks.* Bol. Soc. Ceram. Vidrio, **42**: 69–78 (2003).

4.10   Carbon Products

Henk Nugteren and Mercedes Maroto-Valer
Delft University of Technology, The Netherlands; University of Nottingham, UK

Types of combustion residues involved
The production of activated carbons and carbon artefacts from coal combustion residues is restricted to high-carbon fly ashes and to the concentrates from carbon separation processes, undertaken in the first place to obtain a low-carbon ash residue. For some applications, especially for carbon artefacts, these concentrates need further beneficiation before treatment. However, selective collection of ash from different hoppers in the particulate separation plant may result in ashes having as high a carbon content as 50 % and may bring down the carbon content of the resulting mixed ash. Such ashes are suitable after upgrading for further processing.

Residual carbon from oil-fired fly ash is a possible candidate for similar applications.

State of development
Currently, most of the work is concentrated in the group of Maroto-Valer and co-workers, currently at the University of Nottingham and formerly at the Pennsylvania State University, and it is in the research stage.

4.10.1   Introduction

The combustor design or the operating conditions of the combustor of a coal-fired power station may lead to the incomplete combustion of the coal, which will result in unburned carbon being present in fly ash. Unburned carbon is a constituent of ash that is often seen as an unwanted component, as it reduces the ash quality for many ash applications. In the literature, both the terms 'unburned carbon' and 'fly ash carbon' occur for the same material, and therefore these terms are used here interchangeably. Unburned carbon is known to have a deactivating effect on air-entrainment admixtures in concrete, as these are adsorbed by the carbon particles (see Section 4.2), and therefore allowable upper limits of loss on ignition (LOI) in the order of 3–6 % are set for such applications, depending on standards used in different countries. Since power plants cannot always meet these standards, as the required reduced NO_x emissions inherently increase carbon in ash, unburned carbon is often removed from the ash by separation technologies or by carbon burnout (see Chapter 3). Otherwise, carbon-rich ash, with reported values of up to 50 % carbon content, if taken from the hoppers with the highest carbon values, is either placed in holding ponds or landfilled, a practice by which in the USA millions of tonnes of unburned carbon has been disposed of.[1,2] As an indication, in the USA over 900 million tonnes of coal is used in coal-fired units, to generate about 50 % of the total electricity. This also generates around 70 million tons of fly ash, from which around 43 millions are disposed of.

However, the adsorbent properties of unburned carbon may be turned from a restriction into a profit, if these particles can be separated in such a way that an unburned carbon concentrate is produced. Among the first to investigate the advantages of fly ash carbons as an adsorbent material were researchers from the Center for Applied Energy Research in Lexington, Kentucky,[3] who extended an earlier idea to use the coal waste elutrilithe[4] to the use of fly ash carbon, as it was concentrated from carbon-rich fly ashes without any activation step. Successful application of the unburned carbon fraction in coal combustion residues has considerable environmental benefits, in that it allows the utility industry to meet the strict NO_x emission limits (Clean Air Act Amendments in the USA) without producing residues for which no application is available. It has the further direct environmental benefit of using the activated carbons in environmental technology and the indirect benefit of the continuation and expansion of the use of fly ash as a cement extender.[5,6]

Whereas commercial activated carbon is produced in a two-step process, removal of volatile compounds and activation, unburned carbon from coal combustion only requires the last step, since it has already gone through a devolatilisation step in the combustor. This means a potential lower cost of production and, moreover, that the product has the inherent advantage of being composed of fine particles that are not only rich in micropores (< 2 nm in diameter) but also have a high content of mesopores (2–50 nm).

Recent work has focused on comparing the physical and chemical properties of coal and biomass unburned carbon samples from combustion and gasification processes. It has been reported that wood-based samples have a lower starting point for weight loss (180–300 °C versus 200–400 °C) than the coal-based samples, and their thermal reaction is more complex.[7] Gasification and combustion unburned carbon samples from lignite are typically mesoporous (2–50 nm), while from bituminous coal are typically microporous (< 2 nm).[8]

Carbon is also used for the production of carbon products such as anodes for aluminium smelting and electrodes for steel arc furnaces. Such carbon artefacts are normally produced from calcined petroleum coke, in which the sulfur content is a disadvantage that may be overcome by replacement with unburned carbon from coal combustion.[2]

4.10.2   Forms of Unburned Carbon

Although restrictions on carbon content in ashes is normally given as the upper limit to allowable LOI, one must be careful with this parameter, as LOI often gives only an approximation of the carbon content of an ash.

Petrographic examinations have shown that there are at least three distinct microscopically identifiable carbon types (see Figure 4.39):[2,3,9-13]

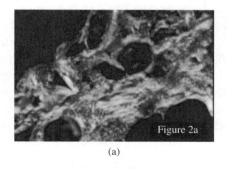

(a)

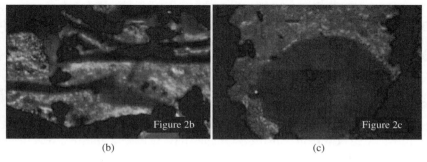

(b)   (c)

Figure 4.39   Optical micrographs of the unburned carbon sample: (a) typical anisotropic mosaic texture; (b) a typical isotropic material trapped in an anisotropic matrix in unburned carbon; and (c) a large particle of inertinite (macrinite) trapped in the anisotropic matrix (from Andrésen *et al.*[2]). Reproduced by permission of M. Maroto-Valer.

- inertite particles, which appear to be nonfused particles;
- isotropic coke, extensively reacted particles that have gone through a molten stage; and
- anisotropic coke, as isotropic coke but with more aligned graphitisable carbon particles.[13]

The density of these particles increases with intensity of reaction, which makes it possible to separate these three fractions by density gradient centrifugation (DGC) using a high-density lithium polytungstate media (max. density 2.85 g cm^{-3}).[14] For this purpose, preliminary triboelectrostatically enriched carbon concentrates were used.

The specific surface area of the carbon particles is relatively low (10–60 m^2 g^{-1}) and increases linearly with density. Pores are 60–90 % in the mesopore range (2–50 nm), with the highest proportions found in the inertite fraction, followed by the anisotropic coke fraction, and lowest in the isotropic coke fraction. Generally, the density and porosity properties seem to be independent of the fly ash sample investigated.[11]

While the carbon particles pass through the boiler, the carbon surfaces undergo changes by breaking and opening of aromatic structures, formation of C-oxygen functional groups, condensed aromatic ring systems, interaction of C=C bonds with binding of oxygen, bridging hydrogen bonds and so on. These processes are far from complete compared to activated carbons, where activation by controlled thermal and chemical treatment aims at developing a microporous structure.[3] This may explain why fly ash carbon has predominantly macro- and mesopores, whereas commercial activated carbons have predominantly micropores.

As the different proportions of the three types of carbon particles may have an impact on the adsorption of air-entraining admixtures, they may also influence a subsequent activation step and/or the application of the products as adsorbents, as was noted, for example, in mercury capture capacities.[11,12] Also, the observation that not all unburned carbon samples are equally suited for activation[13] indicates that the nature of unburned carbon particles may play an important role for further utilisation.

Apart from unburned carbon from coal combustion fly ash, the unburned carbon from oil-fired fly ash may also be a suitable precursor for carbon products. Often, oil-fired fly ash is used for the recovery of its elevated vanadium and nickel contents (see Section 4.11), in which considerable amounts of unburned carbon remain as a by-product. When leached with sulfuric acid, the elemental carbon content of this residue can be as high as 90 %, the remainder being N, H, O and ash.[15]

This carbon has particle sizes between 1 and 100 μm and a median of the order of 50–80μm. The specific surface area is 16–33 m² g⁻¹, the real density 2.05–2.16 and the true density about 0.15 g cm⁻³. The crystalline structure can be classified as poorly crystalline graphitic carbon and the particles appear in two distinct forms, porous and foamed. Pores in the porous particles are mainly in the order of 1–5 nm.[15,16]

The carbon from oil-fired fly ash can also be exploited for flue gas treatment (SO_2 and NO_x).[17]

In general, unburned carbon in class F fly ashes is found in the finer fractions, whereas in class C ashes there are significant large carbon particles present. The class F carbons have significantly lower specific surface areas than the class C carbons.[18]

4.10.3 Separation

Carbon removal from ash can be realised in different ways, as was explained in Chapter 3. The most commonly methods used currently are electrostatic separation, classification, flotation and some other methods as described in detail in Chapter 3. However, the main purpose of this operation is to render an ash residue with a permitted unburned carbon content. Often, the carbon concentrate produced is redirected to the combustor for recovery of its energy content. In order to take advantage of this fraction, the carbon content must be increased further. The LOI value of the carbon concentrates vary enormously from 15 % to even 80 % under extreme circumstances, depending on the methods used, with flotation techniques giving the best concentrates.[19,20] Gray and co-workers treated an industrial fly ash sample by tribo-electrostatic separation, ultrasonic column agglomeration and column flotation, with the objective of recovery of carbon particles for further application.[21] Column flotation turned out to give the best results, with carbon concentrates at a LOI value of 61 % and a carbon recovery of 62 %, with 90 % of the ash reporting to the tails with LOI values < 8 %. It is believed that applying a multi-stage cleaning circuit may significantly improve the cleaning performance.

Following flotation, a further upgrade of the concentrate may be accomplished using an acid digestion step with $HCl/HNO_3/HF$ at 65 °C to reduce the ash content down to less than 4 %.[14,19]

4.10.4 Activation Technology

As noted above, production of activated carbon from an unburned carbon concentrate from coal combustion can be performed in a one-step process, as opposed to the two-step process normally applied in the production of commercial activated carbon products. The activation technology used by Maroto-Valer and co-workers comprises heating in a stainless steel tube reactor under a nitrogen atmosphere to the desired temperature, followed by steam injection while the reactor is heated isothermally for one hour.[13] Preliminary studies show that steam activation provides higher surface areas than either CO_2 or KOH activation.[22] However, this could be due to the lower intrinsic gasification reactions of CO_2 compared to H_2O and to the higher amounts of KOH needed.

   Precursors for activated carbon are typically isotropic, and therefore, unburned carbons with high contents of isotropic coke are most suitable for producing activated carbons.[13] As activation is a gasification reaction and the sample reactivity is related to surface active sites (defects in carbon layer planes, edge atoms and the presence of heteroatoms such as O, S and N), characterisation of unburned carbons from fly ash may predict their suitability for activation. It was indeed found that carbon samples with high isotropic coke content and samples with a relatively high oxygen content showed the highest reactivity during activation.[13] In the same study, activated carbons were produced with a specific surface area as high as $1270 \, m^2 \, g^{-1}$ and a pore volume of $0.815 \, ml \, g^{-1}$, comparable to those reported for commercial activated carbons. Increasing activation time and temperature increased the surface area and porosity, but also resulted in wider pores and lower solid yields. The pore volume distributions for carbon concentrates from six different fly ashes before and after activation are shown in Figure 4.40.[13] This shows that the total pore volume changes by approximately a factor of 10 on activation, and that in general the relative contribution of the micropores to the total pore volume increases.

4.10.5 Fly Ash Carbon Direct Application

The macro- and mesopores of fly ash carbon particles make them suitable for easy access of macromolecular organic species such as dye molecules. This in contrast to microporous active carbons that would either require long diffusion rates to adsorb macromolecules or, due to the small size of the available channel openings, are incapable of such

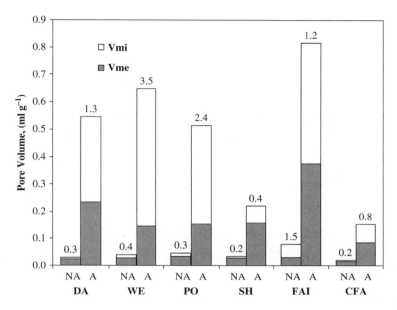

Figure 4.40   Pore volume distribution for unburned carbon concentrates from six different fly ashes (denoted DA, WE, PO, SH, FA1 and CFA), prior and after steam activation at 850 °C. Volumes of micropores and mesopores are depicted as V_{mi} and V_{me}. The numbers given indicate the ratio V_{mi}/V_{me}. Reprinted with permission from reference 13. Copyright © American Chemical Society 2003.

adsorption. With this in mind, Graham *et al.* investigated the adsorption capabilities of dyes for different carbon concentrates from fly ash obtained either by froth flotation or from hydro-classifiers, with carbon contents varying from 31 to 61 %, and compared these to the performance of a commercial activated carbon (F-400, Calgon Carbon Corporation, Pittsburgh, USA).[3] The tests were performed using the reagent-grade dyes Acid Orange 10 and Basic Blue 7 (Victorian blue), as well as effluents containing fibre reactive black and blue dye residues. It was found that in an agitated suspension fly ash carbon adsorbed the dye material very fast compared to the commercial activated carbon (Figure 4.41). After only 10 minutes exposure time, the ash sample with 61 % C had adsorbed 27.2 mg g^{-1} of carbon (74 % of its capacity), whereas the activated carbon reached this level only after more than 10 hours. However, the maximum adsorption capacity of activated carbon was significantly above that of the fly ash carbons, but was only obtained after approximately 24 hours.

The presence of ultra-fine particulate matter derived from dissociated fibre material in the industrial dye residue effluents made it difficult to recover the loaded carbon by filtration. Therefore, the dye effluent was

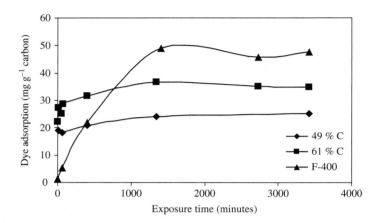

Figure 4.41 Dye adsorption (Victorian blue) on to fly ash carbon samples upgraded by froth flotation to 49 % and 61 % respectively. The F-400 sample is a commercially available activated carbon (98 % C). Reproduced by permission of U.M. Graham.[3]

pumped through a loosely packed fly ash carbon bed that acted as a filter and adsorbent. With this set-up, results similar to those with the reagent-grade dyes were obtained.

The unburned carbon in fly ash was confirmed as the actual adsorbent by Wang et al.,[23] who tested several ashes and carbon-enriched ashes for the adsorption of Methylene Blue and Crystal Violet and found a relation between carbon content and adsorption capacity.

The use of residual carbon from oil-fired fly ash for the decolourisation of dye effluents has also been reported.[15,24] Although the adsorbed amount of dyes per unit weight is only one seventh of that of commercial activated carbon, the fact that the amounts adsorbed per unit surface in these unburned carbons were about 4–5 times greater than that in the activated carbon indicates a high adsorptive ability of unburned carbon.[15]

4.10.6 Applications of Activated Carbons

Mercury Capture

Nearly 90 % of the Hg(II) entering a flue gas desulfurisation unit is removed within this unit, however, none of the Hg^0 is captured. Injection of a carbon adsorbent upstream of the electrostatic precipitator seems a promising technology for the control of mercury emissions in boilers without a FGD system, or with substantial Hg^0 emissions.[25] The cited reference gives an extensive overview of the available mercury control options for coal-fired power plants. As elemental mercury is difficult to capture, a large excess of sorbent is required (mass ratio C : Hg of 10 000 or more), which makes the process expensive. By using carbon originating from the same combustion process, costs may be cut to an acceptable level. In a preliminary study, unburned carbons from coal combustion fly ash were tested along with other sorbents.[26] It was found that, in general, mercury removal was rather poor and variable between large limits, from less than 0.1 to 2.5 mg g^{-1}. Chlorine-promoted activated carbon showed the highest adsorption. This is also supported from commercially activated carbons, treated with zinc chloride ($ZnCl_2$) that produces Cl-containing functional groups such as $[Cl_2-C_nH_xO_y]$ on which mercury can be captured by chemisorption with the formation of $[HgCl]^+$, $[Hg_2Cl_2]^0$ and $[HgCl_4]^{2-}$ complexes.[27] The use of unmodified but concentrated unburned carbon from fly ash or wood ash is covered in a US patent.[28]

Maroto-Valer and co-workers used steam activation to activate carbon particles that were concentrated from fly ash and surprisingly found that the nonactivated de-ashed carbon concentrate was able to capture 1.85 mg g^{-1} of mercury (compared to 2.77 mg g^{-1} for a commercial activated carbon), whereas the activated sample only captured 0.23 mg g^{-1}. Although the specific surface area had increased by a factor of 15 during activation from 53 m^2 g^{-1} to 863 m^2 g^{-1}, the pore sizes had reduced in such a way that mercury capture, a mass-transfer-limited process, was hindered.[19] It is, however, unlikely that the large difference in mercury adsorption capacity is caused by this change in pore volume alone. Therefore, in a follow-up study, XPS analyses were carried out on the same activated and nonactivated carbon samples.[29] It was found that the activation process had removed almost all of F and Cl species and oxygen functional groups from the surface of the sample. So, surface functionality may have an important influence on mercury adsorption capacity as well. This surface functionality was probably enhanced by the de-ashing process carried out with HCl and HF.

However, the work reported could not differentiate between the effect of oxygen functional groups, halogen species and carbon sites, and further studies on the modification of the surface properties of fly ash carbons and their mercury adsorption properties are under way to ascertain the effect of oxygen functionalities, halogen species and carbon sites.

Capture of CO_2

In view of the now generally accepted need for reduction of CO_2 emissions, much research is dedicated to CO_2 sequestration technologies. Successfully applying such technologies may become a prerequisite to guaranteeing the sustainable key role of coal in the 21st century. To sequester CO_2 efficiently, the separation and capture of CO_2 from flue gases is often required. Such processes are normally based on the use of amine solutions.[30] These processes are very expensive, because they are very energy intensive and the amine solutions have only limited lifetimes. It is estimated that the costs of separation and capture of CO_2 is about three-quarters of the total cost of ocean or geological sequestration.[31] An improvement can be made by bonding the amine groups to a solid surface, resulting in an easier regeneration step. Maroto-Valer and co-workers proposed the use of activated carbons produced from unburned carbon from coal power plants as the solid surface instead of expensive commercial molecular sieves and activated carbons.[31,32] They modified their previously activated carbon with some of the amines that are commonly used for CO_2 separation, such as monoethanolamine (MEA), diethanolamine (DEA), methyldiethanolamine (MDEA) and amino-2-methyl-1-propanol (AMP). However, the pore-filling effects of these chemicals on the mesopores and the possible blocking of micropores decreased the pore volume and the specific surface area compared to the unmodified activated carbon. This was reflected in the CO_2 adsorption capacity at low temperatures (30 °C), which was reported as $40\,\mathrm{mg\,g^{-1}}$ for the unmodified and only around $20\,\mathrm{mg\,g^{-1}}$ for the modified activated carbons, except for that modified with MEA, for which the CO_2 adsorption capacity increased to $68.6\,\mathrm{mg\,g^{-1}}$. However, for practical use in flue gas cleaning, the CO_2 adsorption capacity at stack temperature, normally in the range of 100–120 °C, is of more interest. With increasing temperature, the CO_2 adsorption capacity for the unmodified activated carbon quickly decreased from $40.3\,\mathrm{mg\,g^{-1}}$ at 30 °C to $18.5\,\mathrm{mg\,g^{-1}}$ at 70 °C and $7.7\,\mathrm{mg\,g^{-1}}$ at 120 °C. A similar decrease of CO_2 adsorption capacity is also found for MEA- and AMP-loaded carbons. However, for the DEA-modified activated carbon, the highest CO_2 adsorption capacity was measured at 70 °C ($37.1\,\mathrm{mg\,g^{-1}}$) and for the MDEA modified carbon even at 100 °C ($40.6\,\mathrm{mg\,g^{-1}}$). With further increase of temperature, the CO_2 adsorption capacity again decreases, probably due to decomposition of the amines. This means that activated carbons from fly ash loaded with amines have a potential for CO_2

removal from flue gas at stack temperatures, but that the selection of chemicals is critical.[31]

In preliminary experiments, the same researchers found an adsorption capacity for CO_2 as high as $93.6\,mg\,g^{-1}$ for a polyetherimine (PEI) impregnated de-ashed activated carbon concentrate from a different fly ash carbon at an adsorption temperature of $75\,°C$.[31]

By using 3-chloropropylamine-hydrochloride (CPAHCL), CO_2 absorption was reported to be almost double that of the unmodified carbon concentrate, but still gave less than 10 % of CO_2 adsorption compared to commercial sorbents.[33] However, adsorption in this case was investigated at low temperatures ($25\,°C$), which may have had an effect on the results.

4.10.7 Production of Carbon Artefacts

Calcined petroleum coke, selling for $220–250 per tonne, is currently used for the manufacture of carbon bodies for a wide range of applications, including anodes for aluminium smelting and electrodes for steel arc furnaces.[2] Sulfur content of calcined petroleum coke is typically in the range of 3–4 % and this is known to spoil the carbon products used in the aluminium and steel industries. Unburned carbon from coal combustion generally contains less than 0.5 % sulfur, and is therefore seen as a potential low-sulfur replacement for calcined petroleum coke for the production of carbon artefacts. However, in general, unburned carbon from coal combustion has a somewhat higher specific surface area and porosity compared to calcined petroleum coke.

The production of such carbon bodies involves a filler (calcined petroleum coke or, alternatively, the unburned carbon), a binder (coal tar pitch) and additives for the enhancement of certain properties. Three different particle size fractions of petroleum coke are used in formulation and only the finest fraction ($< 75\,\mu m$) may be replaced by unburned carbon.[1] Green carbon pellets are produced from the mixture of ingredients and subjected to controlled heating in a tube furnace to a maximum temperature of $1000\,°C$ for about 24 hours for a complete baking cycle.[2]

Tests in which fine petroleum coke ($< 75\,\mu m$) was partially replaced by unburned carbon resulted in a slight decrease in density of the baked carbon bodies. This is explained by the higher specific surface and porosity of the green carbon, and might be overcome by optimising the formulation, especially the pitch content, as a function of the specific surface area.[2] However, in an earlier study, for which the thermal treatment was only carried out at low temperatures (125 °C), the same authors found an increase in density, due to a synergistic effect between the unburned carbon and the coal tar binder pitch. They thus concluded that unburned carbon could be used successfully as a superior replacement for the fine fraction of petroleum coke in the manufacture of carbon artefacts.[1]

The different forms of carbon in the unburned portion of combustion residues also play a role in the manufacture of carbon bodies. Isotropic and anisotropic carbon is also present in petroleum coke, where a given ratio of anisotropic over isotropic coke is required for different applications.[2] Anisotropic coke is preferred or even required because it has better conducting properties than the more insulating isotropic coke, an important property when it comes to the production of carbon anodes. Moreover, anisotropic coke has a higher density, which is an additional advantage. For carbons from fly ash, this consideration could result in an interesting market diversification: predominantly isotropic carbons for the production of activated carbons and predominantly anisotropic carbons for the manufacturing of carbon bodies.

References

1   Andrésen, J.M., Maroto-Valer, M.M., Andrésen, C.A. and Battista, J.J., *Assessing the potential of unburned carbon as a filler for carbon artifacts.* In Proc. 1999 Fly Ash Utilization Symposium, Lexington, Kentucky, pp. 534–540 (1999).
2   Andrésen, J.M., Zhang, Y. and Maroto-Valer, M.M., *Utilization of unburned carbon as a low sulfur alternative to petroleum coke.* In Proc. 2001 Fly Ash Utilization Symposium, Lexington, Kentucky (2001).
3   Graham, U.M., Robl, T.L., Groppo, J. and McCormick, C.J., *Utilization of carbons from beneficiated high LOI fly ash: adsorption characteristics.* In Proc. 1997 Fly Ash Utilization Symposium, Lexington, Kentucky, pp. 29–36 (1997).
4   Hu, Z. and Vansant, E.F., *Chemical activation of elutrilithe producing carbon-aluminosilicate composite adsorbent.* Carbon, 33: 1293–1300 (1995).
5   Schobert, H.H. and Song, C., *Chemicals and materials from coal in the 21st century.* Fuel, 81: 15–32 (2002).
6   Maroto-Valer, M.M., Lu, Z., Zhang, Y., Shaffer, B.N., Andrésen, J.M. and Schobert, H.H., *Environmental benefits of producing adsorbent materials from unburned carbon.* In Proc. 2001 Fly Ash Utilization Symposium, Lexington, Kentucky (2001).
7   Maroto-Valer, M.M., Zhang, Y. and Miller, B.G., *Development of activated carbons from coal and biomass combustion and gasification chars.* Prepr. Am. Chem. Soc. Div. Fuel Chem., 49: 690–691 (2004).
8   Maroto-Valer, M.M., *Personal communication* (2005).

9   Vleeskens, J.M., Haasteren, T.W.M.B. van, Roos, M. and Gerrits, J., *Behaviour of different char components in fluidized bed combustion: a char petrography study.* Fuel, **67**: 426–430 (1988).

10 Bailey, J.G., Tate, A., Diessel, C.F.K. and Wall, T.F., *A char morphology system with applications to coal combustion.* Fuel, **69**: 225–239 (1990).

11 Maroto-Valer, M.M., Taulbee, D.N. and Hower, J.C., *Characterization of differing forms of unburned carbon present in fly ash separated by density gradient centrifugation.* Fuel, **80**: 795–800 (2001).

12 Hower, J.C., Maroto-Valer, M.M., Taulbee, D.N. and Sakulpitkakphon, T., *Mercury capture by distinct fly ash carbon forms.* Energy & Fuels, **14**: 224–226 (2000).

13 Zhang, Y., Lu, Z., Maroto-Valer, M.M., Andrésen, J.M. and Schobert, H.H., *Comparison of high-unburned-carbon fly ashes from different combustor types and their steam activated products.* Energy & Fuels, **17**: 369–377 (2003).

14 Maroto-Valer, M.M., Taulbee, D.N. and Hower, J.C., *Novel separation of the differing forms of unburned carbon present in fly ash using density gradient centrifugation.* Energy & Fuels, **13**: 947–953 (1999).

15 Ya-Min Hsieh and Min-Shing Tsai, *An investigation of the characteristics of unburned carbon in oil fly ash.* In M.M. Maroto-Valer, C. Song and Y. Soong (eds), *Environmental Challenges and Greenhouse Gas Control for Fossil Fuel Utilization in the 21st Century,* Kluwer Academic/Plenum Publishers: New York, Chapter 27, pp. 387–401 (2002).

16 Hsieh, Y.-M. and Tsai, M.-S., *Physical and chemical analyses of unburned carbon from oil-fired fly ash.* Carbon, **41**: 2317–2324 (2003).

17 Davini, P., *Flue gas treatment by activated carbon obtained from oil-fired fly ash.* Carbon, **40**: 1973–1979 (2002).

18 Külaots, I., Hurt, R.H. and Suuberg, E.M., *Size distribution of unburned carbon in coal fly ash and its implications.* Fuel, **83**: 223–230 (2004).

19 Maroto-Valer, M.M., Zhang, Y., Lu, Z., Granite, E. and Pennline, H., *Development of activated carbons from unburned carbon for mercury capture.* In Proc. 2003 Fly Ash Utilization Symposium, Lexington, Kentucky (2003).

20 Hwang, J.Y., Liu, X. and Zimmer, F.V., *Beneficiation process for fly ash and the utilization of cleaned fly ash for concrete application.* In Proc. 11th Int. Symp. on Use and Management of Coal Combustion By-products, January, Orlando, Fla (1995).

21 Gray, M.L., Champagne, K.J., Soong, Y., Killmeyer, R.P., Maroto-Valer, M.M., Andrésen, J.M., Ciocco, M.V. and Zandhuis, P.H., *Physical cleaning of high carbon fly ash.* Fuel Processing Technology, **76**: 11–21 (2002).

22 Lu, Z., *Characterization and steam activation of unburned carbon in fly ash.* M.Sc. Thesis, Pennsylvania State University (2003).

23 Wang, S., Boyjoo, Y., Choueib, A., Wu, H. and Zu, Z., *Role of unburnt carbon in adsorption of dyes on fly ash.* J. Chem. Tech. Biotechnol., **80**(10): 1204–1209 (2005).

24 Tsai, M.S., Hsieh, Y.M. and Tsai, S.L., *Study on the decolorization of dying effluent by residual carbon from oil-fired fly ash.* in Extended Abstracts, Asian-Pacific Conference on Industrial Waste Minimization and Sustainable Development '97, Taipei International Convention Center, Taiwan, Ministry of Economic Affairs, pp. 249–262 (1997).

25 Pavlish, J.H., Sondreal, E.A., Mann, M.D., Olson, E.S., Galbreath, K.C., Laudal, D.L. and Benson, S.A., *Status review of mercury control options for coal-fired power plants*. Fuel Processing Technology, **82**: 89–165 (2003).

26 Granite, E.J., Pennline, H.W. and Hargis, R.A., *Novel sorbents for mercury removal from flue gas*. Ind. Eng. Chem. Res., **39**: 1020–1029 (2000).

27 Zeng, H., Jin, F. and Guo, J., *Removal of elemental mercury from coal combustion flue gas by chloride-impregnated activated carbon*. Fuel, **83**: 143–146 (2004).

28 Hwang, J. and Li, Z., *Control of mercury emissions using unburned carbon from combustion by-products*. US Patent 6027551 (2000).

29 Maroto-Valer, M.M., Zhang, Y., Granite, E.J., Tang, Z. and Pennline, H.W., *Effect of porous structure and surface functionality on the mercury capacity of a fly ash carbon and its activated sample*. Fuel **84**: 105–108 (2005).

30 Yeh, J.T., Pennline, H.W. and Resnik, K.P., *Study of CO_2 absorption and desorption in a packed column*. Energy & Fuels, **15**: 274–278 (2001).

31 Maroto-Valer, M.M., Andrésen, J.M., Zhang, Y. and Zhe, L., *Development of fly ash derived sorbents to capture CO_2 from flue gas of power plants*. Final technical progress report for DOE, USA, 01-CBRC-E9 (2004).

32 Maroto-Valer, M.M., Lu, Z., Zhang, Y. and Andrésen, J.M., *Using unburned carbon derived sorbents for CO_2 capture*. In Proc. 2003 Fly Ash Utilization Symposium, Lexington, Kentucky (2003).

33 Gray, M.L., Soong, Y., Champagne, K.J., Baltrus, J., Stevens, R.W. Jr, Toochinda, P. and Chuang, S.S.C., *CO_2 capture by amine-enriched fly ash carbon sorbents*. Separation and Purification Technology, **35**: 31–36 (2004).

4.11 Recovery of Values from Combustion Ashes

Michael Cox

University of Hertfordshire, Hatfield, UK

4.11.1 Introduction

Although the major constituents of combustion ashes are silica and alumina, there are many minor elements present that depend on the source of the fuel. Thus the ashes from petroleum and petroleum-derived fuels tend to be high in vanadium oxides and those from municipal waste incineration contain significant amounts of zinc and lead. As mentioned earlier (Chapter 3.3), these minor elements can indicate a potential environmental hazard and so their removal may be important before the ashes can be reused in other products. In addition, some of these minor elements may in themselves represent a value and thus warrant extraction and purification for sale. This section will consider some of the attempts that have been made to extract valuable elements from combustion ash.

4.11.2 Vanadium (Nickel)

Vanadium is the main impurity in fly ash from power stations fuelled by heavy oil combustion. The amounts range from 1 to 7 %, which is comparable with that found naturally in minerals, and so since the 1960s several processes have been described to recover the vanadium. More recently, interest has centred on Orimulsion® ash derived from the combustion of this stable bitumen-in-water emulsion, where the fly ashes can contain up to 12 % V and up to 3 % Ni. The fine and low-density ashes from Orimulsion make landfill disposal difficult and pose an environment problem because of their high leachability in water, with a large part of V and Ni being dissolved.[1] Therefore, to overcome such difficulties, treatment of Orimulsion ash becomes important and the potential recovery of vanadium and nickel provides an economic incentive.

Hydrometallurgical processes have been mainly used to treat both the heavy oil and Orimulsion ashes, although the flow sheets differ in detail because of the significant amount of magnesium salts used as emulsion stabilisers in Orimulsion and the variable amounts of carbon present in the ashes. Thus in the case of Orimulsion ashes it is first necessary to eliminate the magnesium salts with a water wash, to allow the recovery and recycle of the magnesium salts and to decrease the acid consumption in the subsequent leaching step. In cases where the ash contains significant carbonaceous material, preliminary removal of the carbon by heating at 650–1150 °C provides a more concentrated inorganic fraction that leaches more readily.[2] The optimum ignition temperature was about 850 °C, above which the recovery of both vanadium and nickel decreased, presumably due to the formation of refractory vanadium/nickel compounds.[3] Following these preliminary processes, the ashes are leached with sulfuric acid or sodium hydroxide at elevated temperature. Acid leaching dissolves the vanadium, largely as vanadium(IV), nickel and the other metals,[4] whereas alkali leaching will tend to separate vanadium by formation of alkali metal vanadates from the other impurities.[5]

Several different flow sheets have been described to treat this acidic leachate. The simplest provides a number of saleable products containing various amounts of vanadium and nickel. In the process devised by Process Research ORTECH Inc.,[6] the fly ash is partially leached to dissolve 80 % of the feed, leaving a residue containing 30 % vanadium that would be considered as a saleable product. The filtrate is then neutralised to pH 8–10, precipitating a mixed product containing 30 % V and 15 % Ni, also a saleable material. The overall process achieved > 90 % recovery of vanadium and nickel from the ash.

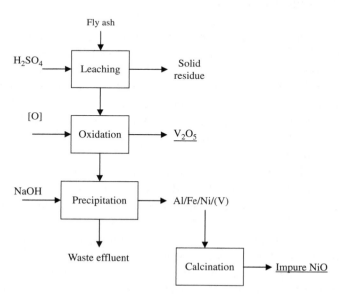

Figure 4.42 The recovery of vanadium and nickel from Orimulsion fly ash (Scheme 1).

The other flow sheets are more complex, as they are designed to separate the elements. The first of these (Figure 4.42) consists of oxidising the vanadium to vanadium(V) with $NaClO_3$, $NaClO$, H_2O_2, O_2, or Cl_2 and precipitation as V_2O_5.[4] The presence of impurities in the V_2O_5 – for example, phosphorus – does not preclude its use in ferrovanadium alloys, but blending with a purer V_2O_5 is required to meet impurity specifications.[2,4] Following removal of the V_2O_5 by filtration, the filtrate was treated with alkali to precipitate residual medals, largely aluminium, iron, nickel and vanadium. Mild calcination of the precipitate gave an impure nickel oxide product that could be used in steel or alloy production. An alternative procedure that provides a purer product is to separate the nickel from the other elements in solution by liquid–liquid extraction, as described in the following flow sheet.

The second flow sheet (Figure 4.43) involves liquid–liquid extraction of the acidic leachate to recover and purify the vanadium, using di-2-ethylhexylphosphoric acid in kerosene in the presence of a modifier.

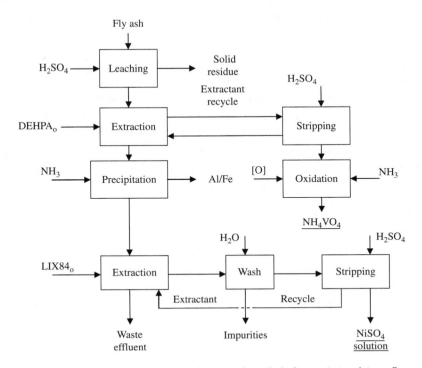

Figure 4.43 The recovery of vanadium and nickel from Orimulsion fly ash (Scheme 2): $DEHPA_o$ = di-2-ethylhexylphosphoric acid in organic diluent; $LIX84_o$ = LIX84 in organic diluent.

The vanadium-loaded organic phase is mixed with dilute sulfuric acid to recover the vanadium, which is then oxidised and precipitated with ammonia to give ammonium polyvanadate (APV). To recover the nickel from the raffinate, the solution is first neutralised with ammonia, precipitating impurities such as iron and aluminium, and forming ammonium nickel sulfate, from which the nickel is extracted by an arylhydroxyoxime (e.g. LIX84, Cognis Inc.). The organic solution is washed with weak sulfuric acid to remove co-extracted impurities, and then nickel is recovered as a strong nickel sulfate solution, with sulfuric acid allowing the organic phase to be recycled in the process. This process was operated commercially for several years at Stenungsund, Sweden.[7] Nickel metal can also be electrochemically deposited from a solution of the double sulfate.[8] It was found that magnesium had little effect on the deposit, but the presence of vanadium did affect the process, with nickel not plating out effectively at $pH = 5$. However, operating at $pH \sim 3-3.5$, nickel could be deposited in the presence of 100 ppm vanadium with a 23 % current efficiency, but at vanadium concentrations of 250 ppm and above the electrodeposition of nickel was inhibited. A commercial process has been based on the results of this EU-funded project and was operating in 1999 in Germany.

A third flow sheet has been described to treat the alkaline leaching of ashes from the burning of heavy oils. Leaching with alkali has been reported to be more efficient than acidic leaching,[5] with 94 % efficiency with $75 \, mol \, dm^{-3}$ NaOH after two hours at $100\,°C$. No oxidant was required and the main products in solution were the V(V) compounds: $Na_4V_2O_7$ and Na_3VO_4, with any V(IV) compounds easily oxidised with H_2O_2. This solution was then adjusted to pH 8 with sulfuric acid to precipitate impurities and the filtrate treated with ammonium chloride to precipitate ammonium metavanadate, which can be dried and roasted to V_2O_5. The residual vanadium in the filtrate was recovered using ion exchange, leaving a raffinate containing $< 1 \, mg \, l^{-1}$ V. The solid residue after leaching still contains up to 15 % Ni, which can be recovered by leaching with sulfuric acid, followed by precipitation of ammonium nickel sulfate.

This solid residue and that from acid leaching (Figures 4.42 and 4.43) often contains a significant amount of carbon and may be further processed to provide an active carbon (Chapter 4.10).

High-temperature chlorination of Orimulsion ash has also been used to separate and recover V, Ni and Mg.[9] Chlorination of the ash with a chlorine/nitrogen mixture was carried out between 400 and $500\,°C$ and the volatilised vanadium chlorides, probably $VOCl_3$, condensed below 80 °C in 99 % yield and > 99 % purity. Separation from iron is achieved as ferric chloride condenses at a higher temperature (~350 °C). Increasing the temperature of chlorination allows the volatilisation of other elements: Fe (500 °C for > 12 h), Ni (600–700 °C condensing > 350 °C, 90 % pure) and Mg (750 °C). The chlorination temperatures of Ni and Mg can be reduced by chlorination in the presence of aluminium chloride at 600 °C. This process produces volatile complexes of the type MAl_2Cl_8 that can be separated by fractional condensation at 410–550 °C ($NiCl_2$) and < 380 °C ($MgCl_2$). Using this process, the nickel yield, relative to direct chlorination, was reduced to 46 % and 78 % pure due to contamination with magnesium.

An alternative pyrometallurgical process to treat petroleum fly ashes to produce ferrovanadium has been developed by Mintek in collaboration with Oxbow Carbon and Minerals.[10] The process consists of initial drying the fly ash at 150 °C, followed by de-carburising and de-sulfurisation pre-treatment steps prior to smelting in a DC arc furnace. Test-work indicated the feasibility of producing a ferrovanadium alloy containing > 15 % V and 6 % Ni, with a V recovery of > 88 %, together with a slag that conformed to the USA EPA safe disposal criteria.

The coking of tar sands to recover petroleum produces a carbonaceous residue that on burning results in a fly ash containing vanadium (~ 3 %) in addition to iron, silicon, aluminium, titanium, some minor metallic elements and unburned carbon. In a flow sheet developed by Process Research ORTECH Inc.,[6] the carbon was initially removed by flotation. The residual ash tailings were then leached under pressure with sodium hydroxide giving, under optimum conditions, a recovery of 93.3 % vanadium as a $5 \, gdm^{-3}$ solution. The vanadium was separated from the other elements by ion exchange with Dow G-55 resin at pH 9.2. Elution of the loaded resin with NH_4Cl/HCl solution allowed the precipitation of ammonium metavanadate, from which vanadium pentoxide could be obtained by calcination.

4.11.3 Gallium

The main source of gallium is as a by-product of the Bayer process for aluminium, but recoverable amounts, up to $120 \, \mu g \, g^{-1}$, also occur in fly ash from coal-burning furnaces. Studies have shown that the gallium content of the ash increases with decreasing particle size, suggesting that the element is concentrated on the surface of the ash.

Once again, hydrometallurgical processes have attracted most attention, with leaching by hydrochloric acid at high liquid/solid ratios being favoured.[11] Following leaching, the solution can be purified by repeated precipitation/dissolution or solvent extraction, where a range of reagents have been used,[12] or ion exchange.[13] A novel process using a solid polyurethane foam as a collector has also been described.[11] Gallium can be eluted from the foam with water, with a small amount of alkali to neutralise any acidity present in the foam. Small amounts of impurities are also carried over and these can be precipitated by slow precipitation with H_2S, and gallium can then be recovered from the treated solution by electrowinning.

Several processes have been described to recover both gallium and vanadium from coal ashes. Two of these follow a hydrometallurgical route, with either dilute sulfuric acid[14] or sodium hydroxide under pressure[15] as leachants. In both cases, the leachate was subjected to ion exchange, using a chelating resin for the acidic solution[14,16] and solvent extraction to concentrate and separate the elements. Another patented process[17] dissolves the ash in a LiCl/KCl eutectic melt at 450 °C in the presence of $AlCl_3$ to form the metal chlorides that are then electrolytically reduced. The patent claims 69 % recovery of gallium, but the vanadium is not recovered.

4.11.4   Germanium

Germanium is another valuable element that occurs in small amounts in coal ash, normally in the form of tetragonal germanium dioxide.

High-temperature chlorination is the favoured route to recovery, with germanium tetrachloride readily formed by roasting a mixture of the ash and ammonium chloride at 400 °C for 90 minutes. The volatile chloride was adsorbed in dilute hydrochloric acid and then hydrolysed to germanium oxide with a recovery of > 80 %.[18] $GeCl_4$ was also the final product of a project to recover germanium from the fly ash following combustion of Czech brown coal.[19] The ash was first leached with either $0.5 \, mol \, dm^{-3}$ sulfuric acid at 80 °C or $5 \, mol \, dm^3$ sodium hydroxide at 110 °C, giving about 50 % and 60 % germanium recoveries, respectively. The leachate was then acidified with hydrochloric acid to give a solution containing $8.5 \, mol \, dm^{-3}$ HCl, from which the $GeCl_4$ could be distilled.

A recent study[20] of the ashes from a coal gasification plant showed that the germanium was present in a different form as water-soluble sulfides and oxides. The amounts of germanium in the ashes depended on the source of the coal, but were generally in the range of 190–320 µg g^{-1}. In addition, gallium (150–220 µg g^{-1}), nickel (1150–2000 µg g^{-1}) and vanadium (3300–5800 µg g^{-1}) were present. Leaching of the germanium with water was dependent upon the calcium content of the ash, and ranged from 45 to 85 % in the case of low-calcium ash and from 33 to 53 % with high-calcium ash. The extractability of the germanium depends on a number of factors, including (as already noted on calcium content) germanium speciation, the water : fly ash ratio, temperature and extraction time. This process is the subject of a patent.[21]

4.11.5 Aluminium

In the 1970s and 1980s, there was a growing concern over a perceived shortage of aluminium from conventional sources that coincided with an increase in the amount of coal fly ash being produced from power stations. This led to a number of flow sheets being proposed for the recovery of aluminium from fly ashes. However, the easing of world tension and the higher costs of processing fly ash have reduced the need for such processes and little recent work has been published on this topic.

The flow sheets generally followed conventional hydrometallurgical practice of leaching, solution purification and recovery, but differed in the need for pre-treatment of the fly ash. Direct leaching with acids provides the simplest route: however, the need to use expensive and corrosive acids such as hydrochloric,[22] or nitric[23] to achieve satisfactory levels of leaching made such flow sheets uneconomic, while leaching even with sulfuric acid under reflux gave very poor leaching efficiency. The reason for this is that a significant proportion of the aluminium in the ash is in the mullite phase and thus is resistant to direct acid leaching.[24] Alkaline leaching is traditionally used for leaching of bauxite in the Bayer process, and it has been found that alkaline pressure leaching of fly ash can simultaneously dissolve silica and alumina. However, separation of silica prior to the precipitation of $Al(OH)_3$ has been identified as a problem.[24]

Pre-treating the ash by sintering at 1000–1200 °C with calcium sulfate : calcium carbonate in a 1:1:1 ratio,[25] or ash : fine coal : calcium oxide (5 : 4 : 1),[26] or addition of 3.75 wt % coal refuse,[27] mainly fixed carbon and iron pyrite, decomposes the silicate phases, forming calcium silicate and calcium aluminate.[28] Following calcination and cooling, the pelletised material can be successfully leached with dilute sulfuric acid (about $2\,mol\,dm^{-3}$), that can be improved by hydrothermal shock by adding hot pellets (350 °C) to the acid.[28] Leaching efficiencies of up to 81 % Al, 70 % Fe and 60 % Ti were generally obtained. The solid residue of leached pellets may be considered as a co-product that could be used as a light aggregate for concrete.[26] Various options are available to purify these acid leachates. The simplest processes – that is, precipitation of the impurities at pH 13 with sodium hydroxide, followed by passing CO_2 through the filtrate to produce $Al(OH)_3$; or the addition of ammonium salts at pH 1.5 to precipitate ammonium aluminium sulfate, followed by calcination to give Al_2O_3 – do not give products that are pure enough for the production of aluminium metal.[26] However, sparging the chloride leachate with gaseous hydrogen chloride provides $AlCl_3.6H_2O$ crystals of > 99 % purity, suitable for further processing.[29] Other procedures to purify the leachate include liquid–liquid extraction with an amine, Primene JM-T(Rohm and Haas), at pH 0.05, or an organophosphoric acid to remove the impurities.[25,26,28] The purified raffinate can then either be concentrated to crystallise aluminium sulfate[28] or treated with ammonium hydroxide to precipitate the aluminium double sulfate.[26] Following calcination, this produces alumina (> 99 % pure) suitable for use in the Hall–Heroult process. If required, stripping the loaded organic phase allows the recovery of iron and titanium.[25,26]

References

1 Dondi, M., Ercolani, G., Guarini, G. and Raimondo, M., *Orimulsion fly ash in clay bricks – part 1, composition and thermal behaviour of ash.* J. European Ceram. Soc., **22**(11): 1729–1735 (2002).

2 Vitolo, S., Seggiani, M. and Falaschi, F., *Recovery of vanadium from a previously burned heavy oil fly ash.* Hydrometallurgy, **62**: 145 (2001).

3 Gardner, H. E., *Recovery of vanadium and nickel from petroleum residues.* US Patent 4,816,236 (1989).

4 Vitolo, S., Seggiani, M., Filippi, S. and Brocchini, C., *Recovery of vanadium from heavy oil and Orimulsion fly ashes,* Hydrometallurgy, **57**: 141 (2000).

5 Chmielewski, A.G., Urbanski, T.S., Migdal, W., *Separation technologies for metals recovery from industrial wastes,* Hydrometallurgy, **45**: 333 (1997).

6 Puvvada, G.V.K., Sridhar, R. and Lakshmanan, V.I., *Recovery of vanadium from fly ash and spent catalysts.* In Vanadium – Geology, Processing and Applications, Proc. Int. Symp. on Vanadium, 2002, CIMM, Montreal, Canada, pp. 171–182 (2002).

7 Ottertun, H. and Strendell, E., *Solvent extraction of vanadium(IV) with di-(2-ethylhexyl)phosphoric acid and tributylphosphate.* In Proc. Int. Conf. Solvent Extraction (ISEC'77), CIMM, **1**: 501 (1979).

8   Sunderland, J.G., Dalymple, I.M., Rodrigues, F., Delmas, F., Nogueira, C. and
    Krummen, N., *Recovery of nickel from used catalysts, Orimulsion fly ash and nickel
    cadmium batteries.* In Proceedings REWAS'99, edited by I. Gaballah, J. Hager and
    R. Solozabal, (eds) TMS/INASMET, **2**: 1161 (1999).

9   Murase, K., Nishikawa, K., Osaki, T., Machida, K., Adachi, G. and Suda, T.,
    *Recovery of vanadium, nickel and magnesium from a fly ash of bitumen-in-water
    emulsion by chlorination and chemical transport.* J. Alloys and Compounds, **264**:
    151–156 (1998).

10 Abdel-Latif, M.A., *Recovery of vanadium and nickel from petroleum flyash.* Min.
    Eng., **15**(11), Suppl. S: 953–961 (2002).

11 Zheng, F. and Gesser, H.D., *Recovery of gallium from coal fly ash.* Hydrometal-
    lurgy, **41**: 187–200 (1996).

12 Gutierrez, B., *Recovery of gallium from coal fly ash by a dual reactive extraction
    process.* Waste Management & Research, **15**(4): 371–382 (1997).

13 Riveros, P.A., *Recovery of gallium from Bayer liquors with an amidoxime resin.*
    Hydrometallurgy, **25**, 1 (1990).

14 Tsuboi, I., Kasai, S., Kunugita, E. and Komasawa, I., *Recovery of gallium and
    vanadium from coal fly ash.* J. Chem. Eng. Japan, **24**, 15 (1991).

15 Lakshmanan, V.I., Vaikuntain, I., Melnbardis, D., Geiser, R.A. and McQueen,
    N.M., US Patent 4,966,761 (1990).

16 Tokuyama, H., Nii, S., Kawaizumi, F. and Takahashi, K., *Separation of V from
    Fe-rich leachant of heavy oil fly ash – application of an ion exchange moving bed.*
    J. Chem. Eng. Japan, **36**(4): 486–492 (2003).

17 Blander, M., Wai, C.M. and Nagy, Z., US Patent 532, 492 (1984).

18 Zhu, G.C., Shi, W.Z. and Tian, J., Rare Metals, **21**(4): 278–281 (2002).

19 Jandová, J., Štefanová, T., Maixner, J. and Mestek, O., *Recovering germanium
    chloride from Czech brown coal.* In Proceedings REWAS'99, edited by: I. Gaballah,
    J. Hager and R. Solozabal (eds), TMS/INASMET, **2**: 1355 (1999).

20 Font, O., Querol, X., López-Soler, A., Chimenos, J.M., Fernández, A.I., Burgos,
    S. and Peña, F.G., *Germanium extraction from gasification fly ash.* Fuel, **84**(11):
    1384–1392 (2005).

21 Burgos Rodr Guez, S., Querol Canceller, X., Font Piqueras, O., Lopez Soler, A., Plana
    Llevat, F., Espiell Alvarez, F., Chimenos Ribera, J.M. and Fernandez Renna, A.I.,
    *Method for recovering metals from the flying ashes generated in an integrated gasifi-
    cation combined cycle-type (IGCC) thermal station.* EP 1 408 127 A1, 6 June 2002.

22 Livingstone, W.R., Rogers, D.A., Chapman, R.J. and Bailey, N.T., *Use of coal
    spoils as feed materials for aluminium recovery by acid-leaching routes, 2. The suit-
    ability and variability of the feed materials on leaching conditions.* Hydrometallurgy,
    10: 79–96 (1983).

23 Duhart, P.M., Ger. Offen. 2,257,521, 7 June 1973.

24 Burnet, G., Murtha, M.J. and Dunker, J.W., *Recovery of metals from coal ash.*
    Ames Laboratory, US DOE Iowa State University, Ames (1984).

25 Seeley, F.G., McDowell, W.J., Felker, L.K., Kelmers, A.D. and Egan, B.Z., *Deter-
    mination of extraction equilibria for several metals in the development of a
    process designed to recover aluminium and other metals from coal combustion ash.*
    Hydrometallurgy, **6**: 277–280 (1981).

26 Matjie, R.H., Bunt, J.R. and van Heerden, J.H.P., *Extraction of alumina from coal
    fly ash generated from a selected low rank bituminous South African coal.* Minerals
    Engineering, **18**(3): 299–310 (2005).

27 Murtha, M.J. and Burnet, G., *Some recent developments in the lime-fly as process for alumina and cement,* Resources and Conservation, 9(1–4): 301–309 (1982).

28 Torma, A.E., *Extraction of aluminium from fly ash.* J. Metall., Berlin, 589–592 (1983).

29 Abbruzzese, C., Marzocchi, G. and Rinelli, G., *Possibility of recovering alumina from ashes.* Ind. Min., 30(6): 323–328 (1979).

5

Novel Products – from Concept to Market

Ian Barnes and Fritz Moedinger (5.6)

Hatterrall Associates, UK; IRSAI srl, Carbonara di Po (MN), Italy

List of Abbreviations and Acronyms

AASHTO	American Association of State Highway and Transportation Officials
ACAA	American Coal Ash Association
AIMCC	L'Association des Industries de Produits de Construction
APC	Air Pollution Control
ASTM	American Society for the Testing of Materials
AUB	Arbeitsgemeinschaft Umweltverträglisches Bauprodukte
BRE	Building Research Establishment (UK)
BS(I)	British Standard (Institute)
CCB	Coal Combustion By-products
CCP	Coal Combustion Product
CCUJ	Centre for Coal Utilisation of Japan
CE	Conformité Européenne
CEN	Comité Européen de Normalisation (European Committee for Standardisation)
CFBC	Circulating Fluidised Bed Combustion
CUAP	Common Understanding of Assessment Procedures
CUR	Civieltechnisch Centrum Uitvoering Research en Regulering (NL)
DIN	Deutsch Institut für Normung

Combustion Residues: Current, Novel and Renewable Applications   Edited by Michael Cox, Henk Nugteren and Mária Janssen-Jurkovičová   © 2008 John Wiley & Sons, Ltd

ECOBA	European Coal Combustion Products Association e.V.
EINECS	European Inventory of Existing Chemical Substances
EMAS	Eco-Management and Audit Scheme
EMS	(integrated) Environmental Management System
EN	European Norm
EPA	Environmental Protection Agency (USA)
EPD	Environmental Product Declaration
ETR	Environmental Tax Reform
EWC	European Waste Catalogue
FBC	Fluidised Bed Combustion
IGCC	Integrated Gasification Combined Cycle
IPP	Integrated Product Policy
ISO	International Standards Organisation
LCA	Life Cycle Analysis/Assessment
LCI	Life Cycle Interpretation
MRPI	Milieu Relevante Product Informatie (environmental performance information (the Netherlands))
MSW	Municipal Solid Waste
OECD	Organisation for Economic Co-operation and Development
OHSAS	Organisational Health and Safety Standard
PCR	Product Category Rule
RTS	Rakennustietosäätiö (environmental performance information (Finland))
SIA	Schweizerischer Ingenieurund Architektenverein (environmental performance information (Switzerland))

5.1 Introduction

Previous chapters have set out the history of the utilisation of combustion ashes and considered different ways in which such residues are currently used and may be used in the future in industrial products. Thus it has been shown that since the time when coal ashes were treated exclusively as waste materials and disposed of in ash mounds and landfills, an increasing percentage has been used, particularly in the construction industry, and pulverised coal-derived fly ash is now firmly established as a valuable secondary raw material in cement formulations, and for the production of lightweight aggregates.

Increasing pressure on the environment, together with increasingly stringent legislation designed to encourage the reuse of materials, makes it vitally important to complete the transformation of combustion ash into a source of valuable products. However, novel products face a number of challenges in moving from an initial laboratory concept to the commercial marketplace. This chapter summarises and discusses some of the legislative and marketing issues that affect products from ash, both for relatively established outlets and for totally new processes and materials.

5.2 Novel Products – Basic Economic Considerations

For a new product to be successful, there must be a market for that product. For ash-derived materials, these can be established markets – as, for example, in the construction industry, in cement formulations and synthetic aggregates – or new and newly developing markets – for example, synthetic zeolites and high-performance mineral-based adhesives.

Any product must be sufficiently attractive to potential customers to be selected over the available alternatives. Thus for an ash product to be successful, it must either offer a clear price and/or performance advantage over the competitive materials, or – for a completely new product – potential customers need to be convinced that the material can be an asset to their business activities.

Novel ash products can be considered as belonging broadly to one of two groups: namely, relatively 'low-tech' or 'high-tech' applications (Figure 5.1). 'Low-tech' applications would be typified by construction products for example, aggregates and so on – and are usually high-volume uses. However, here the ash products are generally competing against low-price naturally occurring alternatives, and the low prices limit the scope for processing and transportation of the new products.

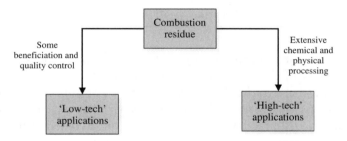

Figure 5.1 Routes to the utilisation of combustion ash residues.

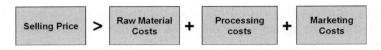

Figure 5.2   Basic economic considerations for a successful product.

The most important consideration for the success of a new product is the cost of bringing that product to market, as summarised in Figure 5.2.

From the earliest stage of development, it is important to place the technical aspects of production within an economic framework. This is relatively straightforward for a product aimed at competing with existing materials, since the market price for competing products is set. For completely new materials, the situation is more difficult and is associated with a greater risk, but the corresponding rewards may be higher if the product is successful. Another important consideration is the market size and potential market share. High-value materials may offer the promise of large profits, but these can only be realised through a significant share of what may be a small market. The high-volume uses tend to be associated with relatively low prices, but in the context of 'avoided disposal cost' can be attractive outlets.

For materials such as combustion ash, the impact of transportation and avoided disposal costs can be considerable, and can effectively limit the 'economic radius' within which a product can be sold. A hypothetical example for a product competing against alternatives selling at €30 per tonne based on that described by Brendel[1] illustrates this effect (Table 5.1).

Table 5.1   Evaluation of economic market radius (after Brendell[1]).

Sales revenue	€30 per tonne
Avoided disposal cost	€10 per tonne
Direct marketing cost	€2 per tonne
Transportation cost	€0.3 tonne/kilometre
Total 'marketing benefit'	€$(30 + 10 - 2) = $ €38 per tonne
Economic 'market radius'	€$38/0.3 = 127$ kilometres

In this case, it is economically viable to sell the product within a 127 km radius of the production plant. As the relative values of *Sales Revenue* and the *Avoided Disposal Cost* increase, the potential market for the product increases. Sales revenue can be improved through the identification of 'added value' uses, while the avoided disposal cost is likely to increase as taxation and legislative forces work to reduce economic routes available for disposal.

For 'high-tech' applications, the potential market may be of a lower volume, but here the products are competing against high-price alternatives, or may be completely new materials with promising prospects. However, the processing costs of manufacturing these materials may be high and may need to be considered carefully in any market assessment.

The above example shows the significance of 'avoided disposal cost' in determining the economic viability for product utilisation. This cost can be considerable, and more importantly, is variable from country to country, depending upon local conditions. This has the effect of making novel applications economically viable in some countries, but not in others. Figure 5.3 sets out the range of nontax-related disposal costs for inert residues such as combustion ashes for 15 European countries.[2] It can be seen that the very high costs of disposal in Luxembourg would make novel products viable that would be uneconomic in Spain or Greece. When landfill taxes are factored in (Section 5.4), disposal costs become even greater, particularly for some countries. If market opportunities for combustion residues are to be maximised, these differences must be recognised and tackled. A report[3] from the European Commission to the Council and the European Parliament recognises that there are still important differences between the member states in the way that they address waste materials with a potential for recovery and reuse, and draws a number of important conclusions, including the need for more 'positive discrimination' to encourage ash utilisation.

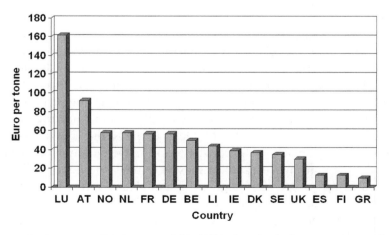

Figure 5.3 Nontax-related costs of landfilling nonhazardous waste in selected countries.

5.3   Specifications and Standards

Specifications for developed uses for combustion residues have been established in many different countries. The aim of these specifications is to set minimum requirements for the performance of ash-derived products tested under laboratory conditions. Specifications have also been set in individual states in the USA and in other countries for the use of fly ash, or other 'waste' material, in structural fill, soil stabilisation, sub-base and base course, and similar applications. Thus for cement applications most of these are based on, or similar to, the ASTM C618 standard in the USA, or the EN450 standard in Europe, and refer to technical rather than environmental standards.

The development of these technical and environmental standards is discussed in Chapter 3, and while it is beyond the scope of this chapter to review such specifications in detail, the large number and their variability demonstrates the plurality of views on the suitability of ash residues for different applications (Table 5.2). While the potential use of ash for a particular purpose is determined primarily by passing the specific tests set out in any relevant standard or specification, it has been emphasised by a number of workers that many of the methods for testing fly ash performance are not relevant or appropriate, and it has been shown that, whilst some ashes may fail standard tests, they can still produce completely acceptable final products. For example, BS EN450 and BS 3892, Part 1 require the loss-on-ignition from ash to be below 7 %, which excludes many ashes that are known through detailed testing to produce completely acceptable cement and concrete products (see Chapter 4.2.1).

Many specifications also concentrate on fly ash from pulverised coal combustion and exclude material derived from the co-combustion of alternative fuels from use in cement and concrete (see also Chapter 4.2). This situation severely restricts the scope for utilising such materials. The harmonisation of the EN450 norm allowed CE marking of PFA with a maximum of 10 % co-firing from January 2006, and made this marking compulsory from 1 January 2007 (at least in the Netherlands). However, progress is being made in this area, and in the Netherlands, Vliegasunie obtained CE marking for applications in cement and concrete for PFA with co-firing percentages higher than 10 %. To this end, through results of the research committee of the Dutch Civil Engineering Centre for Research and Regulations (CUR), a CUAP assessment procedure, 03.01/34 'Fly Ash for Concrete', has been developed. This procedure was approved by the European Organisation for Technical Approvals

Table 5.2 Primary standards for combustion ashes in construction products.

ASTM

ASTM C 618: Standard Specification for Coal Fly Ash and Raw or Calcined Natural Pozzolan for use as a Mineral Admixture in Portland Cement Concrete

ASTM C 311: Standard Test Methods for Sampling and Testing Fly Ash or Natural Pozzolans for Use as a Mineral Admixture in Portland Cement Concrete

ASTM D 5239: Standard Practice for Characterizing Fly Ash for Use in Soil Stabilization

ASTM E 850: Standard Practice for Use of Inorganic Process Wastes as Structural Fill

ASTM E 1861: Standard Guide for Use of Coal Combustion By-products in Structural Fills

ASTM D 5370: Standard Specification for Pozzolanic Blended Materials in Construction Applications

ASTM C 1240: Standard Specification for Silica Fume for Use in Hydraulic-Cement Concrete and Mortar

AASHTO

AASHTO Standard Specifications for Transportation Materials and Methods of Sampling and Testing

European Standards

BS 3892 Part 1 Fly Ash Standard; Part 2 Fly Ash for Use as a Type II Addition

BS EN 450 European Standard for Fly Ash

BS EN 197 European Standard for Multiple Binders (fly ash, cement, silica fume) allowed in Concrete

Australia

AS 3972–1991 Portland Cement

AS 3582.1–1991 Fly Ash

AS3582.2. Ground Granulated Blast Furnace Slag

A.S. 1129 Fly Ash Specification

Canada

CAN/CSA A23.5–97 Canadian Specification for Supplementary Cementing Materials (includes fly ash)

Germany (Deutsches Institut für Normung – Berlin)

DIN 1164–1 German Cement Standard

DIN 1045 Reinforced Concrete Structures; Design and Construction

DIN EN 450 Fly Ash in Concrete – Definition, Demands and Quality Control

ENV 206:1990 (CEN/TC 104) Beton – Eigenschaften, Herstellung, Verarbeitung und Gütenachweis

EN 445:1996 (CEN/TC 104) Einpreßmörtel für Spannglieder – Prüfverfahren

EN 446:1996 (CEN/TC 104) Einpreßmörtel für Spannglieder – Einpreßverfahren

EN 447:1996 (CEN/TC 104) Einpreßmörtel für Spannglieder – Anforderungen für üblichen Einpreßmörtel

EN 450:1994 (CEN/TC 104) Flugasche für Beton – Definitionen, Anforderungen und Güteüberwachung

EN 451–1:1994 (CEN/TC 104) Prüfverfahren für Flugasche – Teil 1: Bestimmung des freien Calciumgehalts

---

Table 5.2   (Continued)

EN 451–2:1994 (CEN/TC 104) Prüfverfahren für Flugasche – Teil 2:
  Bestimmung der Feinheit durch Naßsiebung
The Netherlands
NEN 3550 Dutch cement standard
The United Kingdom (British Standards Institution, London)
BS 3892 PFA as a separate constituent in OPC
BS 6588 Blended cement containing PFA
BS 6610 Pozzolanic pulverised fuel ash cement

(EOTA) in 2004. It resulted in European Technical Approvals (ETA) for co-combustion fly ashes from three different sources in the Netherlands.

A similar problem is seen with the automatic exclusion from some specifications of residues arising from advanced technologies for coal use, such as FBC and IGCC. A survey carried out on the use of CFBC residues in different states in North America provides revealed significant differences in the way in which these residues are treated for potential uses. Thus in Ohio and Texas, CFBC residues may be used in almost everything from cement and concrete products to structural fills, grouting, plastics/paints/metals, concrete blocks and ceramics. In other states, such as Nevada and New Hampshire, no such authorisation exists, and although this does not necessarily mean that CFBC residues may not be used in these products, there may be a presumption against their utilisation.

In addition to CFBC residues being automatically excluded from many specifications, Blondin and Anthony[4] point out that none of the common tests for additives in cement are suitable for determining the potential for using CFBC residues in concrete. Therefore, CFBC residues are failing specifications and tests where they may produce good-quality final products. These authors suggest that new criteria should be set based on the adequate performance of the final product, rather than the exiting 'recipe-based specifications'.

In Japan, the problem with over-rigid standards has been recognised and the Japanese Fly Ash Society is working with electric power companies and the Ministry of International Trade and Industry to amend the Japanese standards for fly ash use in concrete. The amendment aims to promote the use of ash by expanding the applicable area of the current standards. Even fly ashes discarded due to the over-specification of the current standards can be used efficiently if dealt with appropriately, and work is in progress to develop new standards that apply to the different qualities of ashes available.

These latter cases represent ways in which established standards can present a barrier to ash products as they are frequently written around existing materials; the so-called 'recipe-based' standards. Such formulations are based on materials that have been in use for many years and where the components are specified by name; for example, pea gravel: unfortunately, this approach automatically excludes ash products. Specifications based on the performance required of a material – for example, a specific minimum crushing strength after a certain time – are to be preferred, as they do not carry the inherent presumption against ash use present in 'recipe specifications', and ash products can stand or fall on their relative technical merit and cost when compared to alternative materials. The lobbying of technical standards committees and customer user groups by organisations representing the interests of the ash producers can be very successful in breaking down this barrier through ensuring that the ash-derived materials are given 'a fair hearing'.

For novel products, the situation is not necessarily any better, as adherence to accepted standards is crucial for their acceptance. Indeed, for truly novel materials, the relevant standards may not exist, or may only be applicable to a set of similar products in a general sense. Where new standards have to be developed to assist novel ash-derived products into the marketplace, the increasing drive to European Union wide standards requires a pan-European approach, through the CEN standards development committees.

5.4 Legislation and Taxes

Novel ash materials, like all other products, must meet current national, and increasingly international, legislation on health and safety issues, together with relevant environmental regulations. The evaluation of a new material can be a lengthy process, but it is important to ensure that complete and thorough testing is undertaken to meet these requirements.

EU legislation on new products and raw materials is complex, but it can be distilled into two distinct categories:

(1) Manufacturers and importers of chemicals must 'classify' all substances according to the law's description of 'dangerous', as defined in Article 2(2) of 67/548/EEC (as amended), and ranging from explosive to toxic to carcinogenic to flammable. Once classified, chemicals are then required to be packaged and labelled accordingly.

(2) 'New' products must be 'notified' to the member state competent authorities. Products considered 'new' are those that were not included in the European Inventory of Existing Chemical Substances (EINECS), a closed inventory of substances on the EU market as of mid-September 1981. Notification of a new product entails provision of the results of specified testing and the subsequent compilation of a technical dossier, unless it is covered by one of the exemptions granted by Directive 92/32/EEC.

Health and Safety legislation may be regarded as a gatekeeper test of product success. If the material does not comply with health and safety legislation, there is no prospect of the product being successfully marketed.

Of equal importance is the need to demonstrate that any new ash application is environmentally acceptable. In the past, ash-derived materials have sometimes been used as replacements for naturally occurring products without undertaking a thorough environmental assessment. In some cases, these materials have been found to have a deleterious effect on the environment through, for example, the leaching of heavy metals, and this has not only compromised the health of affected persons, but has given ash products a 'bad name' that takes much effort to dispel.

While the systems and procedures for truly novel ash-derived products are daunting, there are also issues with the existing and established outlets for ash residues, and many workers claim that the existing legislation, specifications and guidelines are not working effectively to promote the use of combustion residues, particularly those arising from new technologies such as CFBC and IGCC. In some areas, until the current standards and specifications are amended, ash produced from co-firing waste fuels or from more advanced combustion systems, cannot be sold in the existing marketplace.

In addition to excluding certain types of ash, some legislated standards also limit the quantity of ash that can be used. For example, in Israel the use of ash in cement is limited to 10 %, whereas in many other countries 65 % or more can be used. Many workers are calling for such standards to be updated or replaced with new performance-oriented ash product standards.

Brendel[1] states that the regulations and legislation are also less than ideal in the USA and elsewhere. In many states in the USA, clean coal by-products are regulated as solid wastes, which means that their use requires specific case-by-case approval and long-term environmental monitoring. There is no coherent policy among federal and state agencies regarding the beneficial use of these materials, which has lead to overly conservative requirements. In the highly litigious USA, other legal barriers include the liability of the ash supplier for damages that may occur through ash use.

With increasing focus on the development of sustainable economies with a concomitant reduction of waste, the EU has, and continues to implement a number of articles of legislation in this area. EU environmental legislation can take one of two main forms, both of which take precedence over national legislation:

- *EU regulations.* Once these have been approved and ratified by the European Parliament, regulations are directly applicable and binding on EU member states.
- *EU directives.* Directives set out specific targets and the results to be achieved within a certain timescale. Although the content of directives must be incorporated in to the national legislation within a certain time, the precise form and method of implementation is left to individual member states.

EU directives are the most common form of EU legislation, since the inherent flexibility on how the prescribed goals are achieved allows alignment with local conditions and limitations. However, different interpretation of directives between countries can introduce significant local differences; for example, differing rates of landfill tax. The most important legislation relevant to the utilisation and disposal of combustion ash is summarised below.

As combustion ashes may initially be considered as a waste product of coal combustion, they are subject to various regulations covering their management and disposal. Within the EU, such legislation on waste management consists of the Framework Directive (75/442/EEC, 15 July 1975, as amended 91/156/EEC and 91/692/EEC), which defines general waste management principles, and the Directive on Hazardous Waste (91/659/EEC, 3 December 1991). There are two groups of directives: those that regulate management of certain types of waste and those that determine requirements for waste recycling and disposal.

As a consequence of its low value, the transportation of combustion ash is likely to be limited both for utilisation or disposal. Transportation is governed in particular by the 'Regulation on Supervision of the Shipment of Waste (259/93/EEC, 1 February 1993)', which also covers import and export from the EU.

This regulation can limit or prohibit export, import or transit of waste for various reasons, including: disposal of waste as closely as possible to where it is produced; priority given to the recycling of waste, rather than its disposal, and whether facilities or sufficient capacity for recycling the waste are available in the member state, or the EU.

To ensure more complete implementation of the requirements of this Directive, a reference document, the European Waste Catalogue (EWC 1994), was adopted, to be used as a common definition of wastes throughout the EU.

Where combustion residues arise from more advanced processes, such as CFBC, the ash properties may require that the material falls under a strict control listing and in some cases, the Directive on Hazardous Waste (91/689/EEC, 12 September 1991), the Directive on a List of Hazardous Waste (94/904/EEC, 22 December 1994) and recent successors may apply.

Ashes arising from waste incinerators are governed by the 'Directive on the incineration of hazardous waste (94/67/EEC, 16 December 1994), which contains requirements for the operation of hazardous waste incineration facilities and levels of emissions, including dioxins.

Finally, perhaps the most influential legislative driver for the increased use of combustion ashes is likely to be the Directive on landfills (1999/31/EEC, 26 April 1999), which refers to existing and new landfills, including those landfills on the territory of an industrial company.

This important directive provides information on the registration of a landfill, operational permits for the management of the site, including the types of wastes allowed to be disposed of, monitoring and closure of sites.

The economic consequence of these legislative agents is to drive the relative cost of utilisation versus disposal. Financial incentives that encourage ash utilisation can include the levying of taxes if the ashes are not utilised but are disposed of to landfill. These are the so-called waste taxes, and a number of different types of waste tax exist in the EU. Such taxes can be cost-covering charges, where the revenues are used either to pay for waste disposal services, such as the Dutch household waste tax, or to finance recycling services, such as the Swedish battery charge.

Alternatively, they can be incentive taxes, levied to change environmentally damaging behaviour. Revenues are often used to encourage further behavioural change; for example, the German toxic waste charge, which depends on the potential danger of the waste and cost of treatment. A waste tax can also be used as a fiscal environmental tax, whereby surplus tax revenues can be used to finance budget deficits or shift taxes from labour to resources. This is a way of internalising external costs of a product or service. Such a change in the tax system, which shifts taxes away from labour and capital and on to the use of resources, is known as an environmental tax reform (ETR). An example is the UK landfill tax, which aims to cut waste, encourages reuse or recycling, and boost employment.

Clearly, these three types of taxes are not mutually exclusive; a cost-covering charge may have incentive effects, as might a fiscal tax, or revenues from a fiscal tax may be partially used for environmental purposes.

In those EU member states that have introduced landfill taxes, the charges levied can differ markedly, as shown in Figure 5.4.

When such taxes are considered with the disposal costs presented earlier (Figure 5.3), it is clear that the relative costs of disposal are highly influenced by this taxation component. Therefore, any assessment of the market potential for a new ash product must take these differences into account. To reiterate, applications that are not viable in one country may be viable elsewhere, where local taxation rates are higher.

In addition to making ash disposal more expensive, the promotion of ash use through guidelines and even tax incentives is becoming more common. Countries such as Japan have recycling and reclamation laws or guidelines that promote the effective use of 'waste' materials such as fly ash.[5]

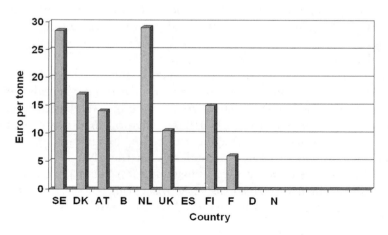

Figure 5.4   Tax-related costs of landfilling nonhazardous waste in selected countries.

In an attempt to enhance the use of fly ash rather than its disposal, the Indian Ministry of Power has collated a series of suggestions, which may translate into official guidelines in the future.[5] These guidelines include:

- banning clay-based brick and other kilns in a 50 km radius of a coal-fired power plant*;
- abolition of excise duty on fly ash based products;
- separate collection of fly ash and bottom ash in power plants;
- installation of dry fly ash collection facilities at power plants;
- concessional prices for land, water and electricity for industries using fly ash near power plants;
- use of standards or quality control evaluation of the fly ash produced;
- introduction of fly ash into codes for civil and construction specifications for materials such as bricks, fillers, and embankments.

Similar policies have already been implemented in China.[5] The Chinese government is considering adjusting tax systems to favour the use of wastes over the extraction of raw materials. It is also aiding the development of large waste utilisation plants and new methods for ash and slag utilisation. Chinese environmental law also now requires that plants be fitted with both wet and dry ash collection systems. Other examples of changes in legislation to promote ash use include a zero tax on some ash applications and reduction of road tolls for the transportation of fly ash within some regions of China.

Vom Berg[6] suggests that although the state can interfere by imposing charges or taking incentive measures, such as tax relief options, it always falls upon the power plant operator to provide the prerequisites for fly ash utilisation, and to develop industrial markets.

Finally, a very powerful driver to the success of new ash products is the decision taken by a country, or a region within a country, mandating the utilisation of wastes such as ash. A good example is the Dutch response to the high oil prices of the early 1980s, when oil-fired power stations switched to coal, with the subsequent increased production of ash. The Dutch government instigated a total ban on the disposal of coal ash to landfill and the Dutch power industry established a joint company, Vliegasunie, which was given the task of securing 100 % utilisation of coal ash: this was successfully achieved within a few years.

---

* This restriction arises from local practice of using fertile top soil to manufacture bricks.

5.5 Customer Acceptance

Despite passing the tests of viable economics, health and safety, environmental legislation and relevant standards, ash products have one final barrier to face: that of customer perception. While potential users continue to consider ash-derived products as at best unconventional and at worst 'waste', producers aiming to market such materials will always face a difficult battle to achieve acceptance. The latter is the hardest to overcome, since the term 'waste' conjures up images of dirty and possibly hazardous substances, a nonuniform material that cannot be produced in a controlled manner, and ultimately an unwanted, technically inferior product. Ironically, through extensive characterisation and utilisation studies around the world, ash products have been shown to be suitable for numerous applications, and often exceed required parameters.

A study[7] in the United States identified the following 11 institutional barriers to increased ash utilisation:

- lack of familiarity with potential ash uses;
- lack of data on environmental and health effects;
- restrictive or prohibitive specifications;
- the belief that fly ash quality and quantity are not consistent;
- lack of fly ash specifications for noncementitious applications, resulting in substitution in such applications by the more restrictive specifications for use of fly ash in cement and concrete;
- the belief that (primary) raw materials are more readily available and more cost-effective;
- the viewpoint of states that EPA procurement guidelines for fly ash in concrete are a rigid ceiling, rather than general guidelines for use;
- actions by environmental agencies that normally support beneficial ash uses in principle, but that frustrate the actual implementation by restrictive regulations;
- restrictive regulation of fly ash as a solid waste in most states;
- lack of state guidelines on beneficial ash use;
- lack of clear federal direction on regulation of beneficial ash use.

These comments are echoed by a survey of the operators of municipal solid waste (MSW) incineration plants (Table 5.3) regarding barriers to increased ash utilisation.[8] Such attitudes are very difficult to overcome, and it is sobering to remember that the use of pulverised coal fly ash in cement formulation took approximately 40 years to achieve widespread acceptance.

Table 5.3   Collated views on the prospects for utilising MSW ash residues.

Country	Comments received
Austria	Very strict national regulations for combustor residue reuse Natural minerals at lower costs and better quality available
Belgium	Due to the lack of incentives, no strong efforts are made for development The law and the regulations are not available for reuse of combustor residues
Canada	Bottom ash: • reluctance of municipalities/contractors to use a 'waste' residue in road construction • variability in product specifications • maximum ratio of bottom ash that can be used is 70 % Treated APC residues: • Reluctance exists on using treated material that was once a special waste
France	Lack of information about long-term behaviour Low cost of natural materials in some regions, cheaper than bottom ash and low cost of land-filling Fear of associations
Hungary	Hungarian regulations and waste categorisation
Japan	Perceived public opinion that reuse products are inferior quality Stable demand is difficult to secure Commercial circulation system of reuse is not established Manufacturing cost of slag is very high compared with natural mine resources
The Netherlands	Different specifications for the treatment of residues and reuse applications Economic viability of treating ash residues for reuse applications
Norway	Cheap raw materials Cheap competitive recycled materials Potential users of the material are sceptical regarding environmental impacts
Spain	Public perception/poor image of waste combustion processes Lack of legal regulations Low cost of raw materials and landfill taxes

Sweden	Lack of standards for combustion residue control and no national regulations mean variations in county administrative board ruling Bottom ash is currently not allowed for civil engineering applications Difficulties in achieving pure metallic fraction from bottom ash
The United Kingdom	End-user concern: • ash properties – variability, different to PFA and potential problems from leachates • potential users reluctance to take a technical risk • may be seen as inferior to products using natural aggregates • transport costs limit markets Regulatory constraints: • Agreement with Environmental Agency (EA) concerning classification of bottom ash • Potential and actual opposition from EA for ash storage and processing schemes as well as use

Where the principal objection to the use of ash products is one of varying quality, much can be achieved with an ash beneficiation process to 'even-out' product quality, or to separate and enrich specific components required for a particular application. Processes for ash beneficiation have been developed and are now increasingly in use to ensure good product quality for ash-derived materials (Chapter 3).

Cornelissen[9] has produced a useful overview of how fly ash has been regarded as waste but, over time and with positive marketing, quality control and upgrading has resulted in ash shifting from a waste to a valuable resource (Figure 5.5).

An important consideration for any customer of a 'waste' material is the potential profit than can be made by its use, and in this context it has been shown that there can be significant profits from using combustion residues. This has been recognised by the establishment of companies set up specifically to market fly ash; for example, Boral Mineral Technologies (USA), ISG Resources Inc. (USA), Wallace Industries (USA), Ash Resources Pty (Republic of South Africa), Pozzolanic Industries (NSW, Australia) and Tanveer Enterprises (New Delhi, India). In countries such as the Netherlands and Belgium, joint companies have been set up by power plant operators and the cement industry and/or power stations to ease the movement between ash production sites and ash markets.

According to Fisher et al.,[10] ash dealers typically buy concrete-quality ash at around $2 per tonne and deliver it to concrete manufacturers at a

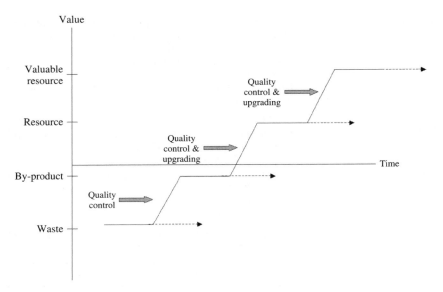

Figure 5.5   Developing ash utilisation through quality enhancements.

cost of $20 per tonne. Once transport and handling costs are removed, the net profit is around $6 per tonne. This is not an insignificant amount, considering that fly ash is still thought of as waste in many areas. It is even more significant when one considers that disposal costs can range from $6 to $27 per tonne.[10] Costs as high as $100 per tonne have been quoted for ash disposal at some power plants in Japan.[11]

Renninger[12] has considered the total market potential for fly ash in the USA. He states that every 'avoided cost' dollar from the use of a by-product as opposed to land-filling is better than a dollar in gross revenue, since there is an associated overhead cost with revenue generation.

In countries such as the USA, with deregulation and the increase in the number of power plants being run privately for profit, there will be a tendency for plants to operate on the lowest costs possible, and this will include accepting the lowest bids for ash disposal or utilisation. According to Bennett,[13] in the long term this will impact unfavourably on the quantities of coal by-products marketed. He warns that unless market growth is stimulated through incentives, utilities will continue to see their disposal expenses climb. He advises that there should be an effort to expand applications and technologies that will ultimately drive disposal volumes and expenses towards zero, while increasing both the demand and price for clean coal by-products.

Callaway[14] suggests that each power plant should consider its own situation and how steps could be taken to turn fly ash into a commodity, and coal-fired power plant operators should start treating fly ash 'disposal' as a business opportunity. This includes all areas of marketing from product evaluation, through storage, transport and delivery, to new product development and customer support. By way of outlining potential markets, Trehan and others[15] have estimated the applications that would be required to use up the ash produced from a 2000 MW$_e$ thermal power station in India. Bearing in mind that Indian coal is very high in ash and assuming that 10 kt of ash were produced every day, this could be used to supply:

- 50 brick plants, producing 5 million bricks per day;
- 20 lightweight aggregate/block plants, producing 13 kt per day; or
- 25 cement plants, producing 75 kt per day.

Although coal-fired power plants in other more developed countries burn coals with lower ash content, thus producing less ash daily, it still represents a significant potential with respect to the construction industry.

Some power plants are already following this route, and are taking the process one step further to the point at which the power plant itself is marketing the ash as a finished product for sale. For example, the construction of a total fly ash utilisation system has been completed at the Abakan power plant in Russia. Between 20 and 30 % of the processed ash and slag is sold to builders, and 70–80 % is used for producing various types of concretes and bricks. By using the waste ash from the plant and taking advantage of energy and steam produced at the plant, the cost of manufacturing products such as bricks and ready-mix concrete is up to 75 % lower than elsewhere.[16]

Countries such as Australia, the Netherlands, the UK and the USA have their own fly ash associations to promote the use of combustion ashes. For example, ECOBA, the European association for use of by-products of coal-fired power stations, was formed in 1990 and has 21 members from 13 member states of the European Union, representing 86 % of all CCPs produced in the EU. These associations also interact on an international level and, since 1993, ECOBA and the American Coal Ash Association (ACAA) have been connected by a 'memorandum of understanding' to allow an increase in the exchange of information. Further, in 1997, the Japanese Centre for Coal Utilisation (CCUJ) joined ECOBA as an affiliate member.

ECOBA provides information to potential users on new applications for fly ash and has the following stated aims:[17]

- the creation of a favourable technical, legislative and regulatory climate for the use of clean coal by-products in Europe;
- the promotion of the recognition and acceptance of clean coal by-products as secondary raw materials;
- the representation of members on technical committees of CEN; and
- the exchange of experience and knowledge between members.

The ACAA has a similar mission 'to advance the management and use of CCBs (clean coal by-products) in ways that are technically sound, commercially competitive, and environmentally safe' and conducts an annual survey and report of clean coal by-products and their use.

In attempting to improve the image and acceptability of combustion residues and their products, there are, broadly, two approaches. In the first, the ash residue is used as a raw material and the product manufactured and marketed without reference to its origins. In the second, the products are marketed with a strong emphasis on the contribution to sustainability and protection of the environment. Such environmental acceptance schemes can be very successful in gaining public support for the recycling of waste products and are gaining in popularity, particularly in the EU.

5.6   Environmental Acceptance Schemes

The degree of eco-efficiency of a production process can be measured by the following principal criteria:

- total energy content – that is, the energy that is required to produce, package, distribute, use and dispose of a specific product;
- consumption of the environment – land for building or mining, forest depletion;
- emissions – greenhouse gases, dust and other chemical and natural substances;
- raw materials – depletion of nonrenewable resources;

- waste generation – packing, production and use;
- recyclability – generation of secondary waste cycles;
- capital – least cost;
- durability – longer periods of use mean less consumption of resources.

The construction sector is one of the frontrunners, not least in the activity taking place in CEN and ISO in providing environmental performance information of its products, but also for a number of reasons, including:

- political pressure;
- major benefits to the sector;
- a level playing field for suppliers;
- an alternative to eco-labelling, which is not suitable for business-to-business communication on construction products;
- elimination of 'blacklists' and 'preference lists' based on varying and questionable methodology.

Businesses can achieve eco-efficiency through:

- optimizing processes and moving from costly end-of-pipe solutions to approaches that prevent pollution at source;
- recycling of wastes, such as using by-products and wastes of one industry as raw materials and resources for another, thus overall creating zero waste;
- eco-innovation, 'smarter' manufacturing using new knowledge and processes to make products more resource-efficient to produce and use;
- new services, such as leasing rather than selling products, thus changing perceptions of companies towards durability and recycling;
- networks and virtual organisations sharing resources to increase the effective use of physical assets.

Eco-efficiency standards actually under development concerning the sustainability and environmental impact of buildings and building products are found in the following ISO and CEN standards supporting the work for the declaration of environmental information:

(1) Standards established in ISO/TC 59:

Standard number	Title of standard
ISO/CD 21930	Environmental declaration of building products
ISO/CE 21931	Framework for the assessment of environmental performance of buildings and constructed assets
ISO/CD 21932	Terminology
ISO/CD 21929	Sustainability indicators
ISO/AWI 15686–6	Buildings and construction assets – Service life planning – Part 6: Guidelines for considering environmental impacts

Some of these documents are fundamental for the work of other standardisation committees such as CEN/TC 350.

(2) Standards under development or revision in ISO/TC 59 SC 17:

Standard number	Title of standard	Comments
ISO/DIS15392	Buildings and constructed assets – Sustainability in building construction – General principles	FDIS ballot initiated on 15th January 2008, 2 months comment period will be reviewed
ISO/TS 21929	Buildings and constructed assets – Sustainability in building construction – Sustainability indicators	Standard published as a technical specification 8th July 2005
ISO/FDIS 21930	Buildings and constructed assets – Sustainability in building construction – Environmental declaration of building products	Published on 1st October 2007. Provides principles and requirements for type III EPDs but does not define requirements for type III programmes and excludes working environment as this is left to national legislation. Seen as a complement to ISO 14025
ISO/TS 21931	Buildings and constructed assets – Sustainability in building construction – Framework for assessment of environmental performance from buildings	Standard published as a Technical Specification on 19th July 2005; To be used in conjunction with ISO 14000 series. International Standard under revision

ISO/AWI TR 21932	Buildings and constructed assets – Sustainability in building construction – Terminology	AHG working with ISO/TC 59 SC 2,14,15,17 and CEN/TC 350 to establish a common terminology document. Working draft study initiated 8th March 2004

(3) Standards that are under development in CEN/TC 350:

Document number	Title of standard	Comment
CEN/TC 350 N 82	Sustainability of construction works – Framework for assessment of integrated buildings performance – Part 1: Environmental, Health and Comfort and Life Cycle Cost Performances	Framework document under discussion

The CEN/TC 350 family of standards will have a substantial impact on the way we built, as it will be possible, once all the standards are issued, to compare buildings from all points of view, not only of environmental impact but also eco-efficiency. At a later stage, non-immediately measurable facts such as health and comfort will also be considered under standards. These standards also focus on communication formats as information flows must be tailored so as to be understood by the intended stakeholders with, if necessary, tailoring reports for different audiences.

Fulfilling these rules and requirements might be made easier for the manufacturer of building products using wastes in substitution for raw materials by adopting a holistic approach. An Integrated Product Policy (IPP) might be the instrument of choice, covering all the impact categories of the product and its manufacturing process, whether the product is a simple brick or a building.

5.6.1 Integrated Product Policy (IPP)

The goal of the IPP is to create a healthier environment for all stakeholders and the ecosystem by promoting a system change, thus extending beyond the production phase into the use of the product. The results and achievements of this policy need to be communicated in a appropriate, standardized way.

The first step in the IPP concept is to establish a function tree (Figure 5.6) in which the following apply:

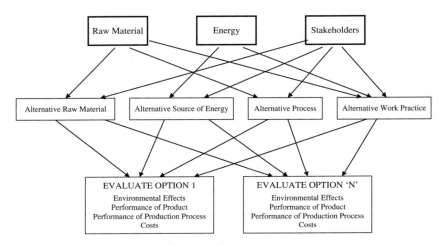

Figure 5.6   IPP functions and their relationships.

- *Raw material* is the primary or secondary (waste) material used in production.
- *Energy* is the energy required to operate the process or equipment *within* the process, but excluding the energy involved in production (extraction) and delivery of materials.
- *Stakeholders* are groups and organizations having an interest or stake in the company, such as the employees, their families, the local community and the society at large, including the media, government, science and technology institutions, NGOs (e.g. Greenpeace), competitors, residents, unions, ecosystem, insurance companies and so on.
- *Environmental effects* are considered to be the consequences of activities of an organisation, products or services that can interact with the environment; that is, flora, fauna, humans, and their interrelations;
- *Performance of product* is considered as the improvement or worsening of a sum of characteristics of the product due to changes in the raw materials, use or type of energy, changes to production process,

or work practice. The term is also intended to include environmental performance.

- *Performance of a production process* is considered as the improvement or worsening of a sum of characteristics of the product due to changes in the raw materials; for example, substitution by 'wastes', use or type of energy, and changes to production process or work practice.

Using the various alternative pathways within this scheme will aid a company in selecting the most appropriate and eco-friendly process. It is incorrect to assume that IPP is only feasible for large companies, and not small or medium-sized enterprises. An example of a small company adopting an IPP system is the Gasser brickworks in Italy, which currently is still the only brick manufacturer in Europe that has an Environmental Product Declaration for its products (SWEDAC and AUB) and ISO 9001 and 14001 certification in place.

At the moment, product and production reporting within an IPP framework is a voluntary exercise. The ability to measure environmental performance must be seen in the light of the trend towards corporate 'leanness', with better indicators to account for, and report on, the financial value created by sustainable development activities.

The tools of an IPP are:

ISO 9001

Implementing an ISO 9001 certified management system has become an almost everyday chore and does certainly not need any further explanation. It must, however, be clearly understood that an ISO 9001 certification does not have any impact on the eco-efficiency of a company or of the products produced. Neither does it inherently impact product quality.

Product Category Rule (PCR)

A PCR is needed as a first step in the process of preparing an Environmental Product Declaration (EPD). The PCR will define the system limits and the amount of information to be included. The intent is to describe product characteristics in terms of specific aspects, to highlight a manufacturing and use perspective as well as the intended normal use of the product in a technical specification. The technical specification will include sufficient information for a customer to assess and evaluate the technical performance and usefulness of the product. The PCR will identify and report the goal and scope of the LCA-based information for the product category and the rules on producing additional environmental information for the product category.

The PCR will also determine the life cycle stages to be included, the parameters to be covered and the way in which the parameters are to be collated and reported.

The PCR:
• describes the selection of parameters to be declared and the way in which the parameters shall be collated and reported;
• includes rules for calculating the LCA, LCI or information modules underlying an EPD, including the specification of the quality of the applied data;
• describes which stages of a product's life cycle are considered;
• includes rules for calculating and reporting additional environmental information for this product category;
• describes what information about the product shall be applied when developing scenarios for the stages that are not covered by the information modules; and
• describes when and how construction products can be compared on the basis of the information provided by EPD.

The principles and procedures set out in ISO 14020, ISO 14025, ISO DIS 21930, ISO 14040, ISO 14044 and ISO/CD 15392 shall also apply in preparing an EPD.

The overall goal of EPD in the construction sector is to encourage the demand for, and supply of, those products that are more eco-efficient through communication of verifiable and accurate information that is not misleading, thereby stimulating the potential for market-driven continuous environmental improvement of products and buildings.

5.6.2 Life Cycle Analysis (LCA)

LCA is a compilation and evaluation based on a specific PCR, according to ISO standards, of the inputs and outputs of material and energy, and the potential environmental impacts associated with the inventory data in each of the selected categories. It consists of a rapidly emerging family of tools and techniques of environmental management systems.

LCA is very important in evaluating the environmental aspects of a product, and hence the production process, through all stages of its life cycle. It allows evaluation of the environmental impact of any changes in the production process; for example, the substitution of raw materials by waste materials. It operates with a number of environmental impact categories, such as 'climate change (global warming)', 'stratospheric ozone depletion', 'human toxicity', 'eco-toxicity', 'photo-oxidant formation', 'acidification', 'nutrification' and 'land use'. The design of the LCA allows for appraisal of the environmental impact at the design stage of a product. Companies will thus be able not only to analyse and monitor their own progress in comparison to others with regard to sustainable production, but with LCA will certainly have a key factor in consumer decision-making in favour of given groups of products or methods of construction (for example, choosing a brick wall in preference to a concrete wall).

The ISO 14000 series of standards are widely accepted as the general framework for life cycle assessment:

- ISO EN 14040 (1997) on principles and framework;
- ISO EN 14041 (1998) on goal and scope definition and inventory analysis;
- ISO EN 14042 (2000) on life cycle impact assessment; and
- ISO EN 14043 (2000) on life cycle interpretation.

These standards, however, do not provide either detailed methodological guidance or tools for the actual performance of life cycle impact assessment (ISO EN 14042).

5.6.3 ISO 14001/EMAS (Eco-Management and Audit Scheme)

Implementing an ISO 9001 certified quality management system has become well-established and documented.* It must, however, be clearly

---

* ISO Standards Compendium: ISO 9000 – Quality management. The Compendium gathers in one volume the 11 published standards and technical reports making up the ISO 9000 family.

understood that ISO 9001 certification does not have any impact on the eco-efficiency of a company or of the products produced.

The implementation of environmental management systems according to ISO 14001 does lead to an increasing need for verification of quantified environmental information requested from raw material suppliers, subcontractors and entrepreneurs. This is of special concern for EMAS-registered companies and sites that are providing information in publicly available environmental statements (as part of the EMAS verification process) that are signed by the managing director or the plant manager, and these need to be checked by the verifiers and guaranteed as trustworthy.

Since previously EMAS focused more on industrial activities, the EC has now developed separate guidelines on how to deal with product issues within EMAS. Environmental Management Systems (EMS) are relevant for all types of organisations – public or private – and can be used to provide a framework for all types of tools, from the greening of procurement to validating green information.

EMS certification by itself does not guarantee a specific environmental product performance, but in the case of EMAS it provides a framework for validating information about such performance by the EMAS verifier.

5.6.4   Environmental Product Declaration (EDP)

For the purpose of this chapter, EPDs will only be considered to be the so-called Type III declarations embedded in a Type III programme, as these are believed to be the only really valuable instrument within an IPP. Table 5.4 lists Type III schemes for the construction industry devised by different European countries.

A Type III EPD is a set of quantified, verifiable and comparable environmental data based on information from a life cycle assessment according to internationally accepted procedures following completed and upcoming ISO standards for LCA. The main advantages of EPDs, for those creating EPDs and providing information to the market, are that they are regarded as being objective. Further, they are nonselective and neutral – because, in contrast to other Environmental Product Declaration schemes, no claims of valuations or predetermined environmental performance levels must be met.

Table 5.4 ISO Type III EPD schemes for the construction industry, devised by different European countries (http://ec.europa.eu/enterprise/construction/internal/essreq/environ/lcarep/lcaselepd.htm) (2008).

		Establishment	Title of scheme	ISO type	Date
1	UK	BRE (Building Research Establishment)	BRE Environmental Profiles of Construction Products	III	1999
2	NL	NVTB (Dutch Construction Product Association)	MRPI (Environmental Relevant Product Information)	III	1999
3	NL	NEN (Dutch Standardisation Organisation)	MEPB (Material Based Environmental Profile for Building) (draft ongoing)	?	2002/3
4	F	AFNOR	XP P01010-1 and XP P01010-2 Environmental Quality of Construction Products: methodology and model (1) and operating data for application in a building (2)	III	2001
5	DK	SBI (Danish Building and Urban Research)	MVDB (Environmental Product Declaration for Building Products) (in progress)	III	2002
6	FIN	VTT / RTS	Environmental Assessment of Building Products/RTS Environmental Declaration	III?	2001
7	S	Swedish Environmental Management Council (Svenska Miljöstyrningsrådet)	Environmental Product Declaration	III	1997
8	S	Ecocycle Council for the Building Sector	Building Product Declaration	II	1997
9	D	German Building Materials Association	Ganzheitliche Bilanzierung von Baustoffe und Gebäude	III	2000
10	D	AUB (Arbeitsgemeinschaft Umweltverträgliches Bauproducte)	Environmental Declaration of Construction Products (in development)	III	2002
11	N	Byggforsk (Norwegian Building Research Institute)	EcoDec	III	2000
12	EU	CEPMC (Council of European Products of Materials for Construction)	Guidance for the Provision of Environmental Information on Construction Products	II	2000

An EPD is also a flexible instrument enabling, after due external review and verification, any changes or improvements of the EPD to be made as required by the company/organisation. EPDs are applicable for all types of products and services within clearly defined product categories. They are designed to meet various information needs within the supply chain and for end-products both in the private and public sectors, as well as for more general purposes in marketing and other information activities.

The EPD is expected to be accurate because the information has to be continuously updated using in-company routines for documentation and follow-up procedures. So an EPD giving accurate impact information about single products will allow environmental impact benchmarking of construction materials and hence finished buildings (Table 5.5).

In the construction sector, information about the environmental performance of materials and products has been collected and published for several years in a number of countries (e.g. AIMCC in France, AUB in Germany, the BRE Environmental Profiles in the UK, MRPI in the Netherlands, RTS Format in Finland and SIA Deklarationsraster in Switzerland).

5.6.5 Organisational Health and Safety Standard (OHSAS)

Within the concept of IPP, workers' safety should receive special attention. Therefore certification of the production facility to an international safety standard such as OHSAS 18001 might be an appropriate way of communicating to the workforce and other stakeholders that workplace safety is a primary concern, despite the use of waste materials as a substitute for natural raw materials. An OHSAS certification can also lead to substantial savings on insurance expenses.

The result of a product manufactured under an IPP can be that such information can be used to facilitate the eco-design of new products and improvements to existing products assisting stakeholders such as architects, contractors and purchasers in their choice of products (for a specific application), and in the use and maintenance of construction works. It can help contractors and recycling companies in the environmentally friendly recovery or safe disposal of waste materials in end-of-life construction work.

Table 5.5 A comparison of information contained in the various European EPD schemes for the construction industry.

Strong points	European EPD schemes (designations as given in Table 5.4)											
	1	2	3	4	5	6	7	8	9	10	11	12
Goal												
Multipurpose use (information, product comparison and input for building calculations)				(x)	x		x		x	x		
Predominant use in environmental evaluation of buildings	x	x	x			x						
Status												
Well-established methodology, several profiles completed	x	x		x		x	x				x	
Will (still) be operational in the future	x	x	(x)	x	(x)	(x)	x			x	x	
Reference unit												
Functional unit including specific technical characteristics of the product in the building		x	n.a.	x	(x)	(x)	x	(x)	(x)			x
Declaration												
Standardised/clear format	x	x		x	(x)	x	(x)	(x)	x	(x)	x	x

Table 5.5 (Continued)

Strong points	European EPD schemes (designations as given in Table 5.4)											
	1	2	3	4	5	6	7	8	9	10	11	12
Quality assurance												
Obligated third party certi-fication/verification/review		x			x		x			x	x	
Hazardous substances and health and safety data												
Information provided on hazardous substances/chemicals				(x)				(x)	x	x	x	
Information provided on indoor environment				(x)		x	(x)				x	
Data quality												
Transparency of data sources and methodology choices used in each profile very high		x	x	x	(x)	(x)	(x)		(x)	(x)	(x)	(x)

5.6.6 An Example of an IPP – the Use of Fly Ash in Heavy Clayware Products

The practice of using waste materials, such as combustion fly ash, as components of the feedstock requires special permits under Italian legislation, making it necessary to introduce methods to safeguard the quality level of the production, and to document any potential hazards that might result from the use of such wastes. This requires the compilation of an extensive environmental performance database, consisting of information on the composition and leaching of the product and on emission data from the stack and the dryer over a period of more than three years. Data on the products are gathered every 40 days, and emission data twice yearly, or more frequently when novel materials are used. Fuel consumption is also recorded and the data broken down to reflect a single unit of product consumption. The possession of such a database makes the start of the IPP project easy and inexpensive.

For the preparation of the LCA, it is likely that some data may be needed to be recorded differently, one example being fuel and electricity use in the quarry. Currently, assumptions have been made based on real summary data that is available (Figure 5.7).

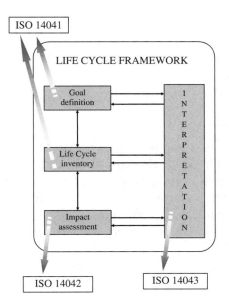

Figure 5.7 The life cycle framework.

The ability to use the standard ISO procedure allows results to be compared and/or benchmarked. Such results can be interpreted in many different ways, but primarily those processes having the highest impact values can be identified and work undertaken to reduce their impact.

For example, Table 5.6 shows the actual impact values and impact values that would be obtained by substituting fossil fuels with fuels from regeneratable sources to fire the kiln. Of course, the total substitution of fuels is not always feasible, but the example illustrates how decisions may be taken on a clear and logical basis.

With the data gathered in the LCA study, it is also possible to decide upon environmental goals as required within an ISO or EMAS certification. Some of the data gathered in the LCA study can also be used in an OHSAS 18001 certification. An argument against the collection of all this data might be the cost. Usually, the cost of such measures is grossly overestimated. The total expense, including the cost of building up the database as set out in Table 5.7, varies according to the level of certification required, but the maximum is €190 000 spread over three years.

5.6.7 Benefits

The potential benefits of an IPP-certified production are manifold:

- a potential increase in market share;
- simplified authorization procedures;
- reduced insurance rates;
- reduced production costs;
- higher profits; and
- a better public image.

In the case of the Gasser brickworks, the use of waste materials as part of the feedstock used in the production of bricks increased from < 5 % to > 15 % in volume, resulting in significant savings in production costs. The search for suitable waste materials is ongoing and a waste materials quota of around 25 % is envisaged. As a result of the environmental certification of the product, the use of waste materials did not lead to any detrimental effect on sales; in fact, the number of environmentally conscious customers have increased, from < 3 % to > 60 % of total turnover.

Table 5.6 Environmental impacts of using alternative fuels.

	Renewable energy ($MJ\,kg^{-1}$)	Nonrenewable energy ($MJ\,kg^{-1}$)	Greenhouse effect, 100 years, CO_2 equiv. ($kg\,kg^{-1}$)	Ozone depletion, R11 equiv.	Photosmog, ethylene equiv.	Acidification, SO_x equiv.	Nutrification, PO_4^{3-} equiv.
Average DACH	0.056	3.55	0.35	0	0.0001	0.001	0.0001
Gasser[a] actual	0.09875	4.8016	0.18975	Negligible	0	0	0
Gasser Biogas	4.90035	0.55	Negligible	Negligible	0	0	0

[a] Gasser Brickworks, Italy.

Table 5.7 Relative costs of gathering data over three years (2002–2004) (Moedinger, private information).

Certification	Cost category	Cost (€)
	Third party	30 000
	Internal	30 000
LCA		
	Third party	12 000
	Internal	5 000
ISO 9.001: 2000		
	Third party	15 000
	Internal	10 000
	Certification	5 000
ISO 14.001: 1996		
	Third party	20 000
	Internal	10 000
	Certification	5 000
EMAS[a]		
	Third party	5 000
	Internal	5 000
	Certification	7 500
OHSAS 18.001		
	Third party	12 000
	Certification	5 000
EPD		
	Third party	9 500
	Internal	Nil
	Certification	3 500
Total costs		190 000

[a] Step only necessary within the EC.

5.6.8 Action Plan

The implementation of an IPP scheme needs to be well planned and requires the involvement of everyone in the company, as employees frequently have the best insight in the details of company operation. A first step is the careful selection of the certification body, and it is important that this body is familiar with the industry and the product being certified. A pre-certification visit by the body is relatively cheap and can be undertaken over two to three days. Based on this inspection, a decision can be taken by the company on whether it needs to involve external consultants. In many cases, a large part of the necessary data and information can be gathered by company personnel (Figures 5.7 and 5.8).

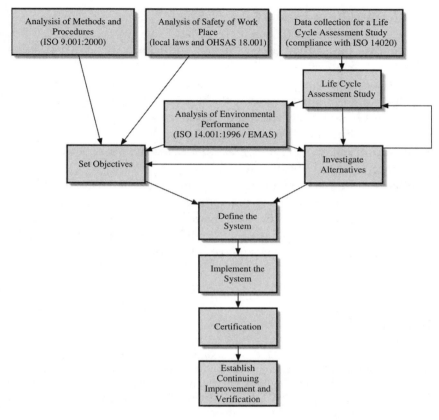

Figure 5.8 Stages in developing an IPP system.

At the core of the IPP process is the life cycle assessment. This can be prepared from generic, or company-specific, data, but the use of generic data is generally discouraged. Company-specific data allow the study to be used as an instrument to fine tune the production process from the point of view of eco-efficiency. The collection of specific data should take place over a sufficiently long period to allow for the elimination of spurious data from periods of abnormal operation. It is preferable to collect data over several years, and this can be facilitated if the records system of the company is tailored to supply such information routinely and automatically.

5.7 Concluding Remarks

This chapter has summarised some of the legislative and marketing issues faced by novel ash products.

This is an exciting time for such products, since the traditional barriers to their use are changing rapidly with the concerted drive to reduce waste within the EU and elsewhere. The consequent EU legislation drivers – for example, the Landfill Directive – and taxation – for example, landfill and depository taxes – are now making it easier than ever before to gain acceptance for ash-derived materials.

Perhaps the greatest challenge that remains is the customer perception of these materials. Here, the activities of bodies representing ash producers (e.g. ECOBA, and their national equivalents) and targeted workshops such as those previously organised under the PROGES Network can do much to raise the profile of this valuable resource.

References

1 Brendel, G.F., *Removing institutional, legal, and regulatory barriers to coal combustion by-product utilisation*. In Proceedings of the 1997 International Ash Utilisation Symposium: Pushing the Envelope, Lexington, Kentucky, 20–22 October 1997.
2 European Topic Centre on Waste and Material Flows (Topic Centre of European Environment Agency); http://waste.eionet.eu.int/
3 Report from the Commission to the Council and the European Parliament on the Implementation of Community Waste Legislation for the Period 1998–2000. Commission of the European Communities Brussels, 19.5.2003 COM(2003) 250 final.
4 Blondin, J. and Anthony, E.J., *A selective hydration treatment to enhance the utilisation of CFBC ash in concrete*. In Proceedings of 49th Annual Purdue University Industrial Waste Conference, West Lafayette, Indiana, 9–11 May 1994.
5 Barnes, D.I., personal communication (2005).
6 Vom Berg, W., *Commercial use of coal combustion by-products in Europe*. In Proceedings of Indo-European Seminar Clean Coal Technology and Power Plant Upgrading. National Thermal Power Corporation Ltd, India, 7 pp. (1997).
7 Pflughoeft-Hassett, D. and Hassett, D., *Developing beneficial use rules for coal combustion by-products*. In Proceedings of University of Kentucky Center for Applied Energy Research International Ash Utilization Symposium, 2001.
8 Barnes, D.I. and Laughlin, K.M., *The management of residues from thermal processes*. Energy from Waste Foundation (2001).
9 Cornelissen, H.A.W., *Micronized fly ash: a valuable resource for concrete*. In Proceedings of 12th International Symposium on Coal Combustion By-product Management and Use, Orlando, Fla, 26–30 January 1997.

10   Fisher, B.C., Blackstock, T. and Hauke, D., *Fly ash beneficiation using an ammonia stripping process*. In Proceedings of 12th International Symposium on Coal Combustion By-product Management and Use, Orlando, Fla, 26–30 January 1997.

11   Carpenter, A.M., *Switching to cheaper coals for power generation*. In IEA CCC/01. IEA Coal Research: London, 87 pp. (1998).

12   Renninger, S., *An overview of the United States Department of Energy's coal combustion by-product utilisation programme*. In Proceedings of 25th Anniversary and 15th National Meeting of the American Society for Surface Mining and Reclamation, St Louis, Missouri, 17–21 May 1998.

13   Bennett, B.H., *The changing face of the utility industry and its impact on coal combustion by-product management*. In Proceedings of 12th International Symposium on Coal Combustion By-product Management and Use, Orlando, Fla, 26–30 January 1997.

14   Callaway, D.W., *CCB management: a business perspective for utility managers*. In Proceedings of 12th International Symposium on Coal Combustion By-product Management and Use, Orlando, Fla, 26–30 January 1997.

15   Trehan, A., Krishnamurthy, R. and Kumar, A., *Utilization of coal ash in India*. In Proceedings of the 1997 International Ash Utilization Symposium: Pushing the Envelope, Lexington, Kentucky, 20–22 October 1997, pp. 615–622.

16   Pavlenko, S., Shishkanow, A. and Bazhenov, Y., *Technological complex for utilisation of high-calcium ash and slag for Abakan Thermal Power Plant*. In Fly Ash, Silica Fume, Slag and Natural Pozzolans in Concrete: Proceedings of 6th CANMET/ACI International Conference, Bangkok, Thailand, 1998.

17   Dietz, S., *Utilisation of CCPs in Europe*. In ECOBA/ACAA Joint Workshop, Toronto, Canada, June 1998.

Index

Combustion Residues: Current, Novel and Renewable Applications Edited by Michael Cox,
Henk Nugteren and Mária Janssen-Jurkovičová © 2008 John Wiley & Sons, Ltd